VOLUME 3

2009 IBC®

SEAOC STRUCTURAL/SEISMIC DESIGN MANUAL

BUILDING DESIGN EXAMPLES FOR
STEEL AND CONCRETE

STRUCTURAL ENGINEERS ASSOCIATION CALIFORNIA

STRUCTURAL ENGINEERS ASSOCIATIONS NATIONAL COUNCIL

ICC INTERNATIONAL CODE COUNCIL®

Copyright

Publisher

Structural Engineers Association of California (SEAOC)

1400 K Street, Suite 212
Sacramento, CA 95814
Telephone: (916) 447-1198; Fax: (916) 444-1501
E-mail: dschinske@seaoc.org; Web address: www.seaoc.org

The Structural Engineers Association of California (SEAOC) is a professional association of four regional member organizations (Southern California, Northern California, San Diego, and Central California). SEAOC represents the structural engineering community in California. This document is published in keeping with SEAOC's stated mission: "to advance the structural engineering profession; to provide the public with structures of dependable performance through the application of state-of-the-art structural engineering principles; to assist the public in obtaining professional structural engineering services; to promote natural hazard mitigation; to provide continuing education and encourage research; to provide structural engineers with the most current information and tools to improve their practice; and to maintain the honor and dignity of the profession."

Editor

International Code Council

Disclaimer

Practice documents produced by the Structural Engineers Association of California (SEAOC) and/or its member organizations are published as part of our association's educational program. While the information presented in this document is believed to be correct, neither SEAOC nor its member organizations, committees, writers, editors, or individuals who have contributed to this publication make any warranty, expressed or implied, or assume any legal liability or responsibility for the use, application of, and/or reference to opinions, findings, conclusions, or recommendations included in this publication. The material presented in this publication should not be used for any specific application without competent examination and verification of its accuracy, suitability, and applicability by qualified professionals. Users of information from this publication assume all liability arising from such use.

First Printing: September 2012

Table of Contents

Preface

This document is the third volume in the three-volume 2009 *IBC Structural/Seismic Design Manual*. It has been developed by the Structural Engineers Association of California (SEAOC) with funding provided by SEAOC. Its purpose is to provide guidance on the interpretation and use of the seismic requirements in the 2009 *International Building Code* (IBC), published by the International Code Council, Inc., and SEAOC's 2005 *Recommended Lateral Force Requirements and Commentary* (also called the Blue Book).

The 2009 *IBC Structural/Seismic Design Manual* was developed to fill a void that exists between the recommendations of the Blue Book, which explains the basis for the code provisions, and everyday structural engineering design practice. The 2009 *IBC Structural/Seismic Design Manual* illustrates how the provisions of the code are used. *Volume I: Code Application Examples,* provides step-by-step examples for using individual code provisions, such as computing base shear or building period. *Volumes II and III: Building Design Examples,* furnish examples of seismic design of common types of buildings. In Volumes II and III, important aspects of whole buildings are designed to show, calculation-by-calculation, how the various seismic requirements of the code are implemented in a realistic design.

The examples in the 2009 *IBC Structural/Seismic Design Manual* do not necessarily illustrate the only appropriate methods of design and analysis. Proper engineering judgment should always be exercised when applying these examples to real projects. The 2009 *IBC Structural/Seismic Design Manual* is not meant to establish a minimum standard of care but, instead, presents reasonable approaches to solving problems typically encountered in structural/seismic design.

The example problem numbers used in the prior manual—2006 *IBC Structural/Seismic Design Manual*—have been retained herein to provide easy reference to compare revised code requirements.

SEAOC and ICC intend to update the *IBC Structural/Seismic Design Manual* with each edition of the building code.

Jon P. Kiland and Rafael Sabelli
Project Managers

Acknowledgments

Authors

The 2009 *IBC Structural/Seismic Design Manual* was written by a group of highly qualified structural engineers. They were selected by a steering committee set up by the SEAOC Board of Directors and were chosen for their knowledge and experience with structural engineering practice and seismic design. The consultants for Volumes I, II, and III are:

Jon P. Kiland, *Co-Project Manager*
Rafael Sabelli, *Co-Project Manager*
Matt Eatherton
Stephen Kerr
John W. Lawson
Joe Maffei
Kevin Moore
Karl Telleen
Douglas S. Thompson
Dan Wendowatz

Reviewers

A number of SEAOC members and other structural engineers helped check the examples in this volume. During its development, drafts of the examples were sent to these individuals. Their help was sought in review of code interpretations, as well as detailed checking of the numerical computations.

Seismology Committee

Close collaboration with the SEAOC Seismology Committee was maintained during the development of the document. The Seismology Committee has reviewed the document and provided many helpful comments and suggestions. Their assistance is gratefully acknowledged.

Production and Art

International Code Council

Suggestions for Improvement

In keeping with SEAOC's Mission Statement: "to advance the structural engineering profession" and "to provide structural engineers with the most current information and tools to improve their practice," SEAOC plans to update this document as structural/seismic requirements change and new research and better understanding of building performance in earthquakes becomes available.

Comments and suggestions for improvements are welcome and should be sent to the following:

Structural Engineers Association of California (SEAOC)
Attention: Executive Director
1020 12th Street, Suite 303
Sacramento, California 95814
Telephone: (916) 447-1198; Fax: (916) 444-1501
E-mail: lee@seaoc.org; Web address: www.seaoc.org

Errata Notification

SEAOC has made a substantial effort to ensure that the information in this document is accurate. In the event that corrections or clarifications are needed, these will be posted on the SEAOC web site at *http://www.seaoc.org* or on the ICC web site at *http://www.iccsafe.org*. SEAOC, at its sole discretion, may or may not issue written errata.

Introduction

The 2009 *IBC Structural/Seismic Design Manual* is intended to help the reader understand and correctly use the IBC structural/seismic provisions and to provide clear, concise, and graphic guidance on the application of specific provisions of the code. It primarily addresses the major structural/seismic provisions of the IBC, with interpretation of specific provisions and examples highlighting their proper application.

The 2009 IBC has had structural provisions removed from its text and has referenced several national standards documents for structural design provisions. The primary referenced document is ASCE/SEI 7-05, which contains the "Minimum Design Loads for Buildings and Other Structures." ASCE/SEI 7-05 is referenced for load and deformation design demands on structural elements, national material design standards (such as ACI, AISC, MSJC and NDS) are then referenced to take the structural load demands from ASCE/SEI 7-05 and perform specific material designs.

The complete 2009 *IBC Structural/Seismic Design Manual* has three volumes. Volume 1 illustrates the application of specific provisions of ASCE 7 and the IBC.Volumes 2 and 3 provide a series of structural/seismic design examples for buildings illustrating the seismic design of key parts of common building types such as a large three-story wood frame building, a tilt-up warehouse, a braced steel frame building, and a concrete shear wall building.

While the 2009 *IBC Structural/Seismic Design Manual* is based on the 2009 IBC, there are some provisions of SEAOC's *Recommended Lateral Force Provisions and Commentary* (Blue Book) that are applicable. When differences between the IBC and Blue Book are significant, they are brought to the attention of the reader.

The 2009 *IBC Structural/Seismic Design Manual* is intended for use by practicing structural engineers and structural designers, building departments, other plan review agencies, and structural engineering students.

How to Use This Document

The 2009 *IBC Structural/Seismic Design Manual*—Volume 3 is based on the 2009 IBC, unless otherwise indicated. Such indication is to be found in the "Code Reference" column at page right. ASCE/SEI 7-05 notation is generally used throughout the document. Code references to the 2009 IBC are enclosed in parentheses. Occasional references to other codes and standards are specifically identified as such (e.g., ACI 318-08, AISC-360-05, etc.)

Abbreviations used in the "Code Reference" column are:

§ – Section	T – Table
F – Figure	Eq – Equation

Generally, each design example is presented in the following format. First, there is an "Overview" of the example: a description of the building to be designed. This is followed by an "Outline" indicating the tasks or steps to be illustrated in each example. Next, "Given Information" provides the basic design information, including plans and sketches given as the starting point for the design. This is followed by "Calculations and Discussion," which provides the solution to the example. Some examples have a subsequent section designated "Commentary" that is intended to provide a better understanding of aspects of the example and/or to offer guidance to the reader on use of the information generated in the example. Finally, references and suggested reading are given under "References." Some examples also have a "Foreword" and/or "Factors Influencing Design" section that contains remarks on salient points about the design.

Notation

The following notation is used in this document. These are generally consistent with those used in ASCE/SEI 7-05 and other standards such as ACI and AISC. Some new notations have been added. The reader is cautioned that the same notation may be used more than once and may carry entirely different meanings in different situations. For example, E can mean the tabulated elastic modulus under the AISC definition (steel) or it can mean the earthquake load under §12.4.2 of ASCE/SEI 7-05.

A = total cross-sectional area of member

A_b = cross-sectional area of a horizontal boundary element (HBE)

A_c = cross-sectional area of a vertical boundary element (VBE)

A_{ch} = cross-sectional area of a structural member measured out-to-out of transverse reinforcement

A_{cv} = gross area of concrete section bounded by web thickness and length of section in the direction of shear force considered

A_{cw} = area of concrete section of an individual pier, horizontal wall segment, or coupling beam resisting shear

A_e = effective net area

A_f = flange area

A_g = gross area of member

A_g = gross area of concrete section. For a hollow section, A_g is the area of the concrete only and does not include the area of the void(s)

A_n = net area of member

A_{nt} = net area subject to tension

A_{nv} = net area subject to shear

A_s = area of nonprestressed longitudinal tension reinforcement

A_{sc} = area of the yielding segment of steel core

A_{sh} = total cross-sectional area of transverse reinforcement (including crossties) within spacing s and perpendicular to dimension b_c

$A_{s,\min}$ = minimum area of flexural reinforcement

A_{st} = area of link stiffener

A_t = area of one leg of a closed stirrup resisting torsion within spacing s

A_{tr} = total cross-sectional area of all transverse reinforcement within spacing s that crosses the potential plane of splitting through the reinforcement being developed

A_{vf} = area of shear-friction reinforcement

A_w = link web area (excluding flange depth)

A_w = web area, the overall depth times the web thickness, dt_w

A_w = effective area of the weld

B = overall width of rectangular HSS member, measured 90 degrees to the plane of the connection

B_1, B_2 = factors used in determining M_u for combined bending and axial forces when first-order analysis is employed

C_a = ratio of required strength to available strength

C_b = lateral-torsional buckling modification factor for nonuniform moment diagrams when both ends of the unsupported segment are braced

C_d = coefficient relating relative brace stiffness and curvature

C_d = deflection amplification factor

C_m = coefficient assuming no lateral translation of the frame

C_{pr} = factor to account for peak connection strength, including strain hardening, local restraint, additional reinforcement, and other connection conditions

C_r = parameter used for determining the approximate fundamental period

C_v = web shear coefficient

D = outside diameter of round HSS member

D = dead load due to the weight of the structural elements and permanent features on the building

E = earthquake load

E = modulus of elasticity of steel, $E = 29{,}000$ ksi (200,000 Mpa)

F_{cr} = critical stress

F_e = elastic critical buckling stress

F_{ex} = elastic flexural buckling stress about the major axis

F_{EXX} = electrode classification number

F_u = specified minimum tensile strength

F_w = nominal strength of the weld metal per unit area

F_y = specified minimum yield stress of the type of steel to be used, ksi (MPa). As used in the *Specification*, "yield stress" denotes either the minimum specified yield point (for those steels that have a yield point) or the specified yield strength (for those steels that do not have a yield point)

F_{yb} = F_y of a beam

F_{yc} = F_y of a column

F_{ysc} = specified minimum yield stress of the steel core, or actual yield stress of the steel core as determined from a coupon test

H = height of story, which may be taken as the distance between the centerline of floor framing at each of the levels above and below, or the distance between the top of floor slabs at each of the levels above and below

I = moment of inertia

I_c = moment of inertia of a vertical boundary element (VBE) taken perpendicular to the direction of the web plate line

K = effective length factor for prismatic member

K_{tr} = transverse reinforcement index

L = live load due to occupancy and moveable equipment

L = distance between VBE centerlines

L = story height, inches

L = length of the member

L' = distance between plastic hinges

L_b = length between points that are either braced against lateral displacement of compression flange or braced against twist of the cross section

L_{cf} = clear distance between VBE flanges

L_p = limiting laterally unbraced length for full plastic flexural strength, uniform moment case

L_r = limiting laterally unbraced length for the limit state of inelastic lateral-torsional buckling

M_f = maximum moment expected at face of column

M_{lt} = first-order moment under LRFD or ASD load combinations caused by lateral translation of the frame only

$M_{\max}$ = absolute value of maximum moment in the unbraced segment

M_n = nominal flexural strength

M_{nt} = first-order moment using LRFD or ASD load combinations assuming there is no lateral translation of the frame

M_p = plastic bending moment

M_{pa} = nominal plastic flexural strength modified by axial load

M_{pe} = plastic moment of beam based on expected yield stress

M_{pr} = probable maximum moment at plastic hinge (steel)

M_{pr} = probable flexural strength of members, with or without axial load, determined using the properties of the member at the joint faces assuming a tensile stress in the longitudinal bars of at least $1.25f_y$ and a strength reduction factor, ϕ, (concrete)

M_u = required flexural strength

M_1 = smaller moment, calculated from a first-order analysis, at the ends of that portion of the member unbraced in the plane of bending under consideration

M_2 = larger moment, calculated from a first-order analysis, at the ends of that portion of the member unbraced in the plane of bending under consideration

N = length of bearing (not less than k for end beam reactions)

P_b = required strength of lateral brace at ends of the link

P_{br} = required brace strength

P_{e1}, P_{e2} = elastic critical buckling load for braced and unbraced frame, respectively

P_n = nominal axial strength

P_u = required axial strength

P_y = nominal axial yield strength of a member, equal to $F_y A_g$

P_{ysc} = axial yield strength of steel core

Q_b = maximum unbalanced vertical load effect applied to a beam by the braces

Q_1 = axial forces and moments generated by at least 1.25 times the expected nominal shear strength of the link

R = seismic response modification coefficient

R_n = nominal strength

R_t = ratio of the expected tensile strength to the specific minimum tensile strength F_u, as related to overstrength in material yield stress R_y

R_u = required strength

R_v = panel zone nominal shear strength

R_y = ratio of the expected yield stress to the specified minimum yield strength, F_y

R_{yb} = ratio of expected yield stress to specified minimum yield stress F_y, for a beam

R_{ye} = ratio of expected yield stress to specified minimum yield stress F_y, for a column

S = snow load

S_h = distance from the face of a column to a plastic hinge, in

U = required strength to resist factored loads or related internal moments and forces, ACI-318

U = shear lag factor

U_{bs} = reduction coefficient, used in calculating block shear rupture

V_c = nominal shear strength provided by concrete, ACI-318

V_e = design shear force corresponding to the development of the probable moment strength of the member

V_{gravity} = beam shear force resulting from $1.2D + f_1L + 0.2S$, IBC

V_n = nominal shear strength of a member

V_p = nominal shear strength of a link

V_{pa} = nominal shear strength of a link modified by the axial load magnitude

V_{RBS} = larger of the two values of shear force at the center of the reduced beam section at each end of a beam

V'_{RBS} = smaller of the two values of shear force at the center of the reduced beam section at each end of a beam

V_s	=	nominal shear strength provided by shear reinforcement, ACI-318
V_u	=	required shear strength
Z	=	plastic section modulus of a member
Z_b	=	plastic section modulus of the beam
Z_c	=	plastic section modulus of the column
Z_e	=	effective plastic modulus of a section (or connection) at the location of a plastic hinge, Z_{RBS}
Z_x	=	plastic section modulus x-axis
Z_{RBS}	=	minimum plastic section modulus at the reduced beam section
a	=	horizontal distance between a column flange and the start of an RBS cut,
b	=	width of compression element as defined in the AISC *Specification*
b	=	length of an RBS cut
b_{bf}	=	width of beam flange
c	=	depth of cut at the center of the reduced beam section
d	=	nominal fastener diameter
d	=	overall member depth
d_z	=	overall panel zone depth between continuity plates
e	=	EBF link length
f_e'	=	specified compressive strength of concrete
f_y	=	specified yield strength of reinforcement, ACI-318
f_1	=	load factor determined by the applicable building code for live loads but not less than 0.5
h	=	clear distance between flanges less the fillet or corner radius for rolled shapes
h	=	distance between horizontal boundary elements' centerlines
h_o	=	distance between flange centroids
k_c	=	distance from outer face of a column flange to web toe of fillet (design value) or fillet weld

l = unbraced length of compression or bracing member

ℓ_n = length of clear span measured face-to-face of supports, ACI-318

ℓ_u = unsupported length of compression member, ACI-318

ℓ_w = length of entire wall or length of segment of wall considered in direction of shear force, ACI-318

n = number of items; such as strength tests, bars, wires, monostrand anchorage devices, anchors, or shearhead arms, ACI-318

r = governing radius of gyration

r_y = radius of gyration about y-axis

s = center-to-center spacing of items; such as longitudinal reinforcement, transverse reinforcement, prestressing tendons, wires, or anchors, ACI-318

t = thickness of element

t_{bf} = thickness of beam flange

t_{bw} = thickness of beam web

t_{cf} = thickness of column flange

t_{cw} = thickness of column web

t_f = thickness of flange

t_p = thickness of plate or panel zone including doubler plates

t_w = thickness of web

w = uniform beam gravity load

w_z = width of panel zone between column flanges

x = parameter used for determining the approximate fundamental period

$\bar{x}$ = connection eccentricity

Δ = first-order interstory drift due to the design loads

Δ_b = deformation quantity used to control loading of test specimen (total brace end rotation for the subassemblage test specimen; total brace axial deformation for the brace test specimen)

Δ_{bm} = value of deformation quantity, Δ_b, corresponding to the design story drift

Δ_{by} = value of deformation quantity, Δ_b, at first significant yield of test specimen

Δ_H = first-order interstory drift due to lateral forces

Ω_o = system overstrength factor

α = angle defining the orientation of reinforcement, ACI-318

α = angle of web yielding in radians, as measured relative to the vertical

α_c = coefficient defining the relative contribution of concrete strength to nominal wall shear strength, ACI-318

β = compression strength adjustment factor

β_1 = factor relating depth of equivalent rectangular compressive stress block to neutral axis depth, ACI-318

β_{br} = required brace stiffness

γ_{total} = link rotation angle

δ_u = design displacement, ACI-318

ε_t = net tensile strain in extreme layer of longitudinal tension steel at nominal strength, excluding strains due to effective prestress, creep, shrinkage, and temperature, ACI-318

θ = interstory drift angle, radians

λ = modification factor related to unit weight of concrete, ACI-318

λ_p = limiting slenderness parameter for compact element

λ_p, λ_{ps} = limiting slenderness parameter for compact element

ΣM^*_{pb} = moment at the intersection of the beam and column centerlines determined by projecting the beam maximum developed moments from the column face. Maximum developed moments shall be determined from test results

ΣM^*_{pc} = moment at beam and column centerline determined by projecting the sum of the nominal column plastic moment strength, reduced by the axial stress P_{uc}/A_g, from the top and bottom of the beam

ϕ = resistance factor

ϕ_b = resistance factor for flexure

ϕ_c = resistance factor for compression

ϕ_d = resistance factor for ductile limit states

ϕ_n = resistance factor for non-ductile limit states

ϕ_t = resistance factor for tension

ϕ_v = resistance factor for shear

ψ_e = factor used to modify development length based on reinforcement coating, ACI-318

ψ_s = factor used to modify development length based on reinforcement size, ACI-318

ψ_t = factor used to modify development length based on reinforcement location, ACI-318

ω = strain hardening adjustment factor

Definitions

Adjusted Brace Strength. Strength of a brace in a buckling-restrained braced frame at deformations corresponding to 2.0 times the design story drift.

Amplification Factor. Multiplier of the results of first-order analysis to reflect second-order effects.

Amplified Seismic Load. Horizontal component of earthquake load E multiplied by Ω_o, where E and the horizontal component of E are specified in the applicable building code.

Backing. Piece of metal or other material, placed at the weld root to facilitate placement of the root pass.

Base. The level at which the horizontal seismic ground motions are considered to be imparted to the structure.

Base Shear. Total design lateral force or shear at the base.

Boundary Elements. Diaphragm and shear wall boundary members to which the diaphragm transfers forces.

Boundary Members. Portions along wall and diaphragm edges strengthened by longitudinal and transverse reinforcement and/or structural steel members.

Buckling-restrained Braced Frame (BRBF). Diagonally braced frame satisfying the requirements of Section 16 of AISC 341 in which all members of the bracing system are subjected primarily to axial forces and in which the limit state of compression buckling of braces is precluded at forces and deformations corresponding to 2.0 times the design story drift.

Charpy V-Notch Impact Test. Standard dynamic test measuring notch toughness of a specimen.

Complete-joint-penetration Groove Weld (CJP). Groove weld in which weld metal extends through the joint thickness, except as permitted for HSS connections.

Concrete. Mixture of portland cement or any other hydraulic cement, fine aggregate, coarse aggregate, and water, with or without admixtures.

Confined Region. The portion of a reinforced concrete component in which the concrete is confined by closely spaced special transverse reinforcement restraining the concrete in directions perpendicular to the applied stress.

Continuity Plates. Column stiffeners at the top and bottom of the panel zone; also known as transverse stiffeners.

Coupling Beam. A beam that is used to connect adjacent concrete wall piers to make them act together as a unit to resist lateral forces.

Dead Loads. The weight of materials of construction incorporated into the building, including but not limited to wall, floors, roofs, ceilings, stairways, built-in partitions, finishes, cladding, and other similarly incorporated architectural and structural items, and fixed service equipment, incuding the weight of cranes.

Design Story Drift. Amplified story drift (drift under the design earthquake, including the effects of inelastic action), determined as specified in the applicable building code.

Design Strength. The product of the nominal strength and a resistance factor (or strength reduction factor).

Design Wall Thickness. HSS wall thickness assumed in the determination of section properties.

Development Length. Length of embedded reinforcement, including pretensioned strand, required to develop the design strength of reinforcement at a critical section.

Doubler. Plate added to, and parallel with, a beam or column web to increase resistance to concentrated forces.

Drift. Lateral deflection of structure.

Ductile Limit State. Ductile limit states include member and connection yielding, bearing deformation at bolt holes, as well as buckling of members that conform to the width-thickness limitations of Table I-8-1 of the *Seismic Provisions*. Fracture of a member or of a connection, or buckling of a connection element, is not a ductile limit state.

Effective Net Area. Net area modified to account for the effect of shear lag.

Embedment Length. Length of embedded reinforcement provided beyond a critical section.

Expected Yield Strength. Yield strength in tension of a member, equal to the expected yield stress multiplied by A_g.

Expected Tensile Strength. Tensile strength of a member, equal to the specified minimum tensile strength, F_u, multiplied by R_t.

Expected Yield Stress. Yield stress of the material, equal to the specified minimum yield stress, F_y, multiplied by R_y.

Factored Load. The product of a load factor and the nominal load.

Filler Metal. Metal or alloy to be added in making a welded joint.

Fillet Weld. Weld of generally triangular cross section made between intersecting surfaces of elements.

Fillet Weld Reinforcement. Fillet welds added to groove welds.

Flat Width. Nominal width of a rectangular HSS minus twice the outside corner radius. In absence of knowledge of the corner radius, the flat width may be taken as the total section width minus three times the thickness.

Frame.

Braced frame. An essentially vertical truss, or its equivalent, of the concentric or eccentric type that is provided in a building frame system or dual frame system to resist shear.

Concentrically braced frame (CBF). A braced frame in which the members are subjected primarily to axial forces.

Eccentrically braced frame (EBF). A diagonally braced steel frame meeting the requirements of Section 15 of AISC 341 that has at least one end of each bracing member connected to a beam a short distance from another beam-to-brace connection or a beam-to-column connection.

Ordinary concentrically braced frame (OCBF). A diagonally braced steel frame meeting the requirements of Section 14 of the *Seismic Provisions* in which all members of the bracing system are subjected primarily to axial forces. Members and connections are designed for moderate ductility.

Special concentrically braced frame (SCBF). A diagonally braced frame meeting the requirements of Section 13 of the *Seismic Provisions* in which all members of the bracing system are subjected primarily to axial forces.

Frame System.

Building frame system. A structural system with an essentially complete space frame system providing support for vertical loads. Seismic force resistance is provided by shear walls or braced frames.

Dual frame system. A structural system with an essentially complete space frame system providing support for vertical loads. Seismic force resistance is provided by a moment-resisting frame and shear walls or braced frames.

Space frame system. A structural system composed of interconnected members, other than bearing walls, that is capable of supporting vertical loads and that also may provide resistance to seismic forces.

Fully Restrained Moment Connection. Connection capable of transferring moment with negligible rotation between connected members.

Gouge. Relatively smooth surface groove or cavity resulting from plastic deformation or removal of material

Gravity Frame. Portion of the framing system not included in the lateral load resisting system.

Gravity Load (*W*). The total dead load and applicable portions of other loads as defined in ASCE/SEI 7-05, §§12.7.2 and 12.14.8.1.

Gusset Plate. Plate element connecting truss members or a strut or brace to a beam or column.

Importance Factor. A factor assigned to each structure according to its occupancy category as prescribed in ASCE/SEI 7-05, §11.5.1.

Inelastic Analysis. Structural analysis that takes into account inelastic material behavior, including plastic analysis.

Interstory Drift Angle. Interstory displacement divided by story height, radians.

Joint (concrete). A portion of a column bounded by the highest and lowest surfaces of the other members framing into it.

***k*-area.** The region of the web that extends from the tangent point of the web and the flange-web fillet (AISC "*k*" dimension) a distance $1^1/2$ inches (38 mm) into the web beyond the "*k*" dimension.

***K*-braced Frame.** A bracing configuration in which braces connect to a column at a location with no diaphragm or other out-of-plane support.

Limit State. A condition beyond which a structure or member becomes unfit for service and is judged to be no longer useful for its intended function (serviceability limit state) or to be unsafe (strength limit state).

Link. In EBF, the segment of a beam that is located between the ends of two diagonal braces or between the end of a diagonal brace and a column. The length of the link is defined as the clear distance between the ends of two diagonal braces or between the diagonal brace and the column face.

Link Intermediate Web Stiffeners. Vertical web stifffeners placed within the link in EBF.

Link Rotation Angle. Inelastic angle between the link and the beam outside the link when the total story drift is equal to the design story drift.

Link Shear Design Strength. Lesser of the available shear strength of the link developed from the moment or shear strength of the link.

Live Loads. Those loads produced by the use and occupancy of the building or other structure and do not include construction or environmental loads such as wind load, snow load, rain load, earthquake load, flood load, or dead load.

Live Loads (Roof). Those loads produced 1) during maintenance by workers, equipment, and materials; and 2) during the life of the structure by movable objects such as planters and by people.

Load and Resistance Factor Design (LRFD). A method of proportioning structural members and their connections using load and resistance factors such that no applicable limit state is reached when the structure is subjected to appropriate load combinations. The term "LRFD" is used in the design of steel and wood structures.

Load Factor. A factor that accounts for deviations of the actual load from the nominal load, for uncertainties in the analysis that transforms the load into a load effect, and for the probability that more than one extreme load will occur simultaneously.

Loads. Forces or other actions that result from the weight of building materials, occupants and their possessions, environmental effect, differential movement, and restrained dimensional changes. Permanent loads are those loads in which variations over time are rare or of small magnitude. Other loads are variable loads (see also "Nominal loads").

Loads Effects. Forces and deformations produced in structural members by the applied loads.

Local Buckling. Limit state of buckling of a compression element within a cross section.

Local Crippling. Limit state of local failure of web plate in the immediate vicinity of a concentrated load or reaction.

Local Yielding. Yielding that occurs in a local area of an element.

LRFD Load Combination. Load combination in the applicable building code intended for strength design (load and resistance factor design).

Maximum Considered Earthquake. The most severe earthquake effects considered by this code.

Moment Connection. Connection that transmits bending moment between connected members.

Moment Frame. Framing system that provides resistance to lateral loads and provides stability to the structural system; members and joints are capable of resisting forces by flexure as well as along their axis.

Net Area. Gross area reduced to account for removed material.

Nominal Load. Magnitude of the load specified by the applicable building code (dead, live, soil, wind, snow, rain, flood, earthquake).

Nominal Strength. Strength of a structure or component (without the resistance factor or safety factor applied) to resist the load effects.

Notch Toughness. Energy absorbed at a specified temperature as measured in the Charpy V-Notch test.

Occupancy Category. Classification assigned to a structure based on its use as specified by the applicable building code.

Overstrength Factor, Ω_o. Factored specified by the applicable building code in order to determine the amplified seismic load, where required by the *Seismic Provisions*.

Panel Zone. Web area of beam-to-column connection delineated by the extension of beam and column flanges through the connection, transmitting moment through a shear panel.

Partial-joint-penetration Groove Weld (PJP). Groove weld in which the penetration is intentionally less than the complete thickness of the connected element.

Percent Elongation. Measure of ductility, determined in a tensile test as the maximum elongation of the gage length divided by the original gage length.

***P*-Delta Effect.** Effect of loads acting on the displaced location of joints or nodes in a structure. In tiered building structures, this is the effect of loads acting on the laterally displaced location of floors and roofs.

***P*-δ Effect.** Effect of loads acting on the deflected shape of a member between joints or nodes.

Plastic Hinge. Yielded zone that forms in a structural member when the plastic moment is attained. The member is assumed to rotate further as if hinged, except that such rotation is restrained by the plastic moment.

Plastic Hinge Location. Location in a beam column assembly where inelastic energy dissipation is assumed to occur through the development of plastic flexural straining.

Plastic Moment. Theoretical resisting moment developed within a fully yielded cross section.

Prequalified Connection. Connection that complies with the requirements of the *Seismic Provisions* or AISC 358.

Probable Maximum Moment at Plastic Hinge. Expected moment developed at a plastic hinge location along a member, considering the probable (mean) value of the material strength for the specified steel and effects of strain hardening.

Protected Zone. Area of members in which limitations apply to fabrication and attachments. See Section 7.4 of the *Seismic Provisions*.

Prototype. The connection or brace design that is to be used in the building (SMF, IMF, EBF, and BRBF).

Quality Assurance. System of shop and field activities and controls implemented by the owner or his/her designated representative to provide confidence to the owner and the building authority that quality requirements are implemented.

Quality Assurance Plan. Written description of qualifications, procedures, quality inspections, resources, and records to be used to provide assurance that the structure complies with the engineer's quality requirements, specifications, and contract documents.

Quality Control. System of shop and field controls implemented by the fabricator and erector to ensure that contract and company fabrication and erection requirements are met.

Reduced Beam Section. Reduction in cross section over a discrete length that promotes a zone of inelasticity in the member.

Reinforcing Fillet. Fillet weld applied to a groove welded "tee joint" to obtain a contour to reduce stress concentrations associated with joint geometry.

Required Strength. Forces, stresses, and deformations produced in a structural component, determined by either structural analysis, LRFD or ASD load combinations or by the *Specification* and *Seismic Provisions*.

Resistance Factor. A factor that accounts for deviations of the actual strength from the nominal strength and the manner and consequences of failure (also called strength reduction factor).

Root. Portion of a multi-pass weld deposited in the first pass of welding.

Root of Joint. Portion of a joint to be welded where the members are closest to each other.

Rotation Capacity. Incremental angular rotation that a given shape can accept prior to excessive load shedding, defined as the ratio of the inelastic rotation attained to the idealized elastic rotation at first yield.

Second-order Analysis. Structural analysis in which equilibrium conditions are formulated on the deformed structure; second-order effects (both P-δ and P-Δ, unless specified otherwise) are included.

Seismic Design Category. A classification assigned to a structure based on its occupany category and the severity of the design earthquake ground motion at the site.

Seismic-force-resisting System. The part of the structural system that has been considered in the design to provide the required resistance to the prescribed seismic forces.

Seismic Forces. The assumed forces related to the response of the structure to earthquake motions, to be used in the design of the structure and its components.

Seismic Load Resisting System (SLRS). Assembly of structural elements in the building that resists seismic loads, including struts, collectors, chords, diaphragms and trusses.

Seismic Provisions. Refers to AISC *Seismic Provisions for Structural Steel Buildings* (ANSI/AISC).

Seismic Response Modification Coefficient, *R*. Factor that reduces seismic load effects to strength level as specified by the applicable building code.

Seismic Response Coefficient. Coefficient C_s, as determined from ASCE/SEI 7-05, §12.8.

Shear Rupture. Limit state of rupture (fracture) due to shear.

Shear Wall. Wall that provides resistance to lateral loads in the plane of the wall and provides stability for the structural system.

Shear Wall-frame Interactive System. A structural system that uses combinations of shear walls and frames designed to resist lateral forces in proportion to their rigidities, considering interaction between shear walls and frames on all levels.

Shear Yielding. Yielding that occurs due to shear.

Simple Connection. Connection that transmits negligible bending moment between connected members.

Site Class. A classification assigned to a site based on the types of soils present and their engineering properties as defined in ASCE/SEI 7-05, §11.4.2.

Site Coefficients. The values of F_a and F_v indicated in ASCE/SEI 7-05, Tables 11.4-1 and 11.4-2, respectively.

Special Plate Shear Wall (SPSW). Plate shear wall systems that meets the requirments of the *Seismic Provisions.*

Special Reinforced Concrete Shear Wall. A cast-in-place wall complying with the requirements of ASCE/SEI 7-05 in addition to the requirements for ordinary reinforced concrete structural walls.

Special Transverse Reinforcement. Reinforcement composed of spirals, closed stirrups or hoops, and supplementary cross-ties provided to restrain the concrete and qualify the portion of the component, where used, as a confined region.

Specification. Refers to the AISC *Specification for Structural Steel Buildings* (ANSI/AISC 360).

Specified Minimum Tensile Strength. Lower limit of tensile strength specified for a material as defined by ASTM.

Specified Minimum Yield Stress. Lower limit of yield stress specified for a material as defined by ASTM.

Splice. Connection between two structural elements joined at their ends to form a single, longer element.

Steel Core. Axial-force-resisting element of braces in BRBF. The steel core contains a yielding segment and connections to transfer its axial force to adjoining elements; it may also contain projectons beyond the casing and transition segments between the projections and yielding segment.

Stiffener. Structural element, usually an angle or plate, attached to a member to distribute load, transfer shear or prevent buckling.

Stiffness. Resistance to deformation of a member or structure, measured by the ratio of the applied force (or moment) to the corresponding displacement (or rotation).

Story Drift Ratio. The story drift divided by the story height.

Strength Design. A method of proportioning structural members such that the computed forces produced in the members by factored loads do not exceed the member design strength (also called load and resistance factor design.) The term "strength design" is used in the design of concrete and masonry structural elements.

Strength, Nominal. The capacity of a structure or member to resist the effects of loads, as determined by computations using specified material strengths and dimensions and formulas derived from accepted principles of structural mechanics or by field tests or laboratory tests of scaled models, allowing for modeling effects and differences between laboratory and field conditions.

Strength Required. Strength of a member, cross section, or connection required to resist factored loads or related internal moments and forces in such combinations as stipulated by these provisions.

Structural Analysis. Determination of load effects on members and connections based on principles of structural mechanics.

Tensile Rupture. Limit state of rupture (fracture) due to tension.

Tensile Yielding. Yielding that occurs due to tension.

Toe of Fillet. Junction of a fillet weld face and base metal. Tangent point of a rolled section fillet.

Torsional Force Distribution. The distribution of horizontal seismic forces through a rigid diaphragm when the center of mass of the structure at the level under consideration does not coincide with the center of rigidity (sometimes referred to as a diaphragm rotation).

Toughness. The ability of a material to absorb energy without losing significant strength.

V-Braced Frame. Concentrically braced frame (SCBF, OCBF or BRBF) in which a pair of diagonal braces located either above or below a beam is connected to a single point within the clear beam span. Where the diagonal braces are below the beam, the system is also referred to as an inverted-V-braced frame.

Wall, Load-bearing. Any wall meeting either of the following classifications:

1. Any metal or wood stud wall that supports more than 100 pounds per linear foot (1459 N/m) of vertical load in addition to its own weight.
2. Any masonry or concrete wall that supports more than 200 pounds per linear foot (2919 N/m) of vertical load in addition to its own weight.

Wall, Nonload-bearing. Any wall that is not a load-bearing wall.

Web Buckling. Limit state of lateral instability of a web.

Web Compression Buckling. Limit state of out-of-plane compression buckling of the web due to a concentrated compression force.

Weld Metal. Portion of a fusion weld that has been completely melted during welding. Weld metal has elements of filler metal and base metal melted in the weld thermal cycle.

Weld Tab. Piece of metal affixed to the end of a welded joint to facilitate the initiation and termination of weld passes outside the structural joint.

X-braced Frame. Concentrically braced frame (OCBF or SCBF) in which a pair of diagonal braces crosses near the mid-length of the braces.

Yield Strength. Stress at which a material exhibits a specified limiting deviation from the proportionality of stress to strain as defined by ASTM.

References

The following codes and standards are referenced in this document. Other reference documents are indicated at the end of each design example.

ACI-318, American Concrete Institute, *Building Code Regulations for Reinforced Concrete*, Farmington Hills, Michigan, 2008.

AISC-360, American Institute of Steel Construction, *Specificaiton for Structural Steel Buildings*, Chicago, Illinois, March 9, 2005.

AISC-341, *Seismic Provisions for Structural Steel Buildings*, March 9, 2005.

IBC, International Code Council, *International Building Code*. Washington, DC, 2009.

SEAOC Blue Book, *Recommended Lateral Force Requirements and Commentary*. Structural Engineers Association of California, Sacramento, California, 1999.

Design Example 1
Steel Concentrically Braced Frames

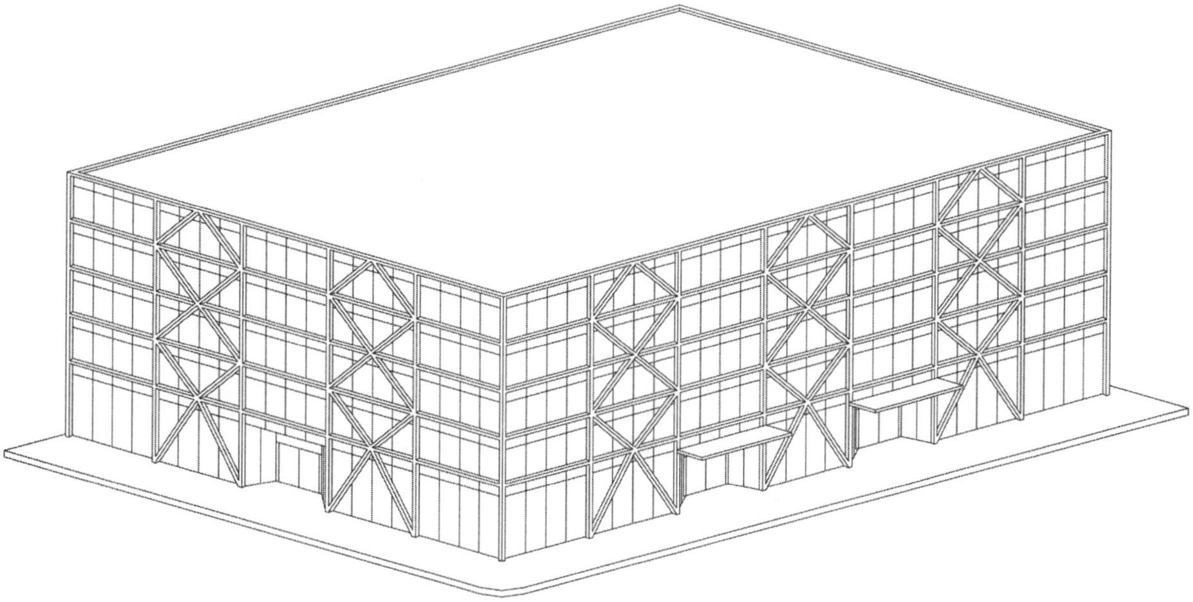

Overview

This example shows procedures for the design of several types of concentrically braced-frame buildings. It is intended to provide specific methods for the design of braced frames that comply with the *International Building Code* and the *AISC Seismic Provisions for Structural Steel Buildings* (AISC 341), guiding designers toward the careful consideration of the performance of concentrically braced frame structures under severe seismic loading. Certain recommendations are given that are different from current practice; nevertheless, the calculation methods illustrated are applicable to a wide range of designs.

Within the broad category of concentrically braced frames, there exist three distinct seismic systems: Special Concentrically Braced Frames, Buckling-Restrained Braced Frames, and Ordinary Concentrically Braced Frames. All three are addressed in this design example. In addition to this variety of systems, there is also a range of brace configurations. Each configuration has special design considerations and procedures. The three most common configurations, the two-story-X, the chevron, and the cross-bracing configurations, are addressed in this design example. The other two configurations, zipper bracing and single-diagonal bracing, are briefly discussed. The configurations permitted for braced frames by AISC 341 are shown in Figure 1-1. Cross-bracing is not used in Buckling-Restrained Braced Frames.

The most common concentrically braced frame system for seismic applications is the special concentrically braced frame (SCBF). SCBFs have moderate ductility and are accordingly designed for moderate strength. The design of an SCBF building is illustrated in Design Example 1A. Design Example 1B describes the special considerations required for the design of SCBF using the chevron configuration; it also describes the design of zipper-braced frames. The two-story X configuration offers better performance (and economy) than the chevron configuration. Design Example 1C describes the special considerations required for the design of SCBF using the cross-bracing configuration, as well as for the single-diagonal configuration.

The Buckling-Restrained Braced Frame (BRBF) is a new type introduced relatively recently into United States design practice. BRBF have high ductility and therefore their required strength is lower than that of SCBF. Design Example 1D shows the differences between BRBF and SCBF design; it is a supplement to Design Example 1A in that the aspects of design that are not specific to BRBF are not repeated.

Ordinary Concentrically Braced Frames (OCBF) are distinguished from SCBF by having fewer seismic-design requirements. The use of the OCBF is severely restricted under the IBC 2009 provisions. OCBF have low ductility and are designed for higher strength than SCBF. Design Example 1E shows the differences between OCBF and SCBF design; like Design Example 1D, Design Example 1E is a supplement to Design Example 1A.

The IBC adopts the AISC 341 (2005) by reference. AISC 341 contains provisions for the design of SCBF, BRBF, and OCBF. AISC 341 (2005 Edition) is the basis of this design.

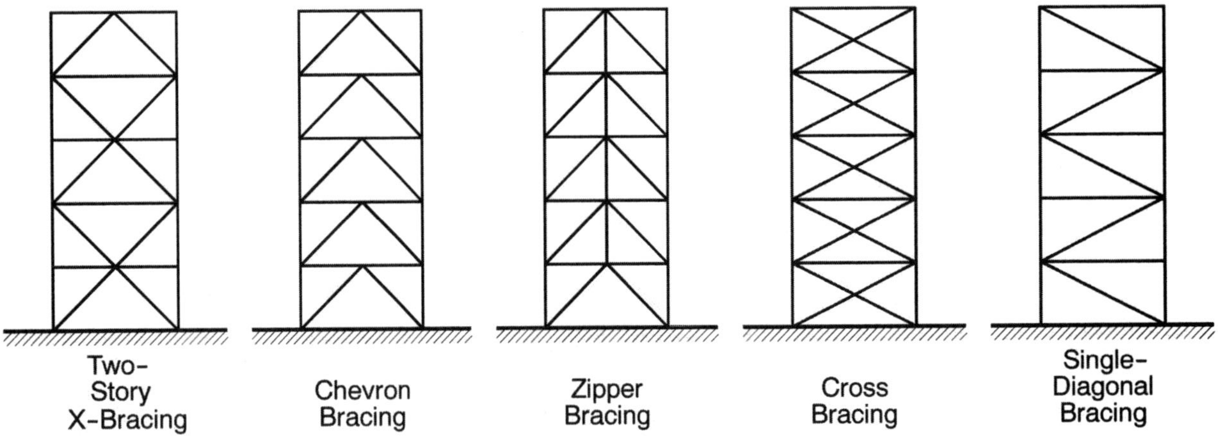

Figure 1-1. Bracing configurations permitted for concentrically braced frames

Outline

Design Example 1A
Special Concentrically Braced Frame

Overview

The category of "Special Concentrically Braced Frame" (SCBF) first appeared as a separate category in the *1994 Uniform Building Code* (ICBO, 1994). This seismic system is distinguished from other concentrically braced frames [now referred to as "Ordinary Concentrically Braced Frames" (OCBF)] by design and detailing requirements specifically directed at maximizing the ductility that braces can provide and preventing brittle limit states. This approach to the design of braced frames grew out of research at the University of Michigan (Astaneh et al., 1985; Hassan and Goel, 1991; Tang and Goel, 1989; Goel, 1992) and the 1992 *AISC Seismic Provisions for Structural Steel Buildings* (AISC LRFD 1992). For a description of the studies, see the SEAOC *Recommended Provisions* (SEAOC, 1999) and Tremblay, 2001.

In their elastic state, SCBF behave as vertical cantilever trusses. As such, columns, beams, and braces are subjected to axial forces. Yield modes involve either axial inelasticity of some of these elements or fracture of the connections. Of these possibilities, buckling and tensile yielding of the braces offer the greatest ductility and least damage to the gravity-load-resisting system, and design provisions for SCBF were created with the intent of confining ductility demands to the braces. Braces are expected to undergo several cycles of buckling without fracture, and to reach strains beyond tension yield.

For this example, the two-story *X* configuration is considered; this configuration is shown in Figure 1-1. The five-story steel frame building shown schematically in Figure 1A-1 is to have special concentrically braced frames for its lateral-force-resisting system. The floor and roof diaphragms consist of lightweight concrete fill over steel decking. A typical plan for the building is shown in Figure 1A-2. An SCBF frame elevation is shown in Figure 1A-3.

For more information, see SEAOC Blue Book article 08.03.050.

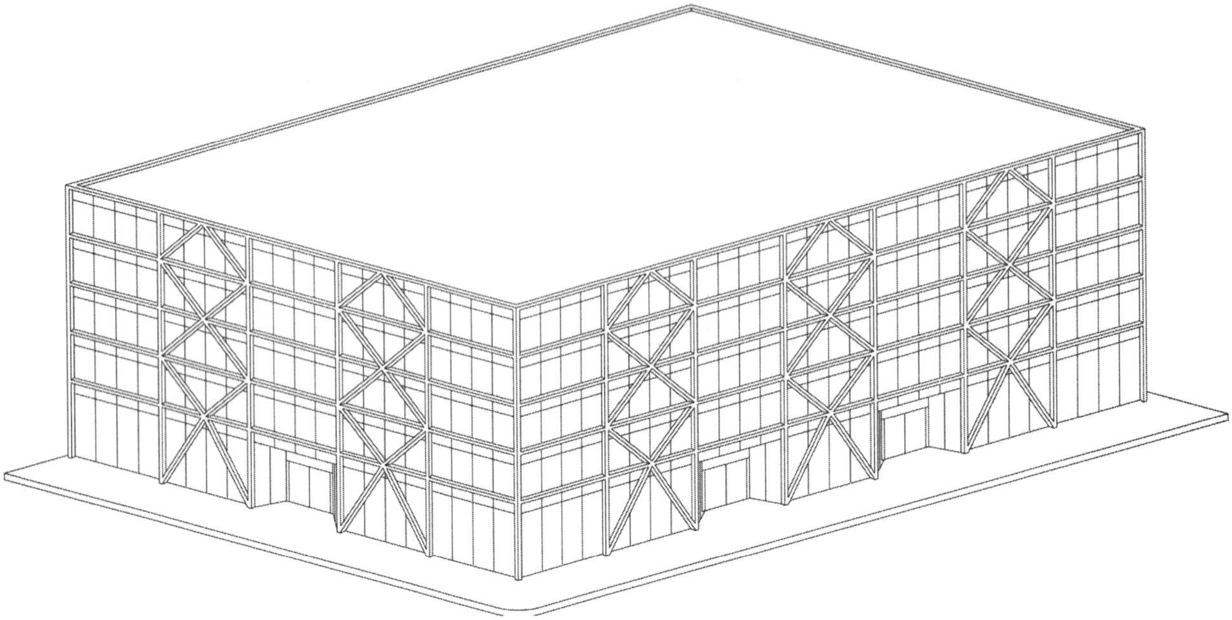

Figure 1A-1. Special concentrically braced frame building

Outline

1. Determination of site ground motion
2. Determination of design base shear coefficient and vertical seismic component
3. Determination of design base shear and vertical distribution of shear
4. Determination of horizontal distribution of shear
5. Determination of reliability/redundancy factor
6. Determination of combined effect of horizontal and vertical earthquake induced forces
7. SCBF member design
8. SCBF connection design
9. Notes on collectors and foundations
10. Additional requirements

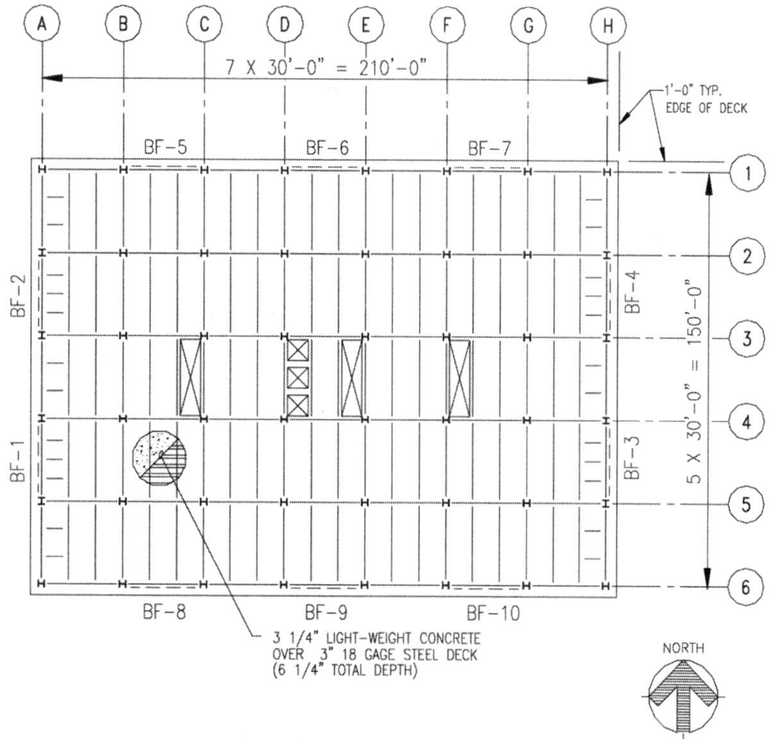

Figure 1A-2. Typical floor and roof framing plan

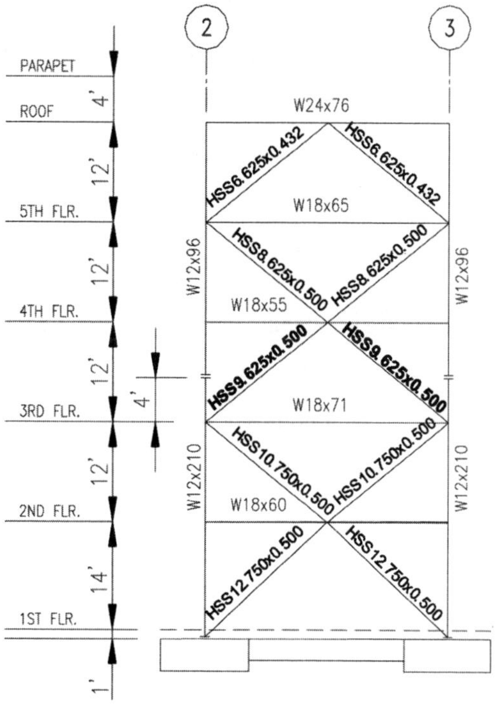

Figure 1A-3. Elevation of frame BF-2

Building Information

Roof weights:

Roofing	6.0 psf
Insulation	3.0
Steel deck + fill	47.0
Roof framing	8.0
Partition walls (10 psf)	5.0 seismic
Ceiling	3.0
Mech/Elec	2.0
Total	74.0 psf

Floor weights:

Floor covering	1.0 psf
Steel deck + fill	47.0
Framing (beams + cols.)	13.0
Partition walls	10.0 seismic (per ASCE §12.7.2)
Ceiling	3.0
Mech/Elec	2.0
Total	76.0 psf

Exterior curtain wall, steel studs, gypsum board, EIFS skin, weight: 20.0 psf

Live loads:

Roof	20 psf
Floor	50 psf

Seismic and site data:

Mapped spectral response accelerations (5% critical damping, Site Class D):

$S_S = 2.1g$ at 0.2-second response §22.0

$S_1 = 0.93g$ at 1.0-second response

$I = 1.0$ (standard occupancy) §11.5.1

Site Class D (§1613.5)

Structural Materials:

Wide flange shapes and plates	ASTM A992, Grade 50 (F_y = 50 ksi)
Hollow Structural Steel (round)	ASTM A500, Grade B (F_y = 42 ksi)
Plate	ASTM A572, Grade 50 (F_y = 50 ksi)
Weld electrodes	E70XX (20 ft-lb at –0° Charpy V-notch toughness for all connections in the seismic-load-resisting system)
Lightweight concrete fill	f'_c = 3000 psi

Calculations and Discussion — **Code Reference**

1. Site ground motion §11.4

The maximum considered earthquake spectral response accelerations for short periods, S_{MS}, and at 1.0-second period, S_{M1}, adjusted for site class effects, are determined by Equations 11.4-2 and 11.4-4.

$F_a = 1.0$ — T 11.4-1

$F_v = 1.5$ — T11.4-2

$S_{MS} = F_a S_S = (1.0)(2.1g) = 2.1g$ — Eq 11.4-1

$S_{M1} = F_v S_1 = (1.5)(0.93g) = 1.4g$ — Eq 11.4-2

The 5-percent-damped design spectral response accelerations at short periods, S_{DS}, and at 1.0-second period, S_{D1}, adjusted for site class effects, are determined by Equations 16-18 and 16-19.

$S_{DS} = 2/3\ S_{MS} = (2/3)(2.1g) = 1.40g$ — Eq 11.4-3

$S_{D1} = 2/3\ S_{M1} = (2/3)(1.4g) = 0.93g$ — Eq 11.4-4

2. Design base shear coefficient and vertical seismic component §§12.2, 12.8

The coefficients for a special concentrically braced framed building are

Response modification coefficient, $R = 6.0$ — T 12.2-1

System overstrength factor, $\Omega_o = 2.0$

Deflection amplification factor, $C_d = 5.0$

The approximate fundamental building period can be computed from the building height (62 feet) and the structural system

$T_a = C_t h_n^{3/4} = 0.020(62^{3/4}) = 0.44$ sec — Eq 12.8-7

The seismic base shear, V, in a given direction is determined in accordance with the following equation

$V = C_s W$ — Eq 12.8-1

The calculation of seismic response coefficient C_s is performed using four equations. These equations, when presented together in a graph of the base shear coefficient versus building period, constitute the response spectrum for the site; this is shown in Figure 1A-4. The first equation defines the coefficient as

$$C_s = \frac{S_{D_S}}{R/I} = \frac{1.40g}{(6.0/1.0)} = 0.233g \qquad \text{Eq 12.8-2}$$

The coefficient need not exceed

$$C_s = \frac{S_{D1}}{(R/I)T} = \frac{0.93g}{(6.0/1.0)0.44\text{ sec}} = 0.352g \qquad \text{Eq 12.8-3}$$

The value of seismic response coefficient, C_s, must be at least equal to the following

$$C_s = 0.044 S_{DS} I > 0.01 \qquad \text{Eq 12.8-5, Supplement No. 2}$$

$$= 0.044(1.4)(1.0) = 0.0616g$$

In addition, for those buildings for which the 1.0-second spectral response, S_1, is equal to or greater than 0.6g, the value of the seismic response coefficient, C_s, must be at least equal to the following value.

$$C_s = \frac{0.5S_1}{(R/I)} = \frac{0.5(0.93g)}{(6.0/1.0)} = 0.078g \qquad \text{Eq 12.8-6}$$

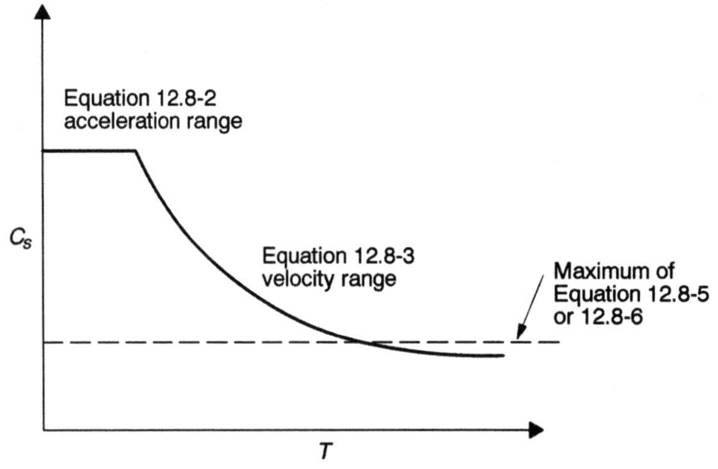

Figure 1A-4. Typical response spectrum

Equation 12.8-2 controls the base shear calculation for a short-period building, and the seismic coefficient is thus

$$V = 0.233\ W$$

The vertical component of seismic acceleration is also computed using the spectrum, although the vertical period is assumed to be very short. The vertical component is defined as part of the seismic load or amplified seismic load

$E = \rho\, Q_E \pm 0.2 S_{DS} D$

$E = \rho\, Q_E \pm 0.2(1.40)D$ Eq 12.4-1, 12.4-2, 12.4-3, 12.4-4

$E = \rho\, Q_E \pm 0.28D$

$E_m = \Omega_o Q_E \pm 0.2 S_{DS} D$ Eq 12.4-1, 12.4-2, 12.4-4, 12.4-7

$E_m = 2.0 Q_E \pm 0.28D$

Q_E = Effect of horizontal seismic forces from base shear V

3. Design base shear and the vertical distribution of shear

First, the building mass is calculated. The floor area at each level is 32,224 square feet. The perimeter of the exterior curtain wall is 728 feet. The roof parapet height is 4 feet. The curtain-wall weights are distributed to each floor by tributary height. (Although this is slightly inaccurate in the case of the top story due to the parapet, the inaccuracy is negligible for the vertical distribution of seismic forces.)

The building mass calculation is shown in Table 1A-1.

Table 1A-1. Building mass calculation

Level	Floor Area (sf)	Floor Weight (psf)	Floor Weight (kips)	Exterior Wall Length (ft)	Exterior Wall Height (ft)	Exterior Wall Weight (psf)	Exterior Wall Weight (kips)	Story Weight W_i (kips)
Roof	32,224	74	2,385	728	10	20	146	2,530
5	32,224	76	2,449	728	12	20	175	2,624
4	32,224	76	2,449	728	12	20	175	2,624
3	32,224	76	2,449	728	12	20	175	2,624
2	32,224	76	2,449	728	13	20	189	2,638
Totals	161,120		12,181				860	13,040

3a. Design base shear §12.8

Using the design base shear coefficient from Part 1, the base shear is

$$V = 0.233\ W = 0.233(13{,}040 \text{ kips}) = 3040 \text{ kips}$$

3b. Vertical distribution of shear §12.8.3

The total lateral force (i.e., design base shear) is distributed over the height of the building in accordance with §12.8.3. The following equations apply.

$$F_x = C_{vx}V \qquad \text{Eq 12.8-11}$$

$$C_{vx} = \frac{w_x h_x^k}{\sum_{i=1}^{n} w_x h_x^k} \qquad \text{Eq 12.8-12}$$

k is a distribution exponent related to building period as follows

$k = 1$ for buildings with a period of 0.5 second or less

$k = 2$ for buildings with a period of 2.5 seconds or greater

$k =$ is interpolated between these values for buildings with period between 0.5 second and 2.5 seconds

Using the building mass tabulated in Table 1A-1, the vertical distribution of shear is determined as shown in Table 1A-2.

Table 1A-2. Vertical distribution of shear

Level	w_x (kips)	Σw (kips)	h_x (ft)	h (ft)	$w_x h_x$ (kip-ft)	$\frac{w_x h_x}{\Sigma w_x h_x}$ (%)	F_x (kips)	ΣV_x (kips)
R	2,530	2,530	62	12	156,860	31.8	967	967
5	2,624	5,154	50	12	131,200	26.6	809	1,776
4	2,624	7,778	38	12	99,712	20.2	614	2,390
3	2,624	10,402	26	12	68,224	13.8	420	2,810
2	2,638	13,040	14	14	36,932	7.6	231	3,041
Totals	13,040				492,928	100	3,041	

4. Horizontal distribution of shear §12.8.4

Although the centers of mass and rigidity coincide, §12.8.4.2 requires inclusion of an accidental torsional eccentricity, e, equal to 5 percent of the building dimension perpendicular to the direction of force.

$e_{ew} = (0.05)(152 \text{ ft}) = 7.6 \text{ ft}$ (east-west)

$e_{ns} = (0.05)\ (212 \text{ ft}) = 10.6 \text{ ft}$ (north-south)

It is assumed that the four frames in the transverse direction are each 25 percent stiffer than the six in the longitudinal direction because they are more heavily loaded. This assumption can be refined in a subsequent analysis, after members have been sized and an elastic analysis has been completed. Many designers estimate the torsional contribution for a symmetric building by adding 5 to 10 percent to the element forces. However, in this example the numerical application of the code provisions will be shown.

Table 1A-3 gives the distribution of direct shear and torsional shear components as percentages of shear force (based on geometry).

Table 1A-3. Calculation of direct shear plus torsion as percentage of story shear in the transverse direction

Frame ID	d (ft)	R_i	$R_i\,d_i$ (ft)	$R_i\,d_i^2$ (ft^2)	$R_{iy}/\Sigma R_y$ (%)	$T_y/\Sigma V_y$ (real) (%)	$\lvert T_y\rvert/\Sigma V_y$ (acc.) (%)	$V_i/\Sigma V_y$ (%)
Transverse Frames								
1	105	1.25	131	13,800	25	0	1.6	26.6
2	105	1.25	131	13,800	25	0	1.6	26.6
3	−105	1.25	−131	13,800	25	0	1.6	26.6
4	−105	1.25	−131	13,800	25	0	1.6	26.6
Longitudinal Frames								
5	75	1	75	5,630	0	0	0.7	0.7
6	75	1	75	5,630	0	0	0.7	0.7
7	75	1	75	5,630	0	0	0.7	0.7
8	−75	1	−75	5,630	0	0	0.7	0.7
9	−75	1	−75	5,630	0	0	0.7	0.7
10	−75	1	−75	5,630	0	0	0.7	0.7
Totals				88,980	100			

Notes:

1. Values d are distances from the center of mass (i.e., the center of the building) to frames in the X (east-west) and Y (north-south) directions, respectively.
2. V_y is direct shear in the Y direction.
3. T_y is shear force on frames due to building torsion caused by Y-direction (north-south) seismic loads.
4. It is convenient to separate accidental torsion from real torsion. Shears due to real torsion are subtracted from or added to the direct shear depending on the direction of the torsional moment. Accidental torsion is always added to direct shear. Accidental torsion is based on a 5% offset of the building mass (§12.8.4.2).

The calculation of direct shear plus torsion for a given frame is based on the following formula.

$$V_{iy} = \Sigma V_y \left(\frac{R_{iy}}{\Sigma R_{iy}} + \frac{e_{real} R_i d_i}{\Sigma R_i d_i^2} + \left| \frac{e_{acc} R_i d_i}{\Sigma R_i d_i^2} \right| \right)$$

Based on the direct and torsional shear values tabulated in Table 1A-3, and on the vertical distribution of shear tabulated in Table 1A-2, the story forces to be used for design of the typical braced frame (BF-2) are as follows.

Table 1A-4. Story shear forces for design of frame BF-2

Level	Story Force F_x (kips)	Story Shear V_x (kips)	BF-2 Force $F_{x,2}$ (kips)	BF-2 Shear $V_{x,2}$ (kips)
R	967	967	257	257
5	809	1,776	215	472
4	614	2,390	164	635
3	420	2,810	112	747
2	231	3,041	61	808

5. Reliability/redundancy factor §12.3.4

Because the design has perimeter bracing of at least two bays per side, ρ may be taken as 1.0.

6. Combined effect of horizontal and vertical earthquake induced forces §12.4.2

For convenience, the seismic design load combinations can be reformulated to combine the vertical component of the seismic motion with the dead load. Also, as the overstrength factor is not considered in combination with the redundancy factor (and the results of the analysis include the redundancy factor), it is sometimes convenient to recalculate the overstrength factor to account for this difference (if $\rho > 1.0$).

$E = \rho\, Q_E \pm 0.28\, D$

$E_m = \Omega_o\, Q_E \pm 0.28\, D$

$\Omega_o = 2.0;\ \rho = 1.0$

$\Omega_o\, Q_E = \Omega_o\, (\rho Q_E)/\rho$

Basic Seismic Load Combinations:

$1.2\, D + 1.0\, E + 0.5L = 1.48\, D + \rho Q_E + 0.5$ §12.4.2.3 (5)

$0.9\, D + 1.0\, E = 0.62\, D + \rho Q_E$ §12.4.3.2 (7)

Special Seismic Load Combinations (Amplified Seismic Load):

$1.2\, D + 1.0\, E_m + 0.5L = 1.48\, D + 2.0\, Q_E + 0.5 = 1.48\, D + 2.0\, Q_E + 0.5$ §12.4.3.2 (5)

$0.9\, D + 1.0\, E_m = 0.62\, D + 2.0\, Q_E = 0.62\, D + 2.0\, Q_E$ §12.4.3.2 (7)

This approach is convenient for using a single set of analytical results for the design of all members. It also provides an illustration of one of the dangers of the redundancy factor. The formulae that include the redundancy factor are used for brace design, and those that do not include it are used for column and beam design. The forces that the framing members are likely to see are greater for a frame with larger braces than for one with smaller ones, and the same overstrength factor should be applied to the two frames. It is apparent from the reformulations that the redundancy factor actually decreases the effective overstrength required of these members and makes unfavorable yield modes more likely. A redundancy factor of 1.3 decreases the effective overstrength factor to 1.54, an extremely low value for a braced frame; use of such a low value in determining column design forces may not be sufficient to preclude column buckling. Designers should not lose sight of the purpose of the required overstrength, even when code provisions permit its reduction.

7. SCBF member design

In seismic design, significant ductility is typically expected of the lateral-load-resisting system. In order for this ductility to be achieved, it is necessary that the system be designed so that non-ductile modes of behavior are precluded. Accordingly, limit-states design is appropriate in the design for seismic loads because design based on elastic analysis may not capture internal forces corresponding to expected behavior. AISC 341 requires that capacity design be used for many calculations (that is, that the required strength of certain elements be determined from the expected strength of other, more ductile, elements, insuring that the more ductile limit state will control the frame behavior). To apply these concepts correctly, it is important that the designer determine the possible yield modes of the frame, and how these affect the force and deformation demands on the frame elements.

7a. Frame analysis

In the elastic range, concentrically braced frames can be readily analyzed as trusses, a computation which can be accomplished by a hand analysis. Although for SCBF buildings (many of which are under five stories), drifts tend to be well below the maximum; calculation of drifts is required, and computer analysis is typically an efficient method. In this example, small eccentricities occur at the connections to facilitate detailing. These eccentricities result in flexural forces in the framing members, as well as larger drifts than those for frames without eccentricities. It is therefore necessary to include the eccentricities in the frame model. As these eccentricities are a function of the framing member depths as well as the framing geometry, it is necessary to establish these at this point. (The calculation of the modified work-point location is given in part 8 of this design example.)

Using the frame forces from Table1A-4, the frame shown in Figure 1A-5 is analyzed. Columns are modeled as fixed at the base and braces as pinned at each end. Although in-plane rotation of the gusseted beam-to-column connection may induce some flexural forces in the brace, these are small at the elastic drift level. Beams are modeled as having fixed connections at their gusseted connections to the columns; these details include beam-to-column joint flange welds.

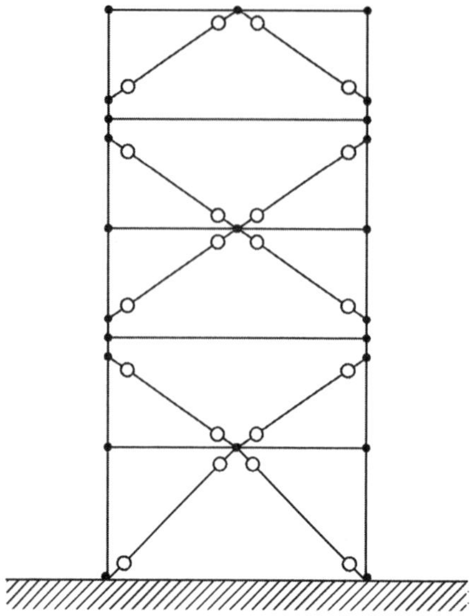

Figure 1A-5. Model of frame BF-2

It is not uncommon for simple beam-end connections such as a shear plate to be used at gusseted beam-column connections. Certain cautions concerning that detailing practice are appropriate. First, bolts should not be used in combination with welds to resist seismic loads (AISC 341, §7.2). As is shown in the design of this connection in the next section, brace forces are transferred through the gusset to the column, an interface that is typically welded; brace forces are also transferred to the column by the beam-to-column connection via the gusset-to-beam weld. Second, these design forces from the brace may be sizable, and a substantial connection may be required. Third, very little is known about the moment-rotation relationship of simple connections to which gussets are added, although Williams observed that they can be considered fully-restrained connections in the elastic range (1996). It would be difficult to demonstrate adequate strength to resist the moments attracted by these stiff connections, especially as such moments are simultaneous with the seismic forces in the connection. Likewise, it would be difficult to demonstrate sufficient inelastic rotation capacity to accommodate rotations due to gravity and seismic loads, as has been done for commonly used simple connections. Detailing of the connection is discussed further in part 8.

Most, if not all, structural analysis programs permit the incorporation of P-Δ effects into the analysis. For this to be done accurately, both the story stiffness and the story gravity load at each level must be reflected in the model. If this is done, the results of this second-order analysis may be used directly in design. Otherwise, forces resulting from lateral translation must be amplified by the story-specific factor, B_2. This includes both axial forces, P_{lt}, and flexural forces, M_{lt} and V_{lt}. In this design example, it is assumed that such a second-order analysis has been performed, and the B_2 amplification factor is not required.

The displacements from the model are converted to design story drifts and compared to the allowable drift from ASCE/SEI 7-05 Table 12.12-1. The story drifts are calculated from the displacements, and divided by the reliability/redundancy factor, ρ (neither this factor, nor the importance factor, *I*, applies to the calculation of drifts). To be compared with the allowable drift, the drifts from the elastic analysis need to be multiplied by the deflection amplification factor, C_d. The design story drift for each level of this frame is less than the allowable drift; the comparison is shown in Table 1A-5. Because the building design is symmetrical and has rigid diaphragms, the average story drift will not exceed the drift of this frame. Thus, the building will meet the drift limitations of Table 12.12-1.

Table 1A-5. Design story drifts for frame BF-2 and allowable story drifts

Story Number	Story Height	Displacement from Model	Drift from Model	Design Story Drift	Allowable Drift
	h_{sx}	δ_{xe}	Δ_{xe}	$\Delta_x = C_d \Delta_{xe}$	$\Delta_{Ax} = 0.020 h_{sx}$
	(ft)	(in)	(in)	(in)	(in)
5	12	2.199	0.503	2.52	2.88
4	12	1.696	0.454	2.27	2.88
3	12	1.242	0.466	2.33	2.88
2	12	0.776	0.428	2.14	2.88
1	14	0.348	0.348	1.74	3.36

These drifts are only in small part due to the eccentricity at the beam-column-brace connections. If that eccentricity were eliminated, the calculated drifts would be 90 to 95 percent of those shown here.

7b. Brace design

Next, the analysis results will be used for the brace design force. Braces in SCBF's must withstand cycles of buckling and tensile yielding. Once brace forces are calculated, a member with sufficient compression capacity must be selected, and in that selection the slenderness and cross-sectional shape chosen are key variables. Several aspects of system behavior depend on these variables, including energy dissipation, susceptibility to local buckling, fracture life, post-buckling stiffness, and the ratio of sliding strength to buckling strength. Square or rectangular cross section HSS is not recommended. Refer to the latest Blue Book article on "Concentric Braced Frames" for more information.

Brace forces from the model are:

$P_D = 8$ kips (compression)

$P_L = 6$ kips (compression)

$\rho P_E = \pm 328$ kips

The critical load combinations for the design of the brace are given by Equations 16-5 and 16-7. Because the gravity loads cause compression in the braces, the maximum brace compression results from Equation 16-5 and the maximum brace tension results from Equation 16-7.

$1.2D + 1.0E + 0.5L = 1.48D + \rho Q_E + 0.5 = 343$ kips (compression)

§12.4 allows live load reduction by 50 percent for live loads less than 100 psf

$0.9D + 1.0E = 0.62D + \rho Q_E = 321$ kips (tension)

The distance from work-point to work-point is 17.9 feet (see §8 for illustrations of the work-point locations). The ductile detailing requirements for SCBF result in the provision of a hinge zone in the connection; it is assumed here that the hinge zone will be at least 9 inches from the work point at each end, resulting in a reduction of the brace length of 18 inches (this assumption must be verified once the connection design is complete). Thus the brace length = 197 inches. HSS 9.625 × 0.500 will be used. Section properties from the 13th Edition LRFD Manual will be used, as these are a more accurate representation of the steel provided by mills today; areas of HSS members tend to be 92 to 93 percent of those given in the 2nd Edition LRFD Manual, or the *1989 Allowable Stress Design Manual.*

Check brace compression strength

$$Kl/r = \frac{1.0(197\text{ in})}{3.24\text{ in}} = 60.1$$

$$F_e = \frac{\pi^2 E}{(Kl/r)^2} = 77.4\text{ ksi} > 0.44F_y$$

$$F_{cr} = [0.658^{F_y/F_e}]F_y = 33.5\text{ ksi} \qquad \text{AISC 360, E3-2}$$

$$\phi P_n = 0.9F_{cr}A = 0.9(33.5\text{ ksi})(13.4\text{ in}^2) = 404\text{ kips} \qquad \text{AISC 360, E3-1}$$

Check brace tension strength

$$\phi P_n = 0.9F_yA_g = 0.9(42\text{ ksi})(13.4\text{ in}^2) = 507\text{ kips} \qquad \text{AISC 360, D2-1}$$

Equation D1-2, which applies to fracture of the net section, is checked in the design of the connection, as additional force levels must be considered in that portion of the design.

$\phi P_n > P_u$ . . . *o.k.*

The HSS 9.625 × 0.500 is adequate.

Check brace slenderness

AISC 341 gives a limit for brace slenderness

$$Kl/r < 4\sqrt{E/F_y} = 105 \qquad \text{AISC 341, §13.2a}$$

For this brace, Kl/r has been calculated as 60.8; the brace complies. (Note that under the exception, braces with $Kl/r \leq 200$ are allowed if capacity design is used for columns. For such slender braces, the Ω_0 may not adequately estimate overstrength.)

Check brace width-to thickness ratio

AISC 341 gives limits for brace compactness; the limit for round sections is

$$D/t \leq 0.044E/F_y = 30.4$$ AISC 341, §13.2d, T I.8.1

The brace complies.

Check lateral distribution of force

AISC 341, §13.2c does not permit a major unbalance between braces working in compression and those working in tension to resist lateral loads. As the frames in this structure are symmetrical and no single-diagonal frames are used, an elastic analysis shows that the seismic force resisted by compression is identical to that resisted by tension, the frame complies with the requirement. Although not directly required by this provision, the result is that the tension strength of braces is similar in each direction, so that even after degradation of the compression strength of braces there is no significant tendency to accumulate drift in one direction. Requirements for frames not complying with this provision (such as frames with single-diagonal bracing) are discussed in Design Example 1C.

7c. Column design

The column at the fourth floor will be designed. The size has been selected as W12×96.

The column design loads are defined by the seismic load combinations of ASCE/SEI 7-05.

$$1.2D + 1.0E_m + 0.5L = 1.48D + 2.0Q_E + 0.5L = 1.48D + 2.0\rho Q_E + 0.5$$ §12.4.3.2 (5)

$$0.9D + 1.0E_m = 0.62D + 2.0Q_E = 0.62D + 2.0\rho Q_E$$ §12.4.3.2 (5)

Compression design of the column will, of course, govern over tension.

The AISC 341, §8c, permits an alternate method of determining the required strength based on 110 percent of the expected strength of the members delivering force to the column. Certain cautions about this are in order. It should be noted that the calculation of expected brace compression strength may be low for three reasons. First, it assumes perfect pins at each end of the brace, when in reality some rotational restraint is provided by the gusset. Second, the calculation of brace strength is based on a generalized column curve that may significantly underestimate the strength of some sections. Third, the computation assumes the maximum allowed out-of-straightness, a condition that does not necessarily occur in the brace.

Additionally, it may not be sufficient to consider only the members directly adjoining the column. Other members may contribute to column forces. A range of possible column forces associated with yield modes is shown in Figure 1A-6. Design Example 1B describes in more detail the column forces from yield of the roof-level braces.

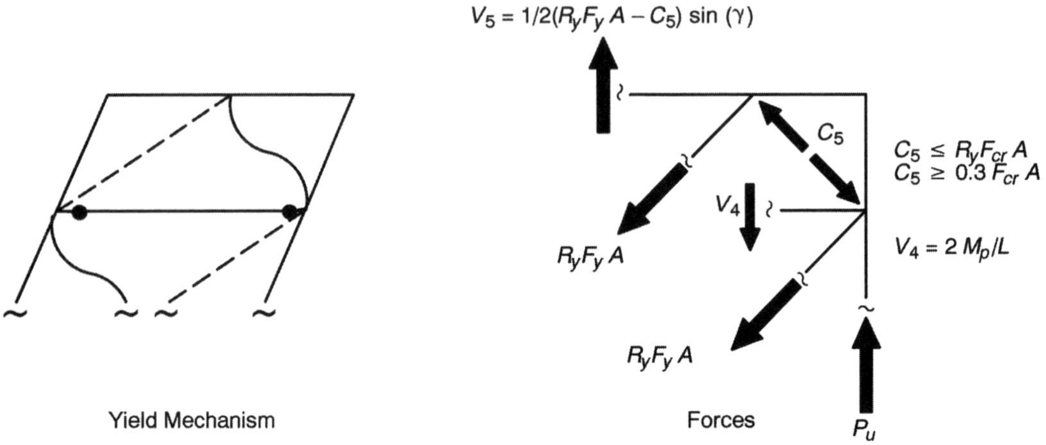

Figure 1A-6. Yield mechanism and corresponding forces acting on column

Figure 1A-6 shows that using only the compression strength of the brace above may not capture the maximum column forces. The maximum column forces can be reliably bounded by the following expression combining the expected strengths of the adjoining beam and braces:

$$P_{max} = 1.1R_y\left(2\frac{M_p}{L}\right) + R_yF_y(A_{4th} + A_{5th})\sin(\gamma)$$

Typically the uplift due to beam flexure is negligible, especially if the adjoining collector beams outside the frame have rigid connections to the column.

$$P_{max} = 1.4(42\text{ ksi})(11.9\text{ in}^2)\sin(37.6°) + 1.4(42\text{ ksi})(7.86\text{ in}^2)\sin(37.6°)$$

$$P_{max} = 427\text{ kips} + 282\text{ kips} = 709\text{ kips}$$

This force is significantly larger than the code-specified seismic force

$$\Omega_o\,Q_E = 2.0Q_E = 2.0(328\text{ kips}) = 656\text{ kips}$$

More research is needed to determine if this type of capacity-design approach is warranted in determining the required strength of columns, especially for columns that can receive forces from braces at multiple levels. As the maximum force is computed on the assumption of simultaneous yield and strain-hardening of the fourth- and fifth-floor braces, it may be unrealistically conservative. However, the code-specified seismic load is less than the force corresponding to the yield of the fourth-floor brace by itself; it is therefore inconsistent with the possible yield mechanisms considered in the design. At a minimum, the column should have sufficient strength to permit the yielding of the brace at one level, or the yield mechanism is likely to be column buckling. Designers must make assumptions as to the forces in the other braces contributing to the column force; while assuming full, strain-hardened yield may be too conservative, assuming no force in those braces is unreasonable. For the purposes of this design example, the full force from fourth floor brace force will be considered in combination with 50 percent of the force from the fifth-floor brace.

$$P_{Em} = 1.4(42\text{ ksi})\left(11.9\text{ in}^2 + \frac{1}{2}7.86\text{ in}^2\right)\sin(37.6°) = 568\text{ kips}$$

The column should be checked for the combined effects of flexural and axial forces. Note that AISC 341 explicitly allows the flexural forces to be ignored when using the amplified load combinations in the design of columns; nevertheless, it is recommended here that flexural forces be considered, as they are simultaneous with the axial forces.

Calculate column axial compression

$$P_u = 1.48D + E_m + 0.5L = 1.48(78\text{ kips}) + 568\text{ kips} + 0.5(44\text{ kips}) = 705\text{ kips}$$

Calculate column major-axis flexure

M_u must be determined considering moment magnification effects. These methods are defined in Chapter C of the AISC 360. Moments are separated into those occurring in the structure without sway, M_{nt}, and those resulting from sway, M_{lt}. Although the effects of sway on moments are most important for moment frames, they are always appropriate to consider. In this design, magnification of lateral-translation forces has been done in the analysis.

Moments in the structure with no translation, M_{nt}, are, in this case, those resulting from gravity loads, while the sway moments, M_{lt}, are those from the seismic load.

$$M_{nt} = 1.48D + 0.5L = 1.48(7.6\text{ kip-ft}) + 0.5(8.2\text{ kip-ft}) = 15.4\text{ kip-ft}$$

$$M_{lt} = 2.0Q_E = 2.0(39.3\text{ kip-ft}) = 78.6\text{ kip-ft}$$

The magnification factor B_1 applies to the M_{nt} moments.

$$B_1 = \frac{C_m}{1 - P_u/P_{E1}} > 1$$ AISC 360, C2-2

$$C_m = 0.6 - 0.4\frac{M_1}{M_2} = 0.6 - 0.4\frac{10.9\text{ kip-ft}}{15.4\text{ kip-ft}} = 0.317$$ AISC 360, C2-4

Note that in reverse curvature a low value of C_m is calculated.

P_{E1} is the Euler buckling load in the plane of the bending moments being considered. P_{E1} can be calculated directly using the Euler equation

$$P_{E1} = \frac{\pi^2 E\, I}{(K\, L)^2} = \frac{\pi^2\, 29{,}000 \text{ ksi}(833 \text{ in}^4)}{\left[1.0\left(12 \text{ ft}\frac{12 \text{ in}}{\text{ft}} - 18 \text{ in}\right)\right]^2} = 15{,}000 \text{ kips}$$

$$B_1 = \frac{0.317}{1-(705 \text{ kips}/15{,}000 \text{ kips})} = 0.33 < 1 \quad B_1 = 1$$

The effective column height is calculated based on end fixity and clear column height.

$$M_u = B_1 M_{nt} + B_2 M_{lt} \qquad \text{C2-1a}$$

$$M_{ux} = 1.0(15.4 \text{ kip-ft}) + (78.6 \text{ kip-ft}) = 94 \text{ kip-ft}$$

Calculate column minor-axis flexure

Additionally, a small minor-axis moment exists due to the eccentricity of the load delivered through the shear plate (this eccentricity is calculated based on Astaneh's method as presented in the AISC Manual.

$M_{nt\text{-}y} = P(e)$ where P is the load delivered through the shear tab factored according to Equation 16-22

$M_{lt\text{-}y} = 0$ kip-ft

B_1 is calculated as for the major-axis moment. For this column, which is subjected to reverse curvature in the minor axis as well, $B_1 = 1$.

$$M_{uy} = 8.9 \text{ kip-ft}$$

Calculate column compression strength

Minor axis buckling governs the compression strength of this column.

$$\frac{K\, L}{r_y} = \frac{1.0(144 \text{ in} - 18 \text{ in})}{3.09 \text{ in}} = 40.8;\ F_e = 172 \text{ ksi}$$

$$F_{cr} = (0.658^{F_y/F_e})F_y = 44.3 \text{ ksi}$$

$$\phi P_n = 0.9\, A_g\, F_{cr} = 0.9(28.2 \text{ in}^2)(44.3 \text{ ksi}) = 1124 \text{ kips} \qquad \text{AISC 360, E3-1}$$

Calculate column major axis flexural strength

The design flexural strength, ϕM_n, may be obtained using the beam-design charts in the AISC Manual. It may also be calculated using the methods in Chapter F of the AISC Specification. From the chart in the manual, the value of ϕM_n corresponding to an unbraced length, L_b, of 12 feet is 546 kip-ft. This value is for a modification factor,

C_b, of 1.0. For the column, C_b may be calculated based on the shape of the moment diagram:

$$C_b = \frac{12.5M_{max}}{2.5M_{max} + 3M_{\frac{1}{4}L} + 4M_{\frac{1}{2}L} + 3M_{\frac{3}{4}L}} \geq 1.0$$ AISC 360, F1-1

C_b may also be taken as 1.0, which is its lower bound. As the moment diagram is linear with almost equal moments at each end, C_b is 2.27. The value of flexural strength from the chart may therefore be increased by a factor of 2.27 up to a limit of ϕM_P, the design plastic moment. This moment may be read from the chart, or may be calculated

$$\phi M_p = \phi Z F_y = 0.9(147 \text{ in}^3)(50 \text{ ksi}) = 551 \text{ kip-ft}$$ AISC 360, F2-1

$$C_b\, \phi\, M_n = 2.27(546 \text{ kip-ft}) = 1239 \text{ kip-ft} > \phi M_p \quad \phi M_n = \phi M_p = 551 \text{ kip-ft}$$ AISC 360, F2-2

Calculate column minor axis flexural strength

The method for calculating this strength is similar to that for the major axis, although the limit state of lateral-torsional buckling does not apply. For this column

$$\phi M_n = \phi M_p = \phi F_y Z = 0.9(50 \text{ ksi})(67.5 \text{ in}^3) = 253 \text{ kip-ft}$$

Check combined effects of flexure and compression on column

$$\frac{P_u}{\phi_c P_n} = \frac{705 \text{ kips}}{1124 \text{ kips}} = 0.63 > 0.2 \qquad \text{Use interaction equation H1-1a}$$

$$\frac{P_u}{\phi_c P_n} + \frac{8}{9}\left(\frac{M_{ux}}{\phi_b M_{nx}} + \frac{M_{uy}}{\phi_b M_{ny}}\right) = \frac{705 \text{ kips}}{1124 \text{ kips}} + \frac{8}{9}\left(\frac{94.0 \text{ kip-ft}}{551 \text{ kip-ft}} + \frac{8.9 \text{ kip-ft}}{253 \text{ kip-ft}}\right) = 0.81 < 1 \quad \ldots \; o.k.$$

AISC 360, H1-1a

The column has adequate strength.

It is worth noting that if the flexural forces were neglected, a smaller column could theoretically be used, although in this case the column size is determined by the compactness requirement. As mentioned earlier, neglecting the flexural forces is not recommended. One reason for this is that the plastic rotation necessary to alleviate the flexural forces would likely occur in the column, as there is no strong-column-weak-beam requirement for SCBF. Also, the stronger columns resulting from the inclusion of flexural forces in the design have a beneficial effect on the performance of braced frames (Tang, 1989). The behavior of columns subjected to high axial forces and rotation demands is currently being studied at the University of California at San Diego.

Check column width-to-thickness ratio

$\lambda < \lambda_p$ AISC 341, §13.5

$$\frac{b}{t} < 0.30\sqrt{E/F_y} = 7.2 \quad \frac{b}{t} = \frac{\frac{1}{2}b}{t_f} = 6.76 \ \ldots \ o.k.$$ AISC 341, T I-8-1

$$\frac{h}{t_w} < 1.12\sqrt{E/F_y}\left(2.33 - \frac{P_u}{\phi_b P_y}\right) = 45.9 \quad \frac{h}{t} = \frac{h}{t_w} = 17.7 \ \ldots \ o.k.$$ AISC 341, T I-8-1

The column complies with the compactness requirements.

7d. Column splice design

The splice shown in Figure 1A-7 will be designed. The location of the splice is mentioned in part 8.4a. This is because the flexural strength of columns (including non-frame columns) is often necessary to prevent story mechanisms as braces degrade over cycles of buckling. (Splice requirements for non-frame columns were introduced for moment-frame buildings in the 2002 provisions and in the 2005 provisions they apply to all buildings.)

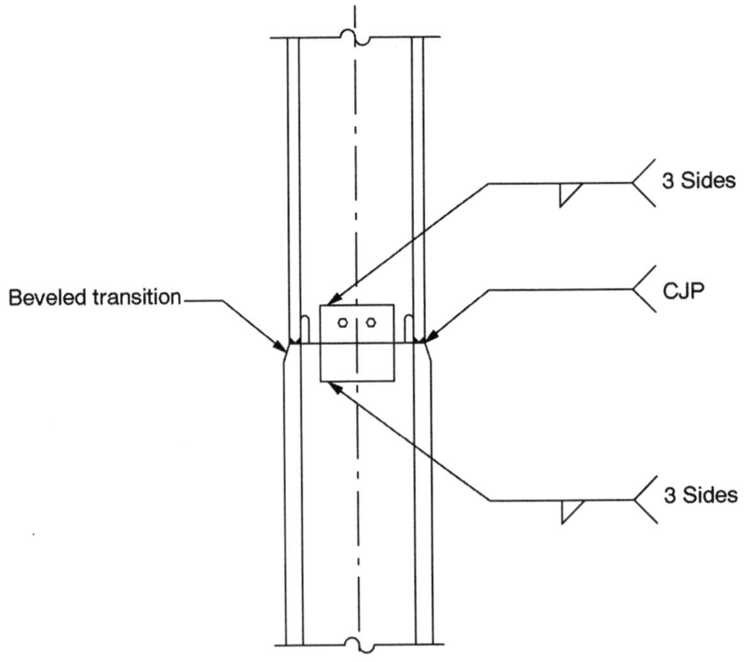

Figure 1A-7. Detail of column splice

Calculate splice demands

Demands are calculated based on the strength of the weaker column section (that is, the column above the splice). Specifically, AISC 341, §13.5 requires that the splice shear strength equal or exceed the maximum force that can develop, which here is taken as the lesser shear strength of the two column sections, and that its flexural strength equal or exceed one half of the lesser flexural strength of the two column sections.

$$V_u = V_n = 0.6\,F_y A_w = 0.6\,F_y(d)t_w = 0.6(50\text{ ksi})[(12.71\text{ in})]\,0.550\text{ in} = 210\text{ kips}$$

$$M_u = \frac{1}{2}M_n = \frac{1}{2}F_y Z = \frac{1}{2}50\text{ ksi}(147\text{ in}^3) = 3675\text{ kip-in} \qquad \text{AISC 341, §13.5}$$

Design shear splice

Add a plate ⅝ × 6½ × 13 inches each side of the column web, with a ⅜ weld on all four edges. The shear demand for each plate is 105 kips.

Check splice-plate shear strength

$$\phi V_n = \phi 0.6 F_y A = 1.0(0.6)(50\text{ ksi})(6\tfrac{1}{2}\text{ in})\tfrac{5}{8}\text{ in} = 122\text{ kips} \ldots \textit{o.k.} \qquad \text{AISC 360, J4-3}$$

Check splice weld strength

Eccentrically loaded welds are most efficiently designed using the AISC Manual. The plate is assumed to have pure shear at its mid-height; this weld is resisted by a C-shaped weld group on the plate edges. Table 8-9 may be used to calculate the required weld size. The angle of loading is 0°.

$C_1 = 1.0$ for E70.

$L = 6\tfrac{1}{2}$ in

$kL = 6\tfrac{1}{2}$ in, $k = 1$ and $x = 0.333$

$L = aL + xL$

$e_x = aL = L - xL = 0.667L$

$a = 0.667$

Interpolating between values for $a = 0.6$ and $a = 0.7$, C is determined to be 4.03.

The required weld throat, D, is calculated to be

$$D_{\min} = \tfrac{1}{16}\text{ in}\,\frac{P_u}{C\,C_1\,L} = \tfrac{1}{16}\text{ in}\,\frac{\frac{1}{2}210\text{ kips}}{(4.03\text{ kip/in})\,1(6.5\text{ in})} = \frac{4.0}{16}\text{ in} < \tfrac{3}{8}\text{ in} \;\ldots\; \textit{o.k.}$$

Design flexural splice

Complete joint penetration welds of both flanges are to be used. Although partial-penetration welds are permissible, the postelastic performance of the frame could be severely limited by a fracture of this connection.

7e. Beam design (Fifth floor)

Beams in braced frames are, of necessity, collector elements. They are therefore designed for the special seismic load combinations of IBC Equations 16-22 and 16-23. In addition, AISC 341 has certain requirements for beams in V-type braced frames.

Calculate beam axial compression

An elastic analysis will show relatively little axial force in the beam at the fifth floor. However, when a brace on either the fourth or fifth floor buckles, forces can redistribute through this beam. It is therefore recommended that the minimum required axial strength of such beams be determined based on consideration of the possible frame yield mechanisms (see Bruneau et al., 1998). Figure 1A-8 shows a yield mechanism involving buckling and tension yielding of braces above and below the beam. This mechanism requires this beam to have significant axial strength. Note that although AISC 341 defines the minimum post-buckling brace compression force as 0.3 P_n, use of that lower bound is in many cases unconservative. For collector design, the expected strength should be used.

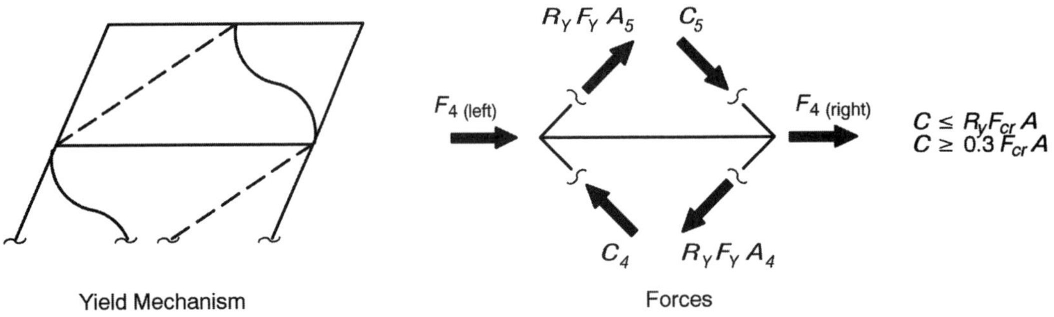

Figure 1A-8. Possible yield mechanisms with different required beam strengths

The beam axial force from the mechanism is computed from the strength of the braces above and below in combination with the story force.

$$P_E \le [R_y F_y A_4 - 0.3 F_{cr} A_5)] \cos\gamma - F_{4\,\mathrm{right}}$$

$$F_{4\,\mathrm{right}} = \frac{1}{2}[(R_y F_y + 0.3F_{cr})A_4 - (R_y F_y + 0.3F_{cr})A_5]\cos\gamma$$

$$= \frac{1}{2}[(1.4 \times 42 \text{ ksi} + 0.3 \times 31.6 \text{ ksi})11.9 \text{ in}^2$$

$$- (1.4 \times 42 \text{ ksi} + 0.3 \times 25.7 \text{ ksi})7.86 \text{ in}^2] \cos 37.6°$$

$$= 115 \text{ kips}$$

$$P_E \leq (1.4 \times 42 \text{ ksi} \times 11.9 \text{ in}^2 - 0.3 \times 25.7 \text{ ksi} \times 7.86 \text{ in}^2) \cos 37.6° - 115 \text{ kips}$$
$$= 391 \text{ kips}$$

This force is a maximum force based on the frame capacity. It should not be amplified using Ω_o. The required axial strength is

$$P_u = 1.48D + Q_{E(\max)} + 0.5L = 1.48(2.1 \text{ kips}) + 391 \text{ kips} + 0.5(1.1 \text{ kips}) = 395 \text{ kips}$$

Calculate beam major-axis flexure

$$M_{nt} = 1.48D + 0.5L = 1.48(11.4 \text{ kip-ft}) + 0.5(4.4 \text{ kip-ft}) = 19.1 \text{ kip-ft}$$
$$M_{lt} = 2.0Q_E = 2.0(71.6 \text{ kip-ft}) = 143 \text{ kip-ft}$$

B_1 is calculated as it was for the column; it is 1.0 for the beam as well. B_2 was calculated in the analysis.

$$M_u = B_1 M_{nt} + B_2 M_{lt} \qquad \text{C2-1a}$$
$$M_u = 1.0(19.1 \text{ kip-ft}) + (143 \text{ kip-ft}) = 162 \text{ kip-ft}$$

Calculate beam axial strength

The W18×65 beam will be checked for both major-axis and minor-axis buckling. (Although the top flange is continuously braced, the section can buckle torsionally; typically, designers perform a minor axis buckling check to address torsional stability). The beam is laterally supported at its midspan and ¼ points in its minor axis. As shown below, the compression strength is governed by the minor-axis slenderness.

$$\frac{KL}{r_y} = \frac{(1.0)87 \text{ in}}{1.69 \text{ in}} = 51.5 \quad \text{(minor axis)}$$
$$\frac{KL}{r_x} = \frac{(1.0)348 \text{ in}}{7.49 \text{ in}} = 46.5 \quad \text{(major axis)}$$
$$F_e = 108 \text{ ksi} > 0.44F_y$$
$$F_{cr} = (0.658^{F_y/F_e})F_y = 41.2 \text{ ksi}$$
$$\phi P_n = 0.9A_g F_{cr} = 0.9(19.1 \text{ in}^2)41.2 \text{ ksi} = 708 \text{ kips}$$

Calculate beam flexural strength

The flexural strength may be determined from the beam design chart in the AISC Manual. For $L_b = 7.5$ feet, $\phi M_n = 477$ kip-ft. The C_b factor for this member may be calculated based on the shape of the moment diagram to be 1.9.

$$\phi M_p = \phi Z F_y = 0.9(133 \text{ in}^3)50 \text{ ksi} = 499 \text{ kip-ft}$$
$$C_b \phi M_n = 1.9(477 \text{ kip-ft}) = 906 \text{ kip-ft} > \phi M_p$$
$$\phi M_n = \phi M_p = 499 \text{ kip-ft}$$

Check combined effects of flexure and compression on beam

$$\frac{P_u}{\phi_c P_n} = \frac{395 \text{ kips}}{708 \text{ kips}} = 0.56 > .02 \quad \text{Use interaction equation H1-1a.}$$

$$\frac{P_u}{\phi_c P_n} + \frac{8}{9}\frac{M_u}{\phi_b M_n} = \frac{395 \text{ kips}}{708 \text{ kips}} + \frac{8}{9}\frac{162 \text{ kip-ft}}{499 \text{ kip-ft}} = 0.85 < 1 \ \ldots \ o.k.$$

7f. Beams at other levels

Check AISC 341 special requirements

AISC 341 has a series of special requirements for beams that are intersected by braces (such as the roof beam and the fourth-floor beam). These requirements are intended to prevent certain unfavorable modes of behavior observed in V-braced frames during past earthquakes. Without these provisions, the yield mechanism of typical stacked chevron-braced frames is likely to include buckling of a brace and flexural yielding of the beam. Such a mechanism can have negative post-buckling stiffness and lead to large drifts (Khatib et al., 1988). The special requirements ensure that the yield mechanism of the frame includes tension yielding of a brace as well as buckling of the complementary brace. This mechanism can provide positive post-buckling stiffness and trilinear hysteresis (Bruneau et al., 1998). See Design Example 1B for a complete treatment of these requirements.

It should be noted that the beam flexural strength requirement of AISC 341, §13.4a item 1 is not applicable to the beam in the bracing configuration in this example. This has been clarified in the commentary in the AISC 341. It has been demonstrated that with this configuration and a beam of low flexural strength a yield mechanism including tension yielding of a brace as well as buckling of the complementary brace is ensured (Sabelli et al., 1998).

AISC 341, §13.4a requires that the beam be continuous between columns, and that it be able to support gravity loads without support by the braces. These requirements are met by the design. The gravity design of the beam will not be presented here; note that composite design is allowed.

Additionally, §13.4a requires consideration of the maximum beam demands based on buckling of one brace and tension yielding of the corresponding one. This requirement applies to traditional chevron-braced frames in which only two braces intersect the beam at the midspan (see Design Example 1B).

Finally, item 2 requires that the midspan be braced for a load calculated based on the beam expected flexural strength using Equations A-6-7 and A-6-8 of AISC 360.

Although the member being braced is actually the diagonal brace, the provision explicitly requires that the beam strength be used to determine the compression force.

$$M_r = R_y F_y Z = 1.1(50 \text{ ksi})(133 \text{ in}^3) = 7315 \text{ in-kip}$$

$$P_{br} = 0.02\, M_r C_d / h_o \qquad \text{Eq A-6-7}$$

$$C_d = 1.0 \text{ for single curvature}$$

$h_o = d - t_f = 17.6$ in

$P_{br} = 8.3$ kip

$$\beta_{br} = \frac{1}{\phi}\left(\frac{10M_r C_d}{L_b h_o}\right) \quad \text{Eq A-6-8}$$

$\phi = 0.75$

$L_b = 15$ feet $= 180$ in

$\beta_{br} = 30.8$ kip/in

Assuming this bracing is provided by means of a diagonal kicker:

$$A_{req} = \frac{\beta_{req} L}{E} = \frac{30.8 \text{ kip/in}(10 \text{ ft})}{29{,}000 \text{ ksi}} = 0.1 \text{ in}^2$$

This requirement is easily met.

Roof beams

Note that the beam at the roof is not part of a two-story X. AISC 341 requires that roof beams be designed to have adequate strength to preclude their flexural yielding prior to tension yielding of a brace below. Studies have shown that if roof beams in chevron-braced frames are not designed for the unbalance load and the braces at that level are the first to yield, a significant concentration of damage can result (Sabelli, 2001). It is prudent to provide some means of preventing this. In this design, the beam has been designed for the unbalance force; the beam's strength for positive flexure has been calculated using composite action with the concrete slab and its strength for negative flexure has been calculated using the slab reinforcement. Alternatively, the braces could be designed to have significantly lower demand-to-capacity ratios than braces at other levels. This is consistent with a ductile yield mechanism of the frame: yielding of another story.

8. Connection Design

SCBF bracing connections differ from those designed using older codes in two fundamental respects: they are designed for forces based on a desired yield mechanism (forces higher than those indicated from an elastic analysis using the design base shear) and they are required to be detailed for ductile post-buckling behavior.

Bracing connections in SCBF are not permitted to be the weak link in the building design. They are required to have sufficient strength to ensure that they will not fracture under forces corresponding to a more desirable yield mechanism of the structure, specifically brace gross-section yielding. The required connection strength is taken as either the expected yield strength of the brace or the "maximum force, indicated by analysis, that can be transferred to the brace by the system" (AISC 341).

Because braces are expected to be subjected to inelastic actions in compression as well as in tension, bracing connections are required to accommodate the effects of brace buckling. Inelastic brace buckling requires three hinges: one at the midpoint and one at each end of the brace. Connections may accommodate the effects of brace buckling in either of two ways: they may provide flexural strength in excess of the expected flexural strength of the brace, thus forcing the end plastic hinge to form in the brace immediately beyond the connection; or they may provide sufficient rotational ductility to accommodate the rotational demands imposed by brace buckling, thus allowing the end plastic hinge to form in the connection.

Designers typically find the latter approach simpler and more convenient. AISC 341 provides a method for providing this ductility based on research by Astaneh (1985). A short length of gusset plate is extended beyond the lines of restraint against folding that the weld to the beam and column provide; the brace does not extend into this zone of the gusset, leaving the plate free to form an inelastic hinge when the brace buckles (see Figure 1A-10). Note that the hinge must form perpendicular to the brace axis. Thus the governing line of restraint may be defined by the gusset corner at the column flange or the gusset corner at the beam flange (or at the gusset edge at the top of the slab). The formation of this hinge relieves the rest of the connection from moments greater than the plate minor-axis flexural strength, which is negligible. Recent work by Roeder (2009) provides an alternative methodology using an elliptical hinge.

Both of the changes in the requirements for bracing connections, the increased design forces and the ductile detailing requirements, lead to larger gussets than those to which most designers (and their clients) are accustomed. These larger gusset plates can be expensive; they can also cause conflicts with other building systems. Additionally, the ductile detailing provisions lead to a series of dimensional requirements that build on one another, so that minor changes in geometry can result in significant increases in gusset plate dimensions. Finally, the restraining effects of these large gusset plates on the beam-to-column connection are not well understood (López, 2001). Because braced frames can see flexural ductility demands when subjected to large ground motions, the behavior of these large haunched connections may be of concern. (There is insufficient information available to make recommendations concerning this aspect of gusset-plate behavior. However, designers are encouraged to consider it, especially for higher performance objectives.)

For these reasons, it is advantageous to design gussets to be as small as possible. Use of A572 material will result in smaller gussets than A36. Using the Uniform Force Method [AISC Manual, (Thornton, 1991) or the Ricker Method, (Thornton, 1991)], gusset plates can be designed to provide an area of plate symmetrical about the brace at its end. This is the most efficient means of transferring the force and results in the smallest required plate area. Although the plate dimensions can be reduced by increasing its thickness, the ability of the relatively thin webs of the framing members to resist the concentrated forces may become a practical limit, and the addition of stiffeners or doublers may not be an economical solution.

This design example uses an eccentric workpoint to facilitate detailing. The workpoint is selected so that the brace axis passes through the intersection of the beam and column flanges; the workpoint is taken as the intersection of this line with the column axis. Additionally, with the modified workpoint, a greater portion of the connection force tends to be transmitted directly to the column, so the beam-to-column connection forces are lower. The Ricker method is used for the calculations, with the forces resolved at the modified workpoint rather than the intersection of the beam and column centerlines. Other methods of analysis are also addressed.

Figure 1A-9 is a schematic of the gusset connection at the beam-column connection. Shown dashed is the outline of a more conventional gusset shape with the traditional workpoint at the intersection of the beam and column centerlines. The modified workpoint results in a somewhat smaller gusset; for the same brace, the conventional approach may require additional stiffening for stability.

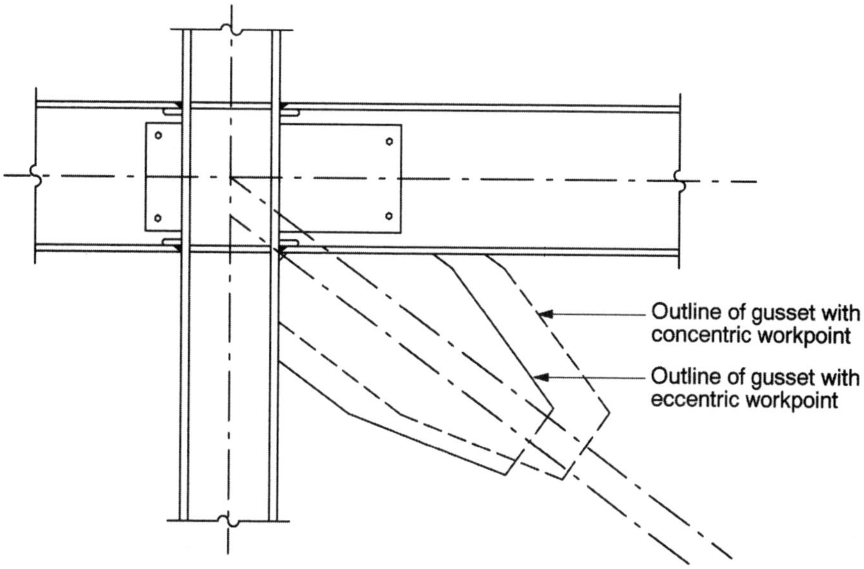

Figure 1A-9. Schematic of connection at beam-column intersection

8a. Brace-beam-column connection design

The connection shown in Figure 1A-10 is to be designed. The connection design is presented beginning with the gusset and continuing to the brace-to-gusset welds. This corresponds to the typical sequence of establishing dimensions in the detailing of the connection for conventional SCBF connection design (the shape of the gusset design presented here lends itself to the opposite sequence). The design of the beam-to-column connection is presented at the end. This is a fundamental part of the connection design and must not be overlooked.

Some of the dimensions used in the design are illustrated in Figure 1A-11.

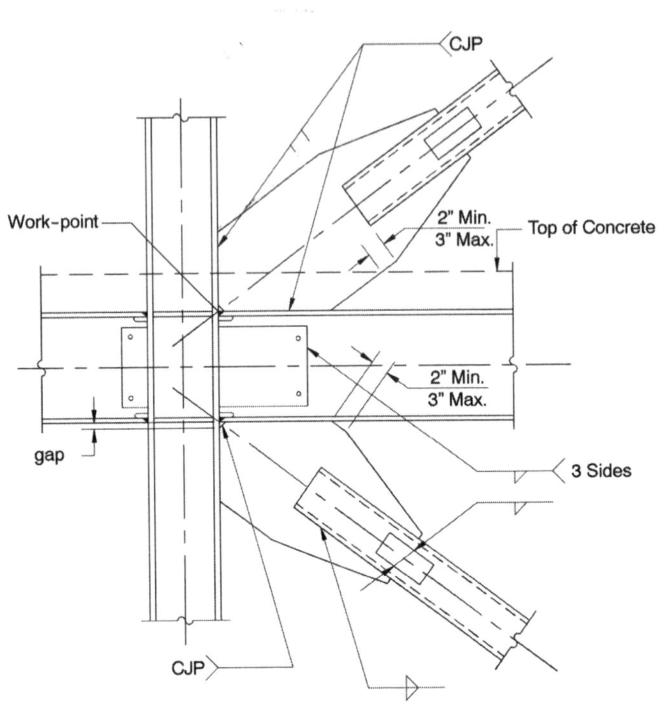

Figure 1A-10. Detail of bracing connection at beam-column intersection

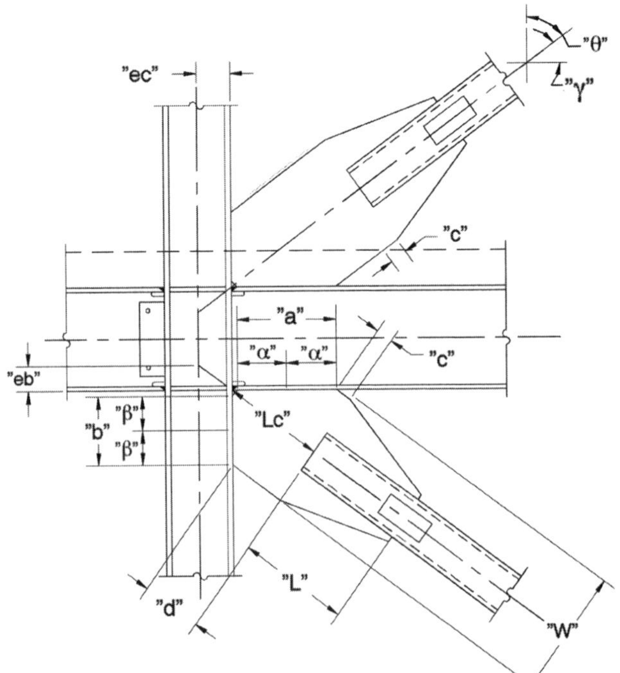

Figure 1A-11. Diagram of dimensions used in the connection design

Compute the bracing connection design force

$$P_u = R_y F_y A_g = 1.4 \times 42 \text{ ksi}(13.4 \text{ in}^2) = 788 \text{ kips}$$ AISC 341, §13.3a

Note that this force is the expected yield strength of the brace; it therefore includes the factor R_y. For A500 steel, AISC 341 gives an R_y value of 1.4. For stability limit states, the demand force can be reduced to the expected compression capacity of the brace. Since the buckling stress is a function of the brace length between hinge zones (the larger the connection, the shorter the brace and thus the higher its buckling strength), which in turn is affected by the connection size, the tension strength can be used as an upper bound for stability checks. However, it is inefficient to alter a connection design based on such a comparison, and explicit calculation of the upper bound of compression strength is recommended when necessary to verify the adequacy of a design.

Design gusset plate using the Ricker Method

The Ricker Method (Thornton, 1991) is a commonly used method of analyzing force distributions in gusset connections. It should be noted that there exists a variety of methods for analyzing these forces, and that an efficient design with sufficient strength, rather than the most accurate analysis, is the purpose of design. So long as strength and ductility are provided, a variety of force distributions in the connection are permissible. The Ricker Method was originally developed for use with simple beam-column connections, typically a bolted interface. This condition requires an analysis method that enforces zero moment at the column face; that requirement is at the root of the Ricker Method distribution of forces. In this design the beam-column connection is fully rigid (for reasons discussed earlier). Therefore, it is not necessary to enforce zero moment at the column face as long as the moment is accounted for in the design.

Determine gusset dimensions

Williams et. al. (1996) conducted a series of finite-element analyses of connections with eccentric and concentric workpoints and observed that as the brace reached its yield stress the stresses at the gusset welds tended to come into alignment with the brace. Additionally, the Ricker Method leads to the most efficient gusset designs; this method will be employed instead of the more traditional Uniform Force Method.

Either force distribution is valid for design (that is, can be used to design a connection that is adequate for the applied load) so long as sufficient strength is provided, as well as moderate ductility to preclude fracture and allow the stresses to distribute to match the design assumption. In either method of determining the force distribution, flexural forces at the beam-column connection must be computed when using an eccentric workpoint.

The design presented here, with the modified work-point and the shaped gussets, allows for a direct calculation of the required gusset plate dimensions. "W" can be calculated directly from the expected brace tensile strength (once a gusset plate thickness is assumed).

$$W_{min} = \frac{R_y F_y A}{\phi F_t t} = \frac{788 \text{ kips}}{0.9 \times 50 \text{ ksi}} = 17.5 \text{ in}^2/t$$

Assume $t = 7/8$ in

$$W_{min} = \frac{17.5 \text{ in}^2}{7/8 \text{ in}} = 20 \text{ in}$$

To ensure that fracture does not govern, the gusset plate section should be checked considering a reduction in area for a "clip" of the corner, which allows flange welding (for unwelded flanges, the clips should still be considered, as there is a gap between flanges).

$$W_{min} = \frac{R_y F_y A}{\phi F_{ut}} + W_{(hole)} = \frac{788 \text{ kips}}{0.75 \times 65 \text{ ksi} \times 7/8 \text{ in}} + 1.5 \text{ in} = 20 \text{ in}$$

Dimensions a and b can be calculated from required width of the gusset

$$a = \frac{W_{min}}{2\cos(\theta)} \quad b = \frac{W_{min}}{2\sin(\theta)}; \quad \begin{array}{l} a = 16.4 \text{ in (use a = 17 in)} \\ b = 12.6 \text{ in (use 13 in)} \end{array}$$

Calculate horizontal and vertical forces to the beam and column

For the design shown, the gusset centerline intersects the beam-to-column flange joint, and equal portions of the brace force can therefore be assumed to be resisted on the two interfaces. The reactions may be computed

$$V_{ub} = V_{uc} = \frac{P_u}{2}\sin(\gamma) = \frac{788 \text{ kips}}{2}\sin(37.6°) = 240 \text{ kips}$$

$$H_{ub} = H_{uc} = \frac{P_u}{2}\cos(\gamma) = \frac{788 \text{ kips}}{2}\cos(37.6°) = 312 \text{ kips}$$

Check vertical section of gusset plate

The vertical section of gusset plate is checked against combined effects of shear and tension using von Mise's yield criterion, assuming a 7/8-inch gusset plate thickness.

$$P_u = \sqrt{H_{uc}^{\ 2} + 3\,V_{uc}^{\ 2}} = \sqrt{(312 \text{ kips})^2 + \frac{(240 \text{ kips})^2}{0.6}} = 507 \text{ kips}$$

$$\phi\, P_n = \phi\, b\, t\, F_y = 0.9(13 \text{ in})(7/8 \text{ in})50 \text{ ksi} = 512 \text{ kips}$$

$\phi\, P_u > P_u$ . . . *o.k.* AISC 360, J4-1

Check horizontal section of gusset plate

$$P_u = \sqrt{V_{ub}^{\ 2} + 3\,H_{ub}^{\ 2}} = \sqrt{(240 \text{ kips})^2 + \frac{(312 \text{ kips})^2}{0.6}} = 572 \text{ kips}$$

$$\phi\, P_n = \phi\, a\, t\, F_y = 0.9(17 \text{ in})\ (7/8 \text{ in})50 \text{ ksi} = 669 \text{ kips}$$

$\phi\, P_n > P_u$ . . . *o.k.*

The welds on these interfaces can also be checked for these forces; a factor of 1.25 on the required strength has been suggested to account for variation in the stress distribution (AISC Manual). This increase is intended to prevent weld fracture from being the governing limit state. It is unclear whether this increase is necessarily sufficient to preclude fracture due to uneven stress distribution from the brace axial force; it does not address any stresses induced by the gusset's behavior as a beam-column-connection haunch. The method followed here is to design the welds to exceed the gusset plate strength and confine any ductility demand to the gusset. Double fillet welds sized for this criterion would need to be at least as thick as the gusset plate for A572 Grade 50 material; a complete-joint-penetration weld is therefore selected.

Check column web for yielding

Beam and column checks are often performed after the gusset design is complete. However, the adequacy of the beam and column webs for the concentrated forces imposed on them is dependant on the gusset design. If a redesign of the gusset is necessary in order to avoid an increase in the beam or column size, it is best to determine this as soon as possible.

$$R_u = H_{uc} = 312 \text{ kips}$$

$$\phi R_n = \phi(b + 5k) F_y t_w = 1.0[13 \text{ in} + 5(1.625 \text{ in})]\ 50 \text{ ksi}(0.550 \text{ in}) = 581 \text{ kips}$$

$$\phi R_n > R_u \ \ldots \ o.k.$$ AISC 360, J10-2

Check column web for crippling at maximum compression force

The maximum compression force can be estimated by

$$P_{max} \le R_y F_{cr} A_{br} = (1.4 \times 33.6 \text{ ksi} \times 13.4 \text{ in}^2) = 630 \text{ kips}$$

Alternatively, it may be calculated by two equations

$$P_{max} \le (0.658^{R_y F_y / F_E}) R_y F_y A_{br} = 578 \text{ kips}$$

$$P_{max} \le \pi^2 EI/L^2 = 1040 \text{ kips};\ P_{max} = 578 \text{ kips}$$

$$R_u = H_{uc} \frac{P_{max}}{R_y F_y A_{br}} = 312 \text{ kips} \frac{578 \text{ kips}}{788 \text{ kips}} = 229 \text{ kips}$$

$$\phi R_n = \phi \left[0.8 t_w^2 \left[1 + 3 \frac{b}{d} \left(\frac{t_w}{t_f} \right)^{\frac{3}{2}} \right] \sqrt{\frac{E F_y t_f}{t_w}} \right]$$ AISC 360, J10-4

$$\phi R_n = 0.75 \left[0.8(0.550 \text{ in})^2 \left[1 + 3 \frac{13 \text{ in}}{12.71 \text{ in}} \left(\frac{0.550 \text{ in}}{0.900 \text{ in}} \right)^{\frac{3}{2}} \right] \sqrt{\frac{29{,}000 \text{ ksi}(50 \text{ ksi})(0.900 \text{ in})}{0.550 \text{ in}}} \right] = 689 \text{ kips}$$

$$\phi R_n > R_u \ \ldots \ o.k.$$

Note that the required compression strength of the gusset is calculated from the expected compression strength of the brace. In this case it is approximately 90 percent of the required tension strength. See §7c for cautions concerning underestimation of the brace capacity.

Check beam web for yielding

$$R_u = V_{ue} = 240 \text{ kips}$$

$$\phi R_n = \phi\,(a + 5\,k)\,F_y t_w = 1.0[17 \text{ in} + 5(1.25)](50 \text{ ksi})0.450 \text{ in} = 511 \text{ kips}$$

$$\phi R_n > R_u \; . . . \; o.k.$$

Check beam web for crippling

$$R_u = V_{ub}\frac{P_{max}}{R_y F_y A_{br}} = 240 \text{ kips}\frac{578 \text{ kips}}{788 \text{ kips}} = 176 \text{ kips}$$

$$N = a = 17 \text{ in}$$

$$N > d = 18.4 \text{ in}$$

$$\phi R_n = \phi\left[0.8 t_w^{\,2}\left[1 + 3\frac{a}{d}\left(\frac{t_w}{t_f}\right)^{\frac{3}{2}}\right]\sqrt{\frac{E F_y t_f}{t_w}}\right]$$

$$\phi R_n = 0.75\left[0.8(0.450 \text{ in})^2\left[1 + 3\frac{17 \text{ in}}{18.4 \text{ in}}\left(\frac{0.450 \text{ in}}{0.750 \text{ in}}\right)^{\frac{3}{2}}\right]\sqrt{\frac{29{,}000 \text{ ksi}(50 \text{ ksi})(0.750 \text{ in})}{0.450 \text{ in}}}\right] = 433 \text{ kips}$$

$$\phi R_n > R_u \; . . . \; o.k.$$

This is at the end of the beam. However, the connection does not allow the web to deform. Equations J10-5a and J10-5b do not apply.

Design brace-to gusset welds

A minimum length for this weld is calculated based on the shear strength of four ½-inch fillet welds

$$L_{min} = \frac{(1/4)P_u}{\phi\left(\frac{\sqrt{2}}{2}\right)(s)0.6\,F_{EXX}} = \frac{(1/4)(788 \text{ kips})}{0.75\left(\frac{\sqrt{2}}{2}\right)(1/2 \text{ in})(0.6)70 \text{ ksi}} = 18 \text{ in} \qquad \text{AISC 360, J2-4}$$

This length should be greater than twice the brace width to reduce the effects of shear lag.

Check critical section of gusset plate in tension

The critical section of the gusset is that area of the section where the tension stress is highest; this is typically at the end of the brace. The maximum gusset plate width on either side of brace centerline is the distance from the gusset corner to the brace centerline. For many designs, this width is reduced by the taper of the plate so that the critical section is significantly less in width than the maximum. In this design, the widths are constant in the critical region. This permits the gusset welds to the beam and column to be sized for their required strength rather than to provide extra width to offset the gusset taper. Furthermore, this design allows the brace to be longer and the gusset to be shorter, because a wider gusset plate provides a point of restraint that is farther from the workpoint. (The line of restraint is defined by the point of contact of the gusset to the beam or column flange; refer to dimension "c" in Figure 1A-11.) Finally, as was described earlier, this design allows for a direct calculation of gusset dimensions based on the expected brace tensile strength.

The gusset width was determined at the start of the design: 10 inches above the brace centerline and 10 inches below. Where the gusset is not proportioned to maximize efficiency, or where the gusset is tapered in this zone, these widths may be very different. One could assume the effective width to be twice the lesser width (a reliable lower bound), or one could analyze the section for tension combined with the flexure resulting from the eccentricity of the brace centerline to the center of the critical section. The effective width for resisting axial force of a rectangular section subjected to flexure can be calculated based on the assumption of a fully plastic section as

$$W_{ef} = \sqrt{4\,e^2 + W^2} - 2\,e$$

For a more thorough explanation of the effective width and eccentricity, see §8b. As there is no eccentricity in this case, the effective width is equal to the full width of the minimum section (the minimum section can be traced using lines perpendicular to the gusset edges and across the brace end).

$$\phi\, P_n = 0.9\, F_y A_g = 0.9(50 \text{ ksi})(\tfrac{7}{8} \text{ in})[(2)(10 \text{ in})] = 788 \text{ kips}$$

$$\phi\, P_n > P_u \;\ldots\; o.k.$$

Note that this width cannot be greater than the "Whitmore" width, which is calculated based on a 30° angle from the brace-to-gusset weld inception point. The minimum brace-to-gusset weld length for which the gusset width can be developed using this 30° angle is therefore

$$L_{\min} = \frac{W_{ef}/2 - d_{br}/2}{\tan(30°)} = \frac{10 \text{ in} - 9.625 \text{ in}/2}{\tan(30°)} = 9 \text{ in}$$

As 18 inches was the designed length of the weld, no reduction in the effective width is required.

Design hinge zone

AISC 341, §13.3b requires that connections be designed to accommodate brace buckling in one of two ways. Connections may be designed for a flexural strength in excess of the brace flexural strength in the plane of buckling, thereby ensuring that inelastic rotations will take place in the brace instead of the connection. Alternatively, a hinge zone in the gusset may be provided to accommodate rotations associated with brace buckling. The second approach is taken in this example.

In the design of this hinge zone, the minimum length of the offset is twice the gusset plate thickness (2 × ⅞ inch = 1¾ inches). The design will specify an offset with a minimum dimension of 2 inches and a maximum of 3 inches to provide construction tolerance. The stability calculations will use 3 inches.

In addition to this offset dimension, the gusset plate should be extended so that the hinge zone occurs above the top of the composite deck where one is present. Others have suggested block-outs in the deck to prevent restraint of the hinge zone.

These two requirements determine the dimension of gusset edge "c" in Figure 1A-11. The hinge zone must be oriented perpendicular to the brace axis. The unrestrained length of plate on the column side of the brace centerline (gusset edge "d") is therefore significantly longer than that on the beam side (gusset edge "c") in this design.

Check gusset-plate edge stability

Based on work by Astaneh (1998), previous editions of this book have included checks for edge buckling. Recent work by Roeder (2009) demonstrates that such checks are not needed.

Check gusset-plate compression strength

AISC 341, §13.3 requires consideration of stability in the design of gusset plates. In the past, the full yield strength of the material was used in the calculation of compression strength. This was never a safe assumption; it is baseless for the typical SCBF connection design, which has unrestrained hinge zones. Several methods have been proposed for calculating the effective length of the gusset plate. The method used here is based on the brace centerline dimension from brace end to beam flange (or column flange, if the brace centerline does not cross the beam flange), which is the longest unsupported length in the modified workpoint design; this length can be determined from the dimensions already established. The effective length factor is typically taken to be 1.2, based on Astaneh's tests and recommendations (1998). Newer studies suggest that this can be reduced to below 1.0 in many cases (Dowswell, 2006).

$$\frac{Kl}{r} = \frac{(1.0)(14.5\text{ in})}{7/8\text{ in}/\sqrt{12}} = 57.4$$

$$\phi F_{cr} = 0.9(0.658^{(Fy/Fe)})Fy = 39.3\text{ ksi}$$

$$\phi P_n = \phi A_g F_{cr} = [(2)(10.0\text{ in}) \times 7/8\text{ in}]\ 39.3\text{ ksi} = 619\text{ kips}$$

$$\phi P_n > P_u \ldots\ o.k.$$

Check gusset plate tear-out

$$\phi R_n = \phi\,(A_v 0.6F_y + U_{bs} A_t F_u)$$

$$= 0.75[2(18\text{ in }(7/8\text{ in})(0.6)50\text{ ksi} + 1.0(7/8\text{ in} \times 9.6\text{ in})\ 65\text{ ksi}]$$

$$= 1118\text{ kips}$$ AISC 360, J4-5

$$\phi R_n > R_u \ldots\ o.k.$$

Check brace tear-out

The slotted connection used resists tear-out primarily in shear. The area effective in tension on either side of the slot is negligible.

$$\phi R_n = \phi A_v 0.6 F_y = 0.9[4(18 \text{ in})(0.465 \text{ in})(0.6)(1.4 \times 42 \text{ ksi})] = 1063 \text{ kips}$$

$$\phi R_n > R_u \ . \ . \ . \ \textit{o.k.}$$

Note that the factor R_y can be used on the brace strength.

Check net section rupture

The requirements of AISC 341 are such that no reduction of the brace gross section is permissible without reinforcement for most steel materials. (The allowable reduction in area is proportionate to the ratio of expected yield strength to ultimate strength; none of the commonly-used brace materials has a sufficiently low ratio to permit any reduction in area). The slotted-tube detail used here requires that the slot extend beyond the gusset edge approximately 1 inch for erection, thus leaving a reduced section. Two reinforcement details are shown in Figure 1A-12. For the detail shown on the right the designer should require that the welds shown be performed in the field so that the plate can be located properly with respect to the hinge zone. (For this reason the field-weld flag is shown here.) Proper quality requirements apply to these connections, and drilled holes at the ends of the slots (or grinding) relieve the likelihood of fracture.

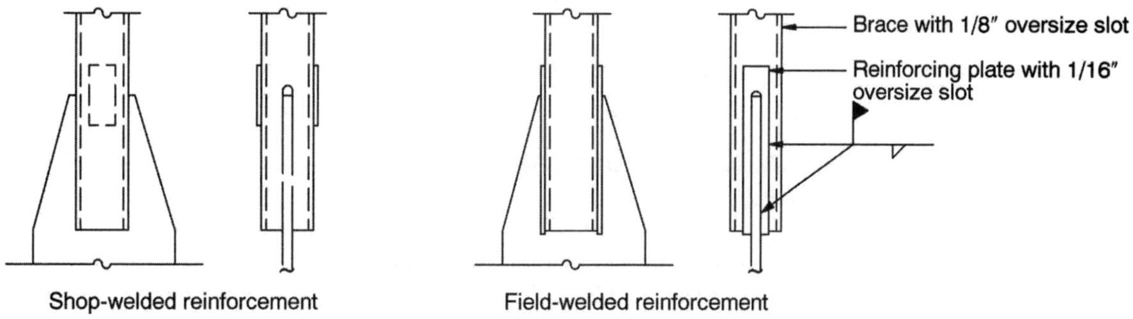

Figure 1A-12. Brace-end reinforcement details

Experimental studies have shown that net-section fracture will be the governing limit state for connection details with reduced sections (Korol, 1996; Cheng, Kulak, and Khoo, 1998).

Reinforcement plates will be added so that the effective area of material in this zone is slightly greater than that of the gross section. A ½ × 4 × 8-inch plate will be added on each unslotted side of the brace at the end of the slot.

$A_e = UA_n$ AISC 360, D3-1

For round sections with L/D > 1.3 no lag reduction is required. The addition of reinforcement requires a reduction, however.

Brace: Subtract width of slot (⅛ in more than gusset)

$$A_n = 13.4 \text{ in}^2 - 2(\tfrac{7}{8} \text{ in} + \tfrac{1}{8} \text{ in})(0.465 \text{ in}) = 12.5 \text{ in}^2$$

Plates:

$$2 \times \tfrac{1}{2} \text{ in} \times 4 \text{ in} = 4 \text{ in}^2$$

$$L = 18 \text{ in}$$

$$\overline{X} = \frac{\Sigma \overline{X}_i A_i}{\Sigma A} = \frac{\dfrac{9.625 \text{ in}}{\pi} 13.4 \text{ in}^2 + \dfrac{9.625 \text{ in} \times 4 \text{ in}^2}{2}}{(13.4 \text{ in}^2 + 4 \text{ in}^2)} = 3.47 \text{ in}$$

$$U = 1 - \frac{3.47}{18} = 0.81$$

$$A_e = 0.81(12.5 + 4) = 13.4 \text{ in}^2 > A_g \; \ldots \; o.k.$$

Welds to develop the reinforcement plate capacity into the plate and back out of it are required. The required weld strength is

$$R_u = b\,t\,F_y = 4 \text{ in } (\tfrac{1}{2} \text{ in})42 \text{ ksi} = 84 \text{ kips}$$

The design weld strength for the weld on half of the plate is

$$\phi\, R_n = \phi\, 0.6 F_{EXX}\,(s\,L)$$

$$\phi\, R_n = 0.75(0.6)70 \text{ ksi}[(\tfrac{5}{16} \text{ in})\; 12 \text{ in} \times \frac{\sqrt{2}}{2} = 84 \text{ kips}$$

Design beam-to-column connection

The beam-to-column connection must be designed considering these forces, in addition to other beam forces determined from the analysis.

In this design, two brace gussets connect to the beam-column joint: one above and one below. It is unlikely that both connecting braces will yield in tension simultaneously. The design will consider the larger yielding in tension in combination with buckling of the smaller, as well as the beam forces from the analysis. The brace above the beam is the smaller brace; the determination of its connection forces will not be presented. Beam forces should be factored using the special seismic load combinations including overstrength.

The vertical force on the connection is the sum of the beam shear and the two vertical forces from the gusset plates

$$V_{\text{con}} = V_{B4} + V_{B5} + V_{\text{beam}} = 240 \text{ kips} + 86 \text{ kips} + 10 \text{ kips} = 336 \text{ kips}$$

The horizontal force is the difference between the two horizontal forces from the gusset plates to the columns (as the horizontal components of the brace forces above and below the connection are opposite), combined with the beam axial (collector) force. This force is at its maximum when the brace below is yielding and the brace above is in its post-buckling state ($0.3P_n$). The force may be reduced considering the unamplified collector force

$$H_{con} = H_{C4} - 0.3H_{C5} - H_{beam} = 312 - 90 - 21 = 201 \text{ kips}$$

The moment at the column face is calculated from the beam-end moment, using the amplified seismic load A_u, in conjunction with the other connection forces. The beam end moment is

$$M_{beam} = 1.48(42 \text{ kip-ft}) + 2.0(85 \text{ kip-ft}) + 0.5(17 \text{ kip-ft}) = 241 \text{ kip-ft}$$

$$M_{con} = 336 \text{ kips}(6.5 \text{ in}) + 241 \text{ kip-ft} - (312 \text{ kips} + 216 \text{ kips})(9.05) = 24.8 \text{ k-ft}$$

This moment need not exceed the expected plastic moment of the beam. The beam moment is always in the same direction as that caused by the vertical components of the gusset forces, V_{B4} and V_{B5}, and opposite that caused by the horizontal components H_{B4} and H_{B5}. The connection moment is therefore

$$M_{con} = V_{con}\alpha + M_{beam} - (H_{B4}\, d/2 + H_{B5}\, d/2)$$

$$M_{con} = 336 \text{ kips}(6.5 \text{ in}) + 241 \text{ kip-ft} - (312 \text{ kips} + 216 \text{ kips})(9.05 \text{ in}) = 24.8 \text{ kip-ft}$$

As the axial forces are sizeable, the beam flanges will be welded to the column to transmit the axial force. The shear will be resisted separately by means of a web plate.

Check shear strength

The shear will be resisted by a 14 × 14-inch plate fillet welded on three sides to the beam web and joined to the column flange with a complete-joint-penetration weld. This plate will fit within the throat of the beam without requiring the welds to intrude into the "*k*" region. Bolts should not be used to resist the required connection forces because they would be sharing forces with the gusset-to-column welds. This is not allowed by AISC 341, §7.2. Erection bolts are acceptable.

The C-shaped weld of the web to the plate is an eccentrically loaded weld group similar to the column splice web plates. Table 8-9 can be used to calculate the required weld size. The angle of loading is 0°.

$C_1 = 1.0$ for E70

$L = 14$ in

$kL = 14$ in, $k = 1$ and $x = 0.333$

$e_x = aL = L - xL = 0.667L$, $a = 0.667$

Interpolating between values for $a = 0.6$ and $a = 0.7$, C is determined to be 3.90.

The required weld throat, D, is calculated to be

$$D_{min} = \frac{1}{16}\frac{P_u}{C\,C_1\,L} = \frac{1}{16}\frac{336}{(4.03)(1)14} = \frac{5.96}{16}$$

A ½-inch fillet weld will be used.

The required shear plate thickness is

$$t_{min} = \frac{P_u}{\phi\, 0.6\, F_y L} = \frac{304 \text{ kips}}{0.75(0.6)(50 \text{ ksi})14 \text{ in}} = 0.97 \text{ in}$$

A 1-inch plate will be used with a complete-joint-penetration weld to the column flange.

Because of the configuration of the shear plate, the full shear does not occur in the beam web at any section. In other connection details, the beam shear strength may be critical. The shear strength of the beam is

$$\phi\, V_n = \phi\, 0.6 F_y A_w = 0.9(0.6)\ (50 \text{ ksi})\ (0.45 \text{ in})18.4 \text{ in} = 224 \text{ kips}$$

This web would be of insufficient strength without the reinforcement provided by the shear plate.

Check flange strength

The demand on the flange welds can be calculated as the combined effect of the flexural and axial demands

$$P_u = \frac{M}{d} + \frac{1}{2} H_{con} = \frac{9.7 \text{ kip-ft}}{18.4 \text{ in} - 0.750 \text{ in}} + \frac{1}{2} 2.43 \text{ kips} = 7.8 \text{ kips}$$

Complete-joint-penetration flange welds will be used.

$$\phi\, P_n = \phi\, F_y b_f t_f = 0.9(50 \text{ ksi})(7.59 \text{ in})0.75 \text{ in} = 256 \text{ kips}$$ AISC 360, T J2.5

$\phi\, P_n > P_u$ . . . *o.k.*

8b. Brace-to-beam mid-span connection design

This connection is very similar to the brace-to-beam connection at the column. All net forces are transmitted through the gusset to the beam. The governing load case corresponds to one brace buckling and the other yielding in tension. The vertical component corresponds to the difference between these two strengths, and the horizontal component corresponds to the sum. Additionally, there is a moment due to the eccentricity of the workpoint to the weld (typically half the beam depth). Astaneh (1998) has recommended reducing this eccentricity in order to reduce the connection size in the chevron configuration. No eccentricity is included in the design of this connection in this example; eccentricity at the beam mid-point may be problematic in the two-story-X configuration. Both the connection and the framing member should be designed to withstand forces corresponding to the simultaneous yield of a brace below and a brace above the beam. The connection is shown in Figure 1A-13.

The same plate thickness (⅞ inch) is used for this connection as was used for the corner gusset. A 70-inch plate length is checked. The length of this gusset is determined by the necessity to extend beyond the braces. If the work point had been

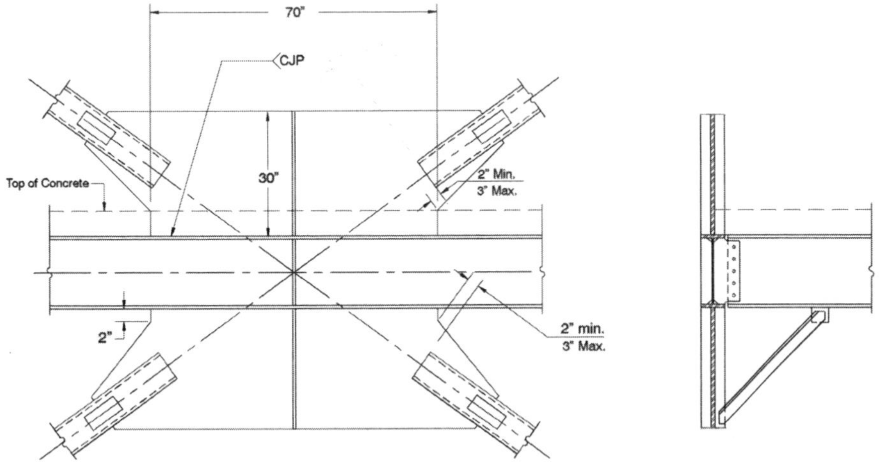

Figure 1A-13. Brace-to-beam connection at midspan

moved up to the top of the beam flange a 48-inch plate would have sufficed. However, significant flexural demands would have required a larger beam. The forces in the plate are calculated at the critical section. For this design, it is at the top of the slab, where the minimum horizontal section is subjected to the maximum forces.

Calculate forces (maximum horizontal force)

The gusset-plate forces are calculated from the expected strengths of the connecting braces.

$$V_{con} = (R_y F_y A_g - P_{max} A_g)\sin(\gamma) = (788\text{ kips} - 578\text{ kips})\sin(37.6°) = 128\text{ kips}$$

$$H_{con} = (R_y F_y A_g + P_{max} A_g)\cos(\gamma) = (788\text{ kips} + 578\text{ kips})\cos(37.6°) = 1082\text{ kips}$$

$$M_{con} = H_{con}\left(t_{slab} + \frac{d_b}{2}\right) = 1082\text{ kips }(6.25\text{ in} + 9\text{ in}) = 16{,}500\text{ in-kips}$$

Calculate eccentricity and effective width

Popov (1952) gives a generalized interaction formula for axial and flexural forces on a steel section in the inelastic range. For the simple case of a rectangular section, the formula can be rewritten to define an effective width. The effective width method is used to convert flexural and axial forces over a section to pure axial force over an effective section. This effective width is located at the center of the section. The axial stress calculated using this width is the same as the flexural stress calculated using the plastic section modulus of the section with the effective width removed from the center. Figure 1A-14 illustrates this concept. See Tamboli (1999) for a similar method.

$$e = \frac{M_{con}}{V_{con}} = \frac{16{,}500\text{ in-kip}}{128\text{ kips}} = 129\text{ in},\ W = 70\text{ in},\ W_{ef} = \sqrt{4e^2 + W^2} - 2e = 9.33\text{ in}$$

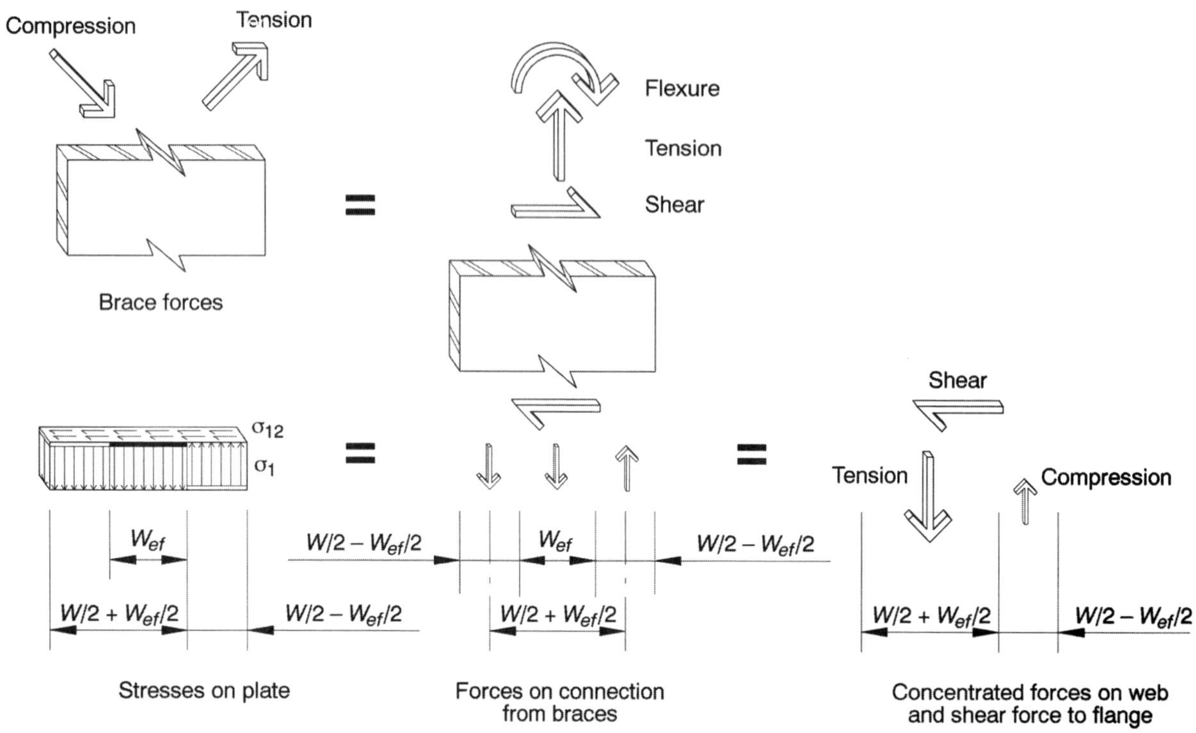

Figure 1A-14. Diagram of stresses in a section using the effective compression width concept

Check combined stresses

The combined stresses are checked against the gusset-plate strengths using von Mise's yield criterion. The principal stress is calculated by dividing the normal force by the reduced effective width, thus accounting for the effect of the flexural force.

$$\sqrt{\left(\frac{V}{\phi F_y L_{ef} t}\right)^2 + 3\left(\frac{H}{\phi F_y L t}\right)^2}$$

$$= \sqrt{\left(\frac{128\text{ kips}}{0.9(50\text{ ksi})(9.33\text{ in})7/8\text{ in}}\right)^2 + 3\left(\frac{1082\text{ kips}}{0.9(50\text{ ksi})(70\text{ in})7/8\text{ in}}\right)^2} = 0.76 \;\ldots\; o.k.$$

Check tension area

From Figure 1A-13 and the brace-to-gusset weld length calculated in part 8a, the width of the critical section can be calculated. It is 10 inches on the beam side of the brace centerline and 18 inches on the opposite side. Thus the total width is 28 inches and the eccentricity is 4 inches The effective width can be calculated from the equation

$$W_{ef} = \sqrt{4e^2 + W^2} - 2e = \sqrt{4(4\text{ in})^2 + (28\text{ in})^2} - 2(4.0)\text{ in} = 21\text{ in}$$

The tension strength is therefore

$$\phi P_n = \phi F_y W_{ef} t = 0.9(50 \text{ ksi})21 \text{ in}(7/8 \text{ in}) = 827 \text{ kips}$$

$$P_u = 788 \text{ kips}$$

$$\phi P_n > P_u \;...\; o.k.$$

Check vertical area in shear

The largest required force to be transferred across the vertical section of plate is the vertical component of the expected brace compression strength:

$$\phi V_n = \phi\, 0.6\, F_y A_w = 0.9(0.6)(50 \text{ ksi})(30 \text{ in})7/8 \text{ in} = 709 \text{ kips}$$

$$V_u = P_{\max} \sin(\gamma) = 578 \text{ kips} \sin(37.6°) = 352 \text{ kips}$$

$$\phi V_n > V_u \;...\; o.k.$$

Check plate compression strength

From the geometry of the plate the length of the gusset plate along the brace centerline is 13.3 inches; the gusset plate is considered restrained by the concrete deck. The effective width is the same as that calculated for the tension strength.

$$\frac{KL}{r} = \frac{(1.0)13.3 \text{ in}}{(7/8 \text{ in}/\sqrt{12})} = 52.7$$

$$F_{cr} = (0.658^{F_y/F_e})\, F_y = 40.8 \text{ ksi}$$

$$\phi P_n = 0.9(7/8 \text{ in})(21 \text{ in})(40.8 \text{ ksi}) = 675 \text{ kips}$$

This is greater than the required compression strength of 578 kips calculated earlier for the brace connection at the beam-column intersection.

Check edge buckling

As in part 8a, the maximum unstiffened length is 18.1 inches. Stiffeners are added at the edges to comply with this limit. These stiffeners must not enter into the hinge zone, as shown in Figure 1A-13.

Check beam web yielding

The mid-span connections are typically quite long. The beam yield modes of web yielding and crippling will only be of concern under very special circumstances, such as beams with very thin webs. The capacities to resist these yield modes are calculated based on the formulae in AISC 60 §J10. To perform these checks, a force perpendicular to the flange needs to be determined, as does a bearing length, N. From Figure 1A-14, the total perpendicular tension force to be resisted by the web yielding strength can be calculated from the normal force and the moment

$$R_u = P_u + \frac{2M_u}{W + W_{ef}} = 128 \text{ kips} + \frac{(2)16{,}500 \text{ in-kip}}{70 \text{ in} + 9.33 \text{ in}} = 544 \text{ kips}$$

The bearing length can also be determined from the figure

$$N = \frac{W}{2} + \frac{W_{ef}}{2} = \frac{1}{2}(70 \text{ in} + 9.33 \text{ in}) = 39.7 \text{ in}$$

The strength to resist web yielding is

$$\phi R_n = \phi (N + 5k) F_y t_w = 1.0(39.7 \text{ in} + 5(1.625 \text{ in})](50 \text{ ksi})0.390 \text{ in} = 932 \text{ kips}$$

Check beam web crippling

The gusset-beam interface at midspan is generally not subject to net compression. There is compression due to flexure, however. That compression force can be determined using Figure 1A-14.

$$R_u = \frac{2M_u}{W + W_{ef}} = \frac{(2)16{,}500 \text{ in-kip}}{70 \text{ in} + 9.33 \text{ in}} = 416 \text{ kips}$$

The bearing length can also be determined from the figure

$$N = \frac{W}{2} - \frac{W_{ef}}{2} = \frac{1}{2}(70 \text{ in} - 9.33 \text{ in}) = 30.3 \text{ in}$$

The strength to resist web crippling is

$$\phi R_n = \phi \left[0.8 t_w^{\,2} \left[1 + 3\frac{N}{d}\left(\frac{t_w}{t_f}\right)^{\frac{3}{2}} \right] \sqrt{\frac{E F_y t_f}{t_w}} \right]$$

$$\phi R_n = 0.75 \left[0.8(0.390 \text{ in})^2 \left[1 + 3\frac{31.8 \text{ in}}{18.1 \text{ in}}\left(\frac{0.390 \text{ in}}{0.630 \text{ in}}\right)^{\frac{3}{2}} \right] \sqrt{\frac{29{,}000 \text{ ksi}(50 \text{ ksi})0.630 \text{ in}}{0.390 \text{ in}}} \right] = 498 \text{ kips}$$

$$\phi R_n > R_u \; . . . \; o.k.$$

Calculate forces (maximum vertical force)

The gusset-plate forces are calculated from the expected tension strength and the residual compression strength of the connecting braces.

$$V_{\text{con}} = (R_y F_y A_g - 0.3F_{cr}A_g) \sin(\gamma) = (788 \text{ kips} - 135 \text{ kips}) \sin(37.6°) = 398 \text{ kips}$$

$$H_{\text{con}} = (F_y A_g + 0.3F_{cr}A_g) \cos(\gamma) = (788 \text{ kip} + 135 \text{ kips}) \cos(37.6°) = 731 \text{ kips}$$

$$M_{\text{con}} = H_{\text{con}}\, t_{\text{slab}} = 731 \text{ kips } (6.25 \text{ in} + 9 \text{ in}) = 11{,}200 \text{ in-kip}$$

Similar procedures should be followed using these forces as were followed with the previous forces based on maximum compression strength. In general, this load case does not govern the design of the connection, as the tension strength of the horizontal section of the gusset plate is greater than its shear strength. However, where very thin beam webs are used, this condition may be the more severe.

9. Notes on collectors and foundations

Collector design in SCBF structures can be accomplished in a variety of ways. Reinforcement in a composite deck can be utilized. The deck must then be treated as a concrete diaphragm complying with the requirements of Chapter 19 of the IBC. If confinement of the concrete is required, this approach is not economical. Alternatively, the steel beam can be designed as a beam-column, with or without composite beam action. This approach can be very efficient; collector connections near the braced frames may need to be significantly more substantial than the simple gravity connections.

10. Base plate design

The anchorage to the foundation must be designed for the amplified seismic load (Equations 16-5 and 16-7 with overstrength). Although this force can be reduced considering the dynamic forces resulting from uplift, use of the footing weight by itself as the limit on member and connection forces is unconservative (SEAOC, 2001). The foundation itself need not be designed to preclude uplift. It is permitted to consider only 75 percent of the overturning moment from the superstructure (§12.13.4) when strength design is used. As discussed earlier, the occurrence of uplift is not likely to eliminate ductility demands in the superstructure and is not an acceptable alternative to connection strength and ductility requirements.

Hohbach et al. (2002) make certain recommendations concerning the detailing of gusset plates at the base connection. In typical building construction, base plates are located at an elevation 12 to 18 inches below the top of a concrete slab. Thus extremely large gussets or block-outs are required if the entire hinge zone of the gusset is required to occur above the top of the concrete slab. Therefore, another approach to these connections is worth consideration. Braces can be extended as close to the base plate as possible; the restraint provided by the surrounding concrete slab can be shown to be sufficient to force the plastic hinge out above the slab, outside of the connection. The design moment can be resolved into a force couple that relies on bearing of the brace and gusset against the concrete. This method can take advantage of the high bearing stress permitted with confined concrete for the lower force in the couple. Alternatively, the shear capacity of the base plate can be considered for the lower force in the couple. Special reinforcing is required where braces are located near the edge of the slab. The end-fixity that this detail provides for the brace has the added benefit of lowering the effective length of the first-floor braces, which are the most heavily loaded, thus permitting smaller sections to be used. (Note that using braces with a lower design tensile strength than those at the level above is not recommended as it could lead to the formation of story mechanisms.)

Designers should always bear in mind that the greatest resistance of SCBF lies in the tension strength of the braces. Although an elastic analysis shows equal forces in opposed braces, the design should address the unequal distribution that results after buckling. In the design of foundations, this can be addressed by tying the footings together to the degree necessary to transfer one half of the horizontal component of the ground-floor brace tensile strength.

11. Additional Requirements

In addition to the requirements for SCBF, AISC 341 gives requirements for the entire seismic-load-resisting system.

The designer should clearly identify the seismic-load-resisting system. Welds that are a part of this system are required to have a Charpy V-Notch toughness of 20 ft-lb at –20°F (Section 7.3b). Additionally, welded, bolted, screwed, and shot-in attachments are not permitted in areas of expected inelasticity of the seismic-load-resisting system (Section 7.4). This is especially important for braces that are to be hidden in walls. Because braces are expected to yield in tension, no attachments should be made to the braces along their entire length; attachments to the gusset plates should also be prohibited in the construction documents. This requirement may necessitate a double wall if braces are not to be exposed.

In the design presented, braces are located adjacent to the exterior curtain wall of the structure. Out-of-plane buckling of the braces could result in a falling hazard if the braces are located very close to the windows. Tremblay (2002) gives a method for estimating the transverse displacement, v, of a brace due to buckling.

$$v = 0.7\sqrt{(2\mu - 1)\,\delta_y L}$$

The ductility, μ, may be estimated using the displacement amplification factor, C_d, and assuming that the elastic displacement is equal to the yield displacement multiplied by the resistance factor, ϕ, for compression. The resulting expression is:

$$v \le 2.5L\sqrt{\frac{F_y}{E}}$$

For this brace, the expected out-of-plane displacement is approximately 15 inches. It may be acceptable to locate the curtain wall within 15 inches of the brace if the glass is resistant to shattering and would not pose a falling hazard.

As mentioned earlier, non-frame columns are instrumental in the resistance of braced-frame buildings to story mechanisms. After buckling of braces, such columns often undergo significant bending, sometimes into the inelastic range (Sabelli, 2001). It is therefore recommended that their splices be designed for a minimum shear strength computed using their flexural strength divided by the clear height. Although this value is half of the maximum shear possible in the columns, it is expected to provide sufficient reliability, considering that the potential story mechanism may not occur at the splice level and may not require the full flexural strength at each end of the column segment.

Finally, Appendix Q of AISC 341 requires that a quality assurance program be developed as part of the contract documents. The quality assurance program should, at a minimum, cover inspection of all bolts and welds of the seismic-load-resisting system, verification of the unrestrained hinge zone in the gusset plate (if one is provided; see part 8 of this example), review of attachments to members of the seismic-load-resisting system, and non-destructive testing of all complete-joint-penetration and partial-joint-penetration welds.

Design Example 1B
Chevron and Zipper Configurations

Overview

This design example supplements Design Example 1A by illustrating the additional requirements that apply if the braced frames employed are of the chevron configuration, rather than the two-story-X (see Figure 1-1). These requirements apply to the beam and column design, as well as to the beam end connection design. Other aspects of the frame design are similar to those illustrated in Design Example 1A.

Traditional chevron-braced (V- and inverted-V-braced) frames have long been known to exhibit poor postelastic behavior. The post-buckling mode of behavior of a traditional chevron-braced frame is characterized by beam flexure (similar to an eccentrically braced frame with a long link connected to one of the columns) rather than the truss action that typically characterizes braced frames. Figure 1B-1 illustrates this mode.

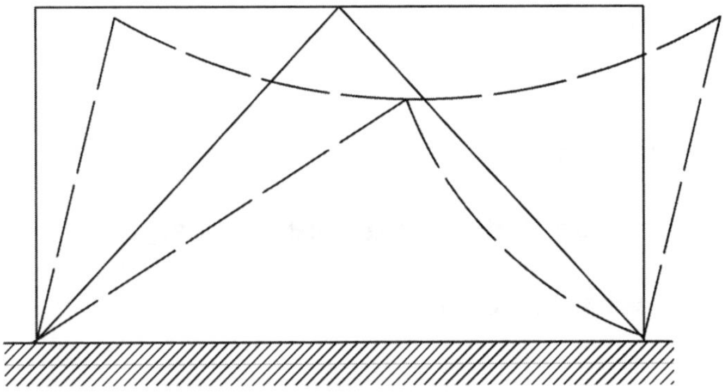

Figure 1B-1. Post-buckling mode of behavior of chevron-braced frames

The resistance of a buckled brace diminishes with additional deformation, sometimes precipitously, so that in the post-buckling range of deformation the vertical component of the force in the tension brace cannot be resisted by the compression brace. The unbalanced vertical component must then be resisted by the beam; flexible or weak beams lead to a negative frame stiffness (Khatib et al., 1988).

SCBF requirements for beams used in this configuration are intended to permit the frame to reach a yield mode based on yielding of the brace in tension prior to flexural yielding of the beam due to the unbalanced force. Figure 1B-2 shows an unfavorable yield mechanism (on the left) that is precluded by the SCBF requirements; it also shows (on the right) the expected, preferred mechanism. This preferred mechanism enables the frame to achieve an overstrength level roughly equivalent to that of other braced frame configurations and, in some cases, a positive post-buckling stiffness (Bruneau et al., 1998). OCBF requirements for chevron braced frames limit the slenderness of braces in order to limit the vertical unbalanced force; this is discussed in Design Example 1E.

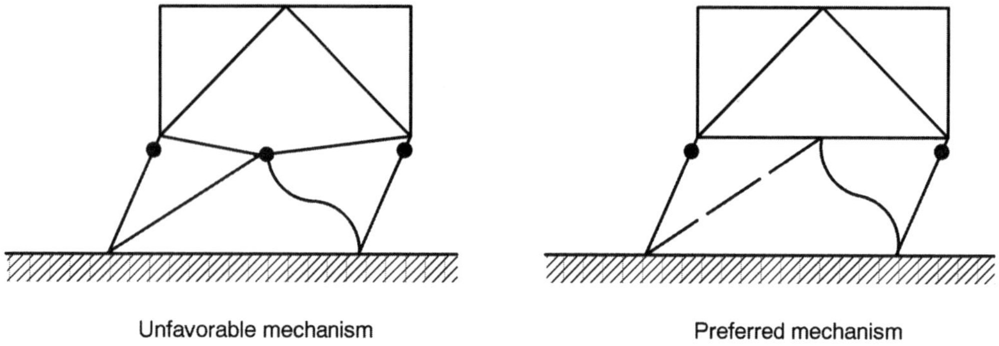

Figure 1B-2. Possible yield modes of chevron-braced frames

The SCBF requirements indirectly lead to stiff beams, but it is nevertheless possible that post-buckling frame stiffness will be very low, even negative (Sabelli et al., 1998). It is therefore recommended that the two-story X configuration be used instead wherever it is feasible.

Outline

1. **Bracing configuration**
2. **Beam design**
3. **Notes on connection and column design**
4. **Zipper configuration**

Calculations and Discussion	Code Reference

1. Bracing configuration

Chevron-braced frames may be either of two varieties: V-bracing or inverted-V-bracing (see Figure 1B-3). In both cases, the buckling of a brace leads to an imbalance in the vertical forces at the beam midpoint. This unbalanced load must be resisted by beam flexure. The direction of the load depends on the bracing configuration: V-bracing will result in an upward load (which is partially offset by gravity effects); inverted-V-bracing will result in a downward force (for which composite beam resistance is advantageous).

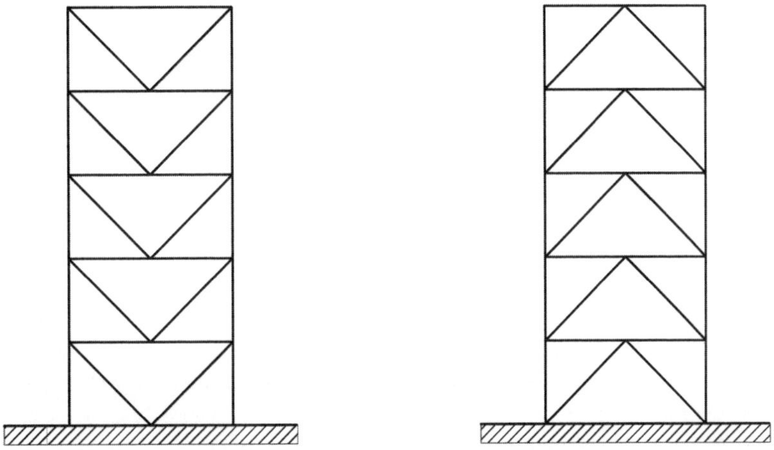

Figure 1B-3. Chevron-bracing configurations

In this design example, the frame will be assumed to be similar to the frame in Design Example 1A. However, the frame will be assumed to be a stacked inverted-V configuration as shown above. Member sizes will be revised as required based on the additional requirements for this configuration.

2. Beam design

In addition to the other requirements for a beam in this configuration (described in Design Example 1A), beams in chevron braced frames must be designed for the vertical unbalanced force. AISC 341,§13.4a requires that the beam have the strength to resist the vertical force, resulting from the simultaneous application of the expected yield strength of one brace, $R_y F_y A$, and the residual compression capacity of the other, $0.3F_{cr}A$.

In this example, the roof beam will be designed. The brace sizes are identical to those determined in Design Example 1A. The resulting vertical unbalanced force is

$$Q_b = (R_y F_y A - 0.3F_{cr}A)\sin(\gamma)$$

$$Q_b = [1.4(42\text{ ksi})7.86\text{ in}^2 - (0.3)34.7\text{ ksi} \times 7.86\text{ in}^2]\sin(37.6°) = 232\text{ kips}$$

AISC 341, §13.4a

The corresponding horizontal force is

$$Q_h = (R_y F_y A + 0.3F_{cr}A)\cos(\gamma)$$

$$Q_h = [1.4(42\text{ ksi})7.86\text{ in}^2 + (0.3)34.7\text{ ksi}(7.86\text{ in}^2)]\cos(37.6°) = 431\text{ kips}$$

To resist these forces a composite beam will be designed. The flexural strength must be determined considering the axial force in the beam. Also, any eccentricity in the connection will result in additional flexural forces. Figure 1B-4 shows this loading.

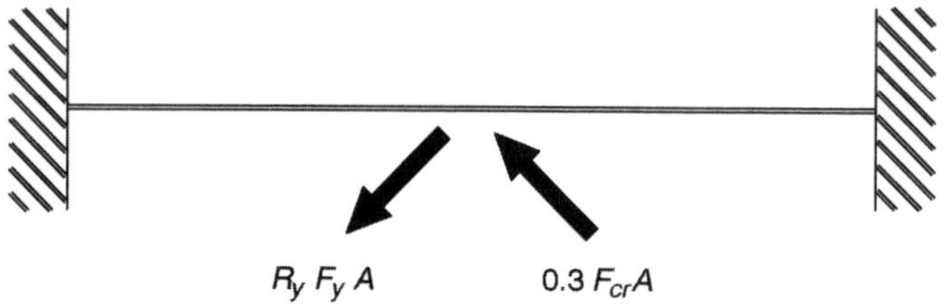

Figure 1B-4. Postelastic loading of a beam in a chevron-braced frame

The sum of the positive and negative beam flexural strengths must be sufficient to resist the effects of the vertical unbalance force, Q_b, in addition to flexure due to gravity loads (assuming no support from the braces). The total effect of the seismic force is $Q_b L/4$, where L is the beam span (see Figure 1B-5). It is also assumed for preliminary analysis that the moment magnification factor B_1 is equal to 1.0. The moment magnification factor B_2 is not applicable to this case because the flexural forces are determined from a capacity analysis rather than from applied loads.

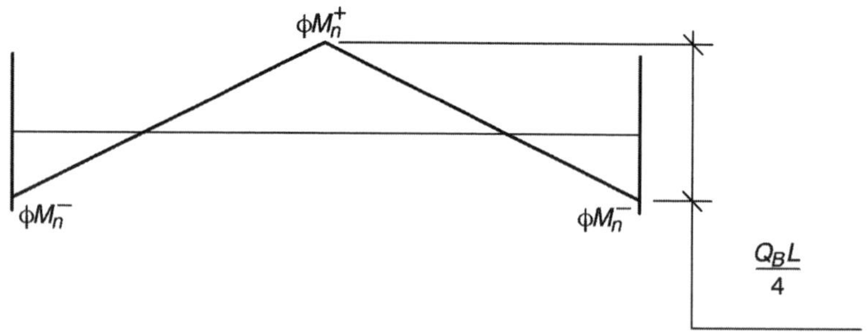

Figure 1B-5. Chevron-beam moment diagram

Since the method of assigning positive and negative moments may result in some flexural yielding prior to the development of the full member capacity, the beam should comply with the slenderness limits of Table I-8-1. Accordingly, a W36×160 beam has been selected.

Calculate axial compression

The axial compression in the beam is due mainly to the seismic loading. Although an elastic analysis may indicate that the story shear arrives to the frame from both sides equally, it is important to consider that the postelastic distribution may be considerably different (due to buckling of a brace above and subsequent loss of its compression strength and stiffness). In this example, the conservative assumption that all load arrives from one side of the frame will be used.

$$P_u = 1.48D + Q_h + 0.5L = 1.48(2.1\text{ kips}) + 431\text{ kips} + 0.5(1.1\text{ kips}) = 435\text{ kips}$$

Calculate flexure

$$M_E = Q_b\, L/4 = 232\text{ kips}(29\text{ ft}/4) = 1682\text{ kip-ft}$$

$$M_{\text{gravity}} = 1.48D + 0.5L = [1.48(696\text{ plf}) + 0.5(300\text{ plf})](29\text{ ft})^2/8 = 124\text{ kip-ft}$$

$$M_{\text{total}} = M_E + M_{\text{gravity}} = 1806\text{ kip-ft} = M_{nt}$$

$$M_{lt} = 0\text{ kip-ft}$$

Calculate axial strength

The W36×160 beam will be checked for both major-axis and minor-axis buckling. The beam is laterally supported at its midspan and ¼ points in its minor axis; it is braced at midspan in its major axis. The additional compression strength provided by the concrete slab will be neglected.

$$\frac{KL}{r_y} = \frac{(1.0)87\text{ in}}{2.50\text{ in}} = 34.8 \quad \text{(minor axis)}$$

$$\frac{KL}{r_x} = \frac{(1.0)174\text{ in}}{14.4\text{ in}} = 12.8 \quad \text{(major axis)}$$

$$F_{cr} = 45.8\text{ ksi}$$

$$\phi P_n = 0.9\, A_g\, F_{cr} = 0.9(47.0\text{ in}^2)45.8\text{ ksi} = 1937\text{ kips}$$

Calculate flexural strength

The flexural strength can be computed using the composite design methods of Chapter I of the AISC 360. For this design, the composite design table of the AISC Manual will be used.

The flexural strength of a composite W36×160 with a 3¼-inch concrete topping over 2-inch deck can be interpolated from the values given

$$Y_2 = 2\text{ in} + \frac{1}{2}3.25\text{ in} = 3.625\text{ in}$$

For a fully composite design, the flexural strength can be as high as 3790 kip-ft. The required flexural strength (such that the combined effects of flexure and compression do not exceed the beam strength) is lower. A partially composite design will be used in which

$$\Sigma Q_n = 1130\text{ kips}$$

Sufficient studs are required to develop this force. Using Equation I3-3, the stud strength is

$$Q_n = 0.5A_{sc}\sqrt{f'_c E_c} < R_g R_p A_{sc} F_u$$ AISC 360, §I3-3

$$Q_n = 17.1 \text{ kip/stud}$$

$$N = \frac{1130 \text{ kip}}{17.1 \text{ kip/stud}} = 66 \text{ studs}$$

These studs are required on each side of the point of maximum moment. Alternatively, channel shear connectors may be employed; these may be more practical for designs requiring such high shear transfer.

The corresponding design flexural strength for this composite design is

$$\phi M_n = 3430 \text{ kip-ft}$$

Check combined effects of flexure and compression

$$\frac{P_u}{\phi_c P_n} = \frac{435 \text{ kips}}{1937 \text{ kips}} = 0.22 > 0.2 \quad \text{Use interaction equation H1-1a}$$

$$\frac{P_u}{\phi_c P_n} + \frac{8}{9}\frac{M_u}{\phi_b M_n} = \frac{435 \text{ kips}}{1937 \text{ kips}} + \frac{8}{9}\frac{1806 \text{ kip-ft}}{3430 \text{ kip-ft}} = 0.69 < 1 \; \ldots \; o.k.$$

A similar check is required using the negative flexural strength and negative moment. Note that without composite design, the required size of a fixed-end beam is W36×182; alternatively, a non-composite, simply-supported W36×210 beam may be used.

3. Notes on column and connection design

Because of the postelastic redistribution of forces, the beam is expected to see large flexural forces not indicated by the elastic analysis. Accordingly, the beam reactions from the elastic analysis are not adequate for the design of the beam support connections, nor for the design of the column. Engineers should be careful to design the entire system so that the desired yield mode is achieved and nonductile modes, such as connection failure or column buckling, are precluded.

In this design, beam negative flexural strength is considered in order to permit use of a lighter beam. The beam end connection and the adjoining members (that is, the column and the beam connecting to the opposite side of the column) must resist this flexural force; if the size of these members must be increased in order to resist the flexural force, the savings from utilizing the beam negative flexural strength in the design will be less significant.

Figure 1B-6 shows the column forces from the elastic design and those derived considering the postelastic mode of behavior expected of inverted V-type chevron-braced frames. As the special seismic load combinations are permitted in lieu of design forces based on concurrent yield of braces on all levels, a corresponding method is proposed here: the seismic portion of the column design force, Q_E, in an inverted V-type chevron-braced frame should be taken as that indicated for the level below using a typical elastic analysis and the seismic load combinations with overstrength factor. For the first floor, the frame reaction should be used.

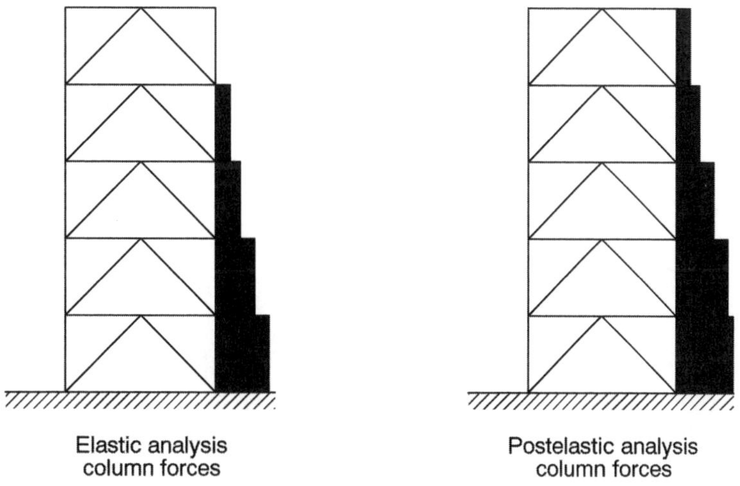

Figure 1B-6. Axial force diagrams for chevron-braced frames

4. Zipper bracing

Although zipper bracing is similar in appearance to chevron bracing, it is in fact a distinct configuration with a different yield mode; it therefore requires its own design approach.

The requirements for chevron-braced frames lead to a structure in which the expected yield mode involves: first, the buckling of a brace; second, the yield of the complementary brace in tension at the same level; and third, formation of plastic hinges in the columns. Beams are designed to withstand the unbalanced force from the braces, so that beam yield is not a part of the mechanism (as illustrated in part 2). This yield mode is superior to one involving flexural yielding of beams, but it is nevertheless a single-story mechanism. Any capacity to spread yielding over several stories is provided by the flexural strength of the building columns.

In contrast, the zipper configuration was developed to reduce the tendency toward story mechanisms by designing for a mechanism entailing beam yielding (Khatib et. al., 1988). Figure 1B-7 shows two possible yield mechanisms for zipper-braced frames. The first one is consistent with the SCBF requirements for chevron-braced frames; it is a story mechanism and thus concentrates ductility demands and involves plastic-hinge rotation in the columns. In the second mechanism, the vertical "zipper braces" transmit the vertical unbalanced forces to adjacent levels until the beams at several levels yield simultaneously. This yield mechanism is not characterized by the post-buckling loss of stiffness that non-SCBF chevron-braced frames exhibit. In fact, its postelastic behavior can be superior to that of chevron-braced SCBF (Sabelli et. al., 1998). The preferred yield mechanism for zipper-braced frames is similar to that of flexure-governed EBF. It is appropriate to use the compactness requirements for EBF beams in the design of these frames.

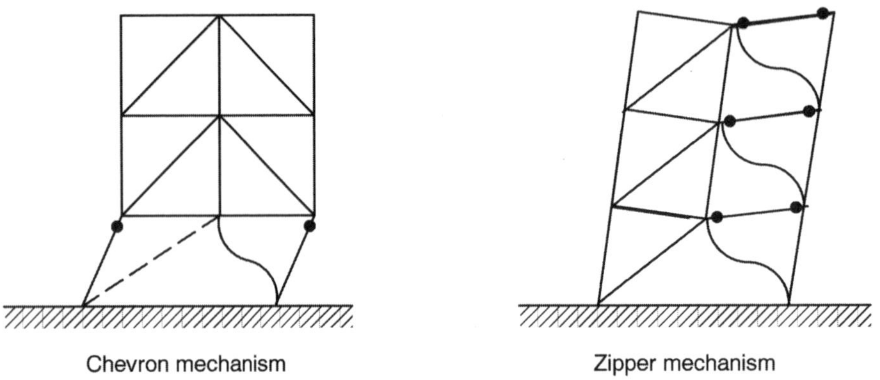

Figure 1B-7. Possible yield mechanisms for zipper-braced frames

No established design criteria exist for these zipper braces. Trial-and-error experimentation with non-linear time-history analysis has led to the recommendation that the required axial strength of the zipper brace be the design strength of the other braces adjoining the connection from the opposite side of the beam (Sabelli, 2001).

Tremblay (2003) has established mechanism-based design guidelines for the required strength of zipper braces. Provisions for zipper-braced frames are currently being developed for AISC 341; these include a capacity-based method for establishing maximum forces in the zipper brace. The draft provisions also include beam-strength requirements intended to favor the "chevron mechanism" over the "zipper mechanism"; the adequacy of the zipper mechanism, which is supported on buckled braces, has not been adequately established for inclusion in design codes. León has developed similar procedures, and has tested an assembly (2003).

Design Example 1C
Cross-Bracing and Single-Diagonal Configurations

Overview

The cross-bracing configuration is sometimes used in special concentrically braced frames. For most buildings, the two-story X configuration offers greater economy because it requires fewer connections than does a cross-braced frame (three for the two-story X compared to five for the cross-braced frame). However, cross-braced frames can offer a distinct advantage: the configuration can significantly reduce the effective brace length, thereby permitting the use of smaller braces. As Design Example 1A illustrates, connection and column size are dependent on brace size; if the brace size can be reduced, savings may be realized throughout the design.

A great deal of analytical and experimental research exists on the appropriate effective length for use in the design of braces in the cross-bracing configuration (Stoman, 1989; Kitipornchai and Finch, 1986; el-Tayem and Goel, 1986; DeWolf and Pelliccione, 1979; Picard and Beaulieu, 1987; Wang, 1992). Most of the analytical studies employ two key concepts: the out-of-plane bracing effect provided to the brace in compression by the high tension forces in the complementary brace; and the out-of-plane critical stiffness dividing first-mode and second-mode buckling as given by Timoshenko (1936).

Most experimental studies of cross-braced frames studied frames with back-to-back angles, in which both braces were continuous without the need of a splice. Tremblay et al. (1995) conducted an experimental study on cross-braced frames with hollow-structural-steel sections. For these sections, a special splice is necessary to provide continuity of the interrupted brace. Tremblay confirmed that the half-length is appropriate for determining the effective length for these frames. The study also demonstrated that the brace hysteretic behavior in cross-braced frames is similar to that of braces in other configurations, although the hysteretic energy dissipated in compression is slightly lower.

The results of these analytical and experimental studies can be used to determine the effective length for most typical connection conditions. The recommended effective length factors are summarized in Table 1C-1. These factors apply to the brace half-length as illustrated in Figure 1C-1. Note that for pin-end braces, the half-length can be measured from the centerline of splice to the hinge zone in the gusset plate. For fixed-end braces, the half-length may be taken conservatively from the column centerline to the centerline of the splice.

Table 1C-1. Effective Length Factors for Cross-Braced Frames

Splice Condition[4]	Fixed In-Plane Fixed Out-of-Plane		Fixed In-Plane Pinned Out-of-Plane			
Brace	Continuous Brace or Spliced Brace		Continuous Brace		Spliced Brace	
End Condition[3]	Theoretical *K* value	Recommended *K* value[5,6]	Theoretical *K* value	Recommended *K* value[5,6]	Theoretical *K* value	Recommended *K* value[5,6]
Fixed Out-of-Plane	0.70[7]	0.8	Note 8	(1.0)[9]	0.70[7]	0.8
Pinned Out-of-Plane	1.00[7]	1.0	Note 8	(2.0)[9]	1.00[7]	1.0
Fixed In-Plane	0.54[10]	0.7	0.54[10]	0.7	0.54[10]	0.7
Pinned In-Plane	0.80[10]	0.9	0.80[10]	0.9	0.80[10]	0.9

Notes:

1. Values apply to frames in which braces are of the same length and cross-section.
2. Values apply to frames in which the brace tension force is approximately equal to the compression force in the opposed brace. Where brace compression due to gravity loads exceeds 15% of the computed buckling strength, values may be unconservative. See applicable references for adjustments.
3. See Figure 1C-2 for examples.
4. See Figure 1C-3 for examples.
5. Values combine theoretical values with increases in effective length recommended in AISC Table C-C2.2.
6. Recommended values are appropriate for brace design; for connection design, the higher theoretical values should be used to determine the required compression strength.
7. See Stoman (1989).
8. See Kitipornchai and Finch (1986); el-Tayem and Goel (1986); Picard and Beaulieu (1987).
9. Recommended values for this splice condition are conservative and are not meant to preclude designers from calculating lower values. See note 8.
10. See Sabelli and Hohbach (1999).

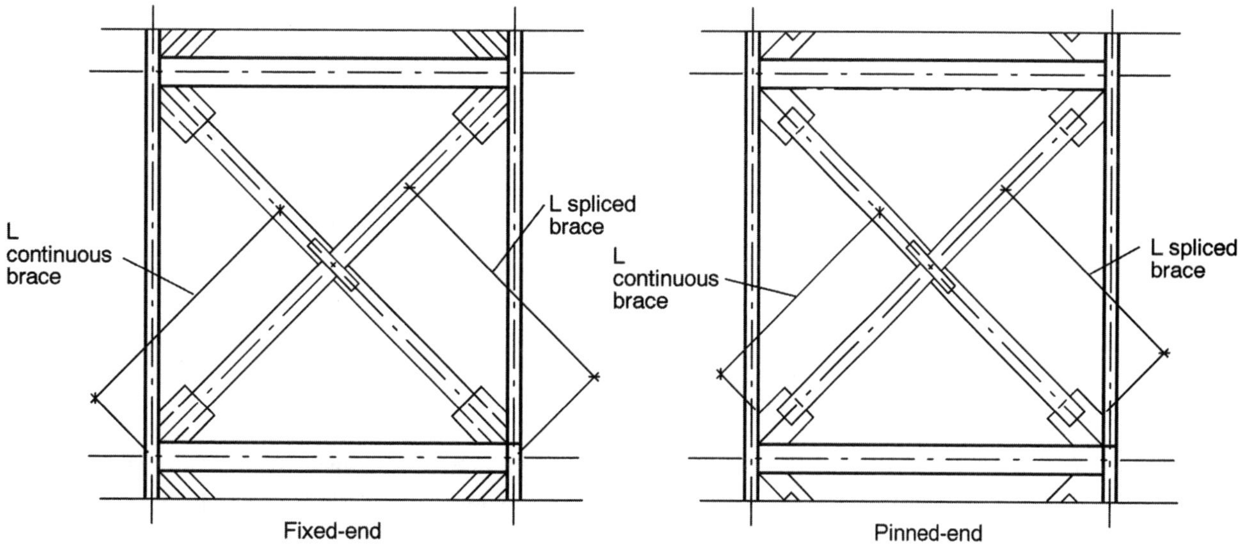

Figure 1C-1. Brace lengths to be used with effective length factor in Table 1C-1

Because of limits on width-thickness ratios, square or round sections are usually preferred where HSS members are used as braces. Because of the difficulty of splicing round HSS, square HSS are more common. See Design Example 1A for a discussion of square HSS. In such cases, the plane of buckling is governed by the effective-length factors, which are determined by the splice and end conditions. Some brace-end connection details are shown in Figure 1C-2. The first detail (detail A) shows a connection that provides both in-plane and out-of-plane fixity via two gussets that sandwich the brace. (The out-of-plane flexural rigidity may not be complete; see footnotes in Table 1C-1 for the recommended effective length factor). This connection is suitable for SCBF if it can provide flexural strength in excess of the expected brace flexural strength, including strain-hardening (see AISC 341, §13.3). Note that the gussets transmit only shear at their connections to the beam and column flanges due to their offset from the webs, resulting in flexibility with respect to forces perpendicular to the flanges. Thought must be given to the construction sequence of such a detail. It has the advantage of obviating the need for brace reinforcement.

The second connection illustrated in Figure 1C-2 (detail B) is a more common SCBF connection that provides a hinge zone. The design of such a connection is illustrated in Design Example 1A. The detail can be considered pinned out-of-plane; in-plane, it provides fixity, but the flexural rigidity may not be complete due to the bending of the walls of the steel-tube brace.

The third connection shown in Figure 1C-2 (detail C) provides a hinge in-plane, and fixity out-of-plane. Again, full flexural rigidity should not be assumed. Where braces are concealed in a wall, this detail requires that special attention be given to the possible in-plane restraint provided by the wall framing and non-structural elements. Although the effect of such restraint is to strengthen the brace, the actual mode of buckling must be accommodated by the end connections so that non-ductile modes of behavior are precluded.

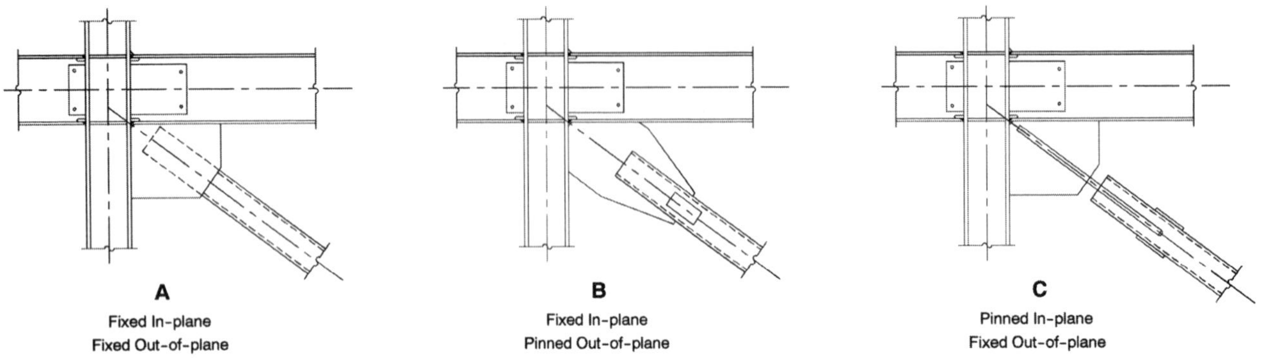

Figure 1C-2. Gusset connections providing differing end-fixity conditions

Two splice connection details are shown in Figure 1C-3. In the first connection, (A), full fixity is provided both in-plane and out-of-plane by means of two gussets that sandwich the spliced brace. This detail adds to the width of a brace, increasing the width of the wall required to conceal the braces. However, it should be noted that the effective length factor appropriate for this detail is lower than that for detail B, so that the brace size required may be significantly smaller.

The second splice connection (detail B) provides a single plate, that passes through a continuous brace and to which segments of the spliced brace are welded. This second connection provides fixity in-plane but not out-of-plane. This connection is not recommended because significant out-of-plane displacement is necessary in order to achieve sufficient restraint of the continuous brace (virtually all of the out-of-plane restraint provided by the discontinuous brace is due to catenary action). This required out-of-plane displacement will increase as the braces elongate under cyclic loading (see Tremblay, 2001). Furthermore, as was illustrated in Design Example 1A, slotted connections in HSS braces require reinforcement of the reduced section, so this connection may not offer any significant cost savings.

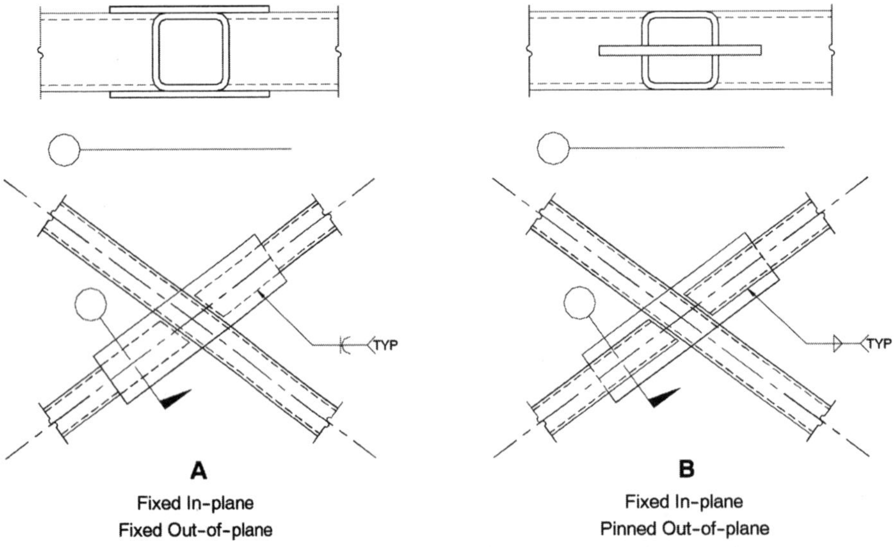

Figure 1C-3. Splice details withdiffering continuity conditions

In this design example, the second end detail (B) from Figure 1C-2 and the first splice detail (A) from Figure 1C-3 are used. Frame loads from Table 1A-4 are used to compute brace forces.

Calculations and Discussion | **Code Reference**

1. Design of brace

Calculate axial force

From Table 1A-4, the sum of the horizontal components of the two braces is 472 kips. The brace angle is 22°. Therefore, the brace axial force is

$$P_u = \frac{472\text{ kips}}{2\cos(22°)} = 255\text{ kips}$$

(assuming that the braces resist no gravity load but all of the lateral seismic load).

Determine brace length

The full diagonal length is 32 feet. Subtracting 18 inches on either end for the connection (an assumption that must be verified after the connection design is complete), the length from gusset hinge to gusset hinge is 29 feet. As discussed earlier, buckling occurs over the half length. The brace half-length is 14.5 feet, and the effective length factor from Table 1C-1 is 1.0 for out-of-plane buckling and 0.7 for in-plane buckling for either brace. (The end connection provides in-plane fixity and a pin out-of-plane; the splice connection provides full continuity in both planes.)

Select brace

Using the column-design tables in the AISC Manual, an HSS 6×6×½ is selected. This brace has a capacity of 260 kips at a length of 15 feet and is significantly smaller than the brace designed in Design Example 1A, and the connection design force, which is based on the brace strength, is correspondingly lower. However, the low brace angle in the cross-bracing configuration may cause the connection to be significantly larger, possibly requiring stiffening.

Check slenderness

$$\frac{K\,L}{r} = \frac{14.5\text{ feet}}{2.23\text{ in}} = 78 < 4\sqrt{E/F_y} = 100 \;\ldots\; o.k.$$

AISC 341, §13.2

Check width-thickness ratio

AISC Seismic gives limits for brace compactness; the limit for square tubes is

$$\frac{b}{t} >= 0.64\sqrt{E/F_y} = 16$$

AISC 341, §13.2, T I-8-1

where the dimension b is defined in AISC 360 (Section B5.1) as the flat width of the tube between the curved corners. Typically, b can be taken as the overall width minus 3 times the thickness, as the corner radius is in the range of 150 to 300 percent of the wall thickness. For the HSS 6×6×½ brace, b can be taken as 6 inches – 3(0.465 in) = 4.61 in. Therefore, b/t = (4.61 in/0.465 in) = 9.9. The brace complies.

2. Design splice connection

To justify the effective length factor used, full continuity of the splice connection must be demonstrated. There are three requirements for demonstrating this. As the stability of the brace against buckling depends on its flexural stiffness, the first requirement is that the splice region have a moment of inertia at least as great as that of the brace. The other two requirements are given by the AISC 341: §13.3 requires that the connection have axial strength at least as great as the expected brace axial strength; and also requires that the connection have a flexural strength at least 110 percent of the expected brace flexural strength ($1.1R_y F_y Z$).

For the splice connection, two plates, ⅞ × 8 inches will be checked. The brace properties are

$$I = 50.5\text{ in}^4$$

$$R_y F_y A = 1.4(46\text{ ksi})9.74\text{ in}^2 = 627\text{ kips}$$ AISC 341, §13.3

$$1.1\, R_y F_y Z = 1.1(1.4)(46\text{ ksi})19.8\text{ in}^3 = 1403\text{ kip-in}$$

The splice connection properties are

$$I_p \;= 2\,t_p\,w\left(\frac{b+t_p}{2}\right)^2 = 2(0.875\text{ in})8\text{ in}\left(\frac{6\text{ in}+0.875\text{ in}}{2}\right)^2 = 165\text{ in}^4$$

$$\phi\, P_n \;= \phi\, 2\, w\, t_p\, F_y = 0.9[2(8\text{ in})](0.875\text{ in})50\text{ ksi} = 630\text{ kips}$$

$$\phi\, M_n = \phi\, Z\, F_y = \phi[t_p\, w(b + t_p)]F_y = 0.9[0.875\text{ in}(8\text{ in})(6\text{ in} + 0.875\text{ in})]50\text{ ksi} = 2166\text{ kip-in}$$

The splice meets the three criteria for continuity.

It should be noted that the flexural strength requirement is not simultaneous with the axial strength requirement. The reason for this is simple: each of these is based on full yielding of the entire cross section, so that no additional forces are possible. Understanding the transition states between uniform compression stress across the brace section to the formation of a full plastic hinge requires an understanding of inelastic buckling. (Inelastic buckling is the only relevant mode to consider, as plastic-hinge formation is consistent with neither elastic buckling nor squashing). At the onset of inelastic buckling, one face of the compression member begins to unload in the plastic-hinge area, eventually reversing the strain and yielding in tension; the compression stress in the opposite face continues to increase until a full plastic hinge is formed. Thus analyzing the connection separately for the two conditions, axial yield and plastic-hinge formation, reliably bounds the demands on the connection.

The splice plates require welds to the spliced brace to develop its axial strength. The required length of four groove welds is

$$L_{\min} = \frac{\frac{1}{4}P_u}{\phi 0.6 F_{\text{EXX}}\left(\frac{5}{8}t\right)} = \frac{\frac{1}{4}627\text{ kips}}{0.75(0.6)70\text{ ksi}\left(\frac{5}{8}\right)0.465\text{ in}} = 17.1\text{ in}$$

The plate length should therefore be approximately 50 inches to permit these welds to be placed on all four sides of both brace segments, leaving 16 inches for the crossing of the continuous brace. This length could be reduced by the addition of slot welds of the plate to the brace wall.

3. Single-diagonal bracing

AISC 341 has special requirements for single-diagonal frames and other frames that do not have a reasonable balance between braces in compression and those in tension. Section 13.2 requires that a line of bracing be checked to ensure that the percentage of force resisted by braces in tension is between 30 and 70 percent. If the line does not comply with this requirement, the braces must be designed for the seismic load combinations with overstrength factor (which include factor Ω_o).

This requirement leads to a system with an effective response reduction coefficient, R, of approximately 3.0. However, the effective overstrength factor applied to columns and collectors is 1.0. This approach is not well suited to conditions in which there is significant uncertainty as to the magnitude of the lateral force; seismic design may be considered such a case. It is therefore not recommended here.

Design Example 1D
Buckling-Restrained Braced Frame

Overview

Buckling-restrained braced frames (BRBF) are a relatively new development in the United States. They are similar to special concentrically braced frames (SCBF) in that a triangulated vertical framework of members makes up the primary seismic-load-resisting system. The chief difference between the two systems is in the braces: buckling-restrained braces achieve significantly higher ductility and energy dissipation by effectively eliminating buckling and the poor hysteretic performance associated with it. Because of this, the tension and compression behavior is very similar, and the requirements for BRBF have been simplified accordingly.

Steel is a very ductile material. Less-than-ductile performance of certain steel systems or elements is often attributable to extreme concentrations of inelastic strain demands. In brace buckling, inelastic hinges form in three segments of the brace-and-gusset system. Axial compression of the brace after buckling results in lateral translation of the brace midpoint and rotation in the inelastic hinges. Local buckling eventually occurs at the midpoint hinge, further magnifying the required strain (Tremblay, 2002).

In buckling-restrained braces, a fairly long yielding segment of steel is provided (typically 50 percent or more of the centerline-to-centerline diagonal length). This segment can yield in axial compression as well as in tension. The segment is part of an axial-force-resisting steel core. Its effective slenderness is extremely low due to the lateral restraint provided by a surrounding casing of steel, concrete, or both.

For buckling to be precluded, this casing must be kept free from axial forces. Several methods of confining the axial force to the steel core have been proposed (Aiken, 2001; Tremblay, 1999; Kalyanaraman et al., 1998); an idealization of one method is shown in Figure 1D-1. Some of these methods are fairly specialized; some are patented. All require significant precision in assembly. Under these circumstances, the design and construction of BRBF in the United States has grown around the specification of proprietary braces provided by a steel fabricator or a specialty manufacturer.

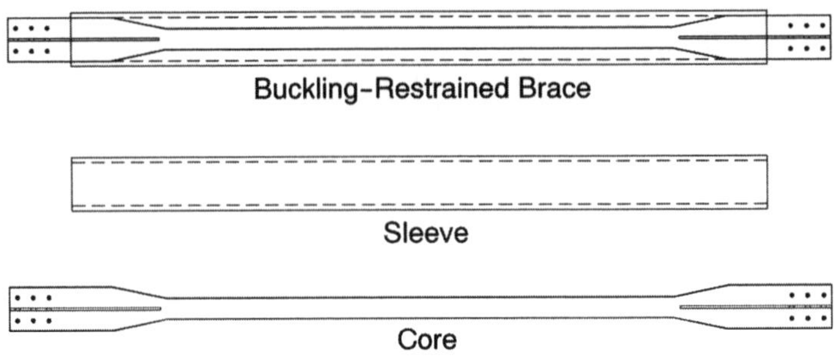

Figure 1D-1. A buckling-restrained brace

As BRBF are a recent development, they were not yet addressed by building codes such as the IBC (ICC, 2000) until the AISC 341 (AISC, 2005). In the first years of their use in the United States, certain conditions constrained the use of BRBF. Many of the building designs specifying buckling-restrained braces were required to undergo peer review. At the same time, designers recognized that the intricacy of buckling-restrained braces requires that some testing be done, but no consensus existed as to the appropriate testing requirements. These conditions effectively limited the use of BRBF to categories of projects that could support, and justify, the effects of special review and testing procedures on project schedule and budget.

In the United States, design approaches for this system have typically followed the same force-based approach that is used for other braced frames, with the buckling-restrained braced frame acting as the primary lateral-load-resisting system for both wind and seismic loadings. Such a procedure was proposed by Clark et al. (1999), based largely on the procedures for designing eccentrically braced frames.

Outline

1. **Determination of base shear**
2. **Determination of member and connection design forces**
3. **Calculation of deflections**
4. **Required testing**

Calculations and Discussion | Code Reference

1. Determination of base shear

The method for computing the base shear is similar to that for the special concentrically braced frame, with differences in the values of certain coefficients. The design will use coefficients appropriate for BRBF with rigid beam-column connections.

BRBF are more flexible than SCBF. Accordingly, their periods are longer, and a larger value of the coefficient C_T is used.

$$T_a = C_T h_n^{3/4} = 0.030(62^{3/4}) = 0.66 \text{ sec}$$ Eq 12.8-7

The seismic base shear, V, in a given direction shall be determined in accordance with the following equation

$$V = C_S W$$ Eq 12.8-1

The calculation of the seismic response coefficient, C_s, is performed using four equations. The first equation defines the coefficient as

$$C_s = \frac{S_{DS}}{R/I} = \frac{1.40g}{(8.0/1.0)} = 0.175g$$ Eq 12.8-2

The coefficient need not exceed

$$C_s = \frac{S_{D1}}{(R/I)T} = \frac{0.93g}{(8.0/1.0)0.66 \text{ sec}} = 0.176g$$ Eq 12.8-3

The value of seismic response coefficient, C_s, must be at least equal to the following

$$C_s = 0.044 S_{DSI} = 0.044(1.4)(1.0) = 0.0616g$$ Eq 12.8-5, Supplement No. 2

In addition, for those buildings for which the 1.0-second spectral response, S_1, is equal to or greater than 0.6g, the value of the seismic response coefficient, C_s, shall not be taken as less than

$$C_s = \frac{0.5S_1}{(R/I)} = \frac{0.5(0.93g)}{(8.0/1.0)} = 0.058g \qquad \text{Eq 12.8-6}$$

Therefore, Equation 12.8-2 controls the base shear calculation for a short period building and the seismic coefficient is thus

$$V = 0.175W$$

Note that the design base shear is lower for BRBF than for SCBF.

Because the period of the building is above 0.5 second, the value of k used in the seismic force distribution is greater than 1.0. It is determined by linear interpolation between periods of 0.5 and 2.5 seconds, with corresponding k-values of 1.0 and 2.0. For this design, the value is 1.07.

2. Member design

To illustrate the design of BRBF, differences between BRBF and SCBF in the design of braces, columns, and beams are discussed below.

2a. Brace design

Buckling-restrained braces are typically a specification item. In this arrangement, required brace strengths are given in the construction drawings and the manufacturer designs the braces to comply, using the material and grade specified. The material grade has a significant effect on the brace stiffness. (The lower the yield stress, the greater the required area, resulting in a stiffer brace.) It is also possible that the brace stiffness can be varied by controlling the yield length of the brace. Manufacturers have limits on the acceptable range of the yield length. It should also be noted that decreasing the yield length concentrates the inelastic strain, reducing the cumulative energy dissipation capacity, although many of the tests of buckling-restrained braces have shown cumulative energy dissipation capacity well in excess of expected demands (López, 2001).

It is very beneficial to the designer to contact the BRB manufacturers to confirm that their material strength assumptions are close to the available material strength. In this design the material will be specified as A36, with additional requirements of a minimum yield strength of 40 ksi and a maximum yield strength of 44 ksi. Required brace yield strengths are shown in Figure 1D-2. Note that the chevron configuration is used. Because the tension and compression strengths of buckling-restrained braces are similar, this configuration is not problematic for BRBF as it is for SCBF. (Section 2d shows the design of the beam connected to the chevron braces.)

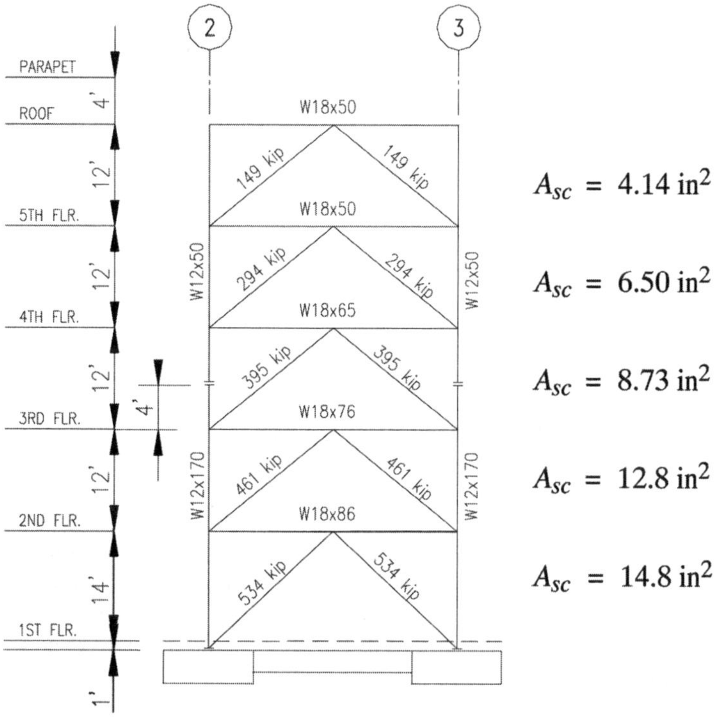

Figure 1D-2. Braced-frame elevation

The steel core area is determined using AISC 341-05 Equation 16-1

$$P_u >= \phi \, P_{ysc} = 0.9 F_{ysc} A_{sc}$$

$$A_{sc} = \frac{P_u}{(0.9 \times F_{ysc})}$$

2b. Brace connection design

With certain simplifications, gusset plates are designed similar to those for SCBF. BRBF gussets are not required to accommodate buckling of the brace; hinge zones are therefore not required, nor must the gussets have flexural strength in excess of that of the brace. Brace manufacturers may have certain size and thickness requirements in order to accommodate brace end connections. Small eccentricities may be permissible in the connection design if the resulting brace rotations are still within the tested limits; in this design, the workpoint is taken as the intersection of the beam and column centerlines.

The maximum connection force is determined using the brace strength and factors determined from testing. These factors are β, which represents the overstrength in compression (buckling-restrained braces tend to be somewhat stronger in compression than in tension), and ω, which represents strain-hardening within the expected deformation range. The factor R_y, representing expected yield strength as compared to nominal yield strength, is assumed not to be applicable, as the final cross-sectional size of a buckling-restrained brace is typically determined considering the material yield strength as determined from coupon tests, so that material deviation from nominal does not result in brace overstrength. The brace yield strength can thus be calculated from the required strength by dividing by the resistance factor.

Values of β and ω are assumed

$$\beta = 1.08$$

$$\omega = 1.25$$

The brace connection force at the fourth floor is

$$Q_E = 1.1\beta\,\omega\,P_{ysc} = 1.1\beta\,\omega\frac{P_u}{\phi} = 1.1(1.08)(1.25)\frac{294\text{ kips}}{0.9} = 485\text{ kips} \qquad \text{AISC 341, §16.3}$$

Note that using design forces based on factors determined in testing requires that measures be taken to ensure that the values used are adequate. Buildings may be designed with a certain brace product in mind, and the manufacturer may have sufficient test data to preclude the need for project-specific testing. Alternatively, the project specifications may require testing that justifies the values used. In either case, the designer must use values that are appropriate for the available brace types.

2c. Column design forces

Column design forces are determined using an explicit consideration of the capacity of all connecting braces.

For the design of the fourth-floor column, the maximum compression force corresponds to the yielding of the fifth-floor brace. The design force for that brace is 149 kips; the yield force is greater due to brace compression overstrength and strain-hardening (as well as the resistance factor used in design).

For BRBF, a more appropriate formulation is

$$Q_E = \beta\,\omega R_y F_{ysc} A_{sc}$$

Thus the maximum seismic axial force on the column is the vertical component of the strain-hardened brace compression strength from the fifth-floor level

$$Q_E = 1.08(1.25)1.1\,\sin(37.6°)40\text{ ksi}(4.14\text{ in}^2) = 150\text{ kips}$$

For this design, as the yield stress of the core is limited to the range of 40-44 ksi, R_y = 44 ksi/40 ksi = 1.1.

This value is significantly lower than that calculated using the overstrength factor

$$\Omega_o Q_E = 2.5(93\text{ kips}) = 233\text{ kips}$$

The required strength at the third floor can be determined similarly, using the vertical component of the strain-hardened brace compression strengths from the fourth-floor and fifth-floor levels.

2d. Beam design forces

Beam design forces are determined considering the maximum unbalance load resulting from unequal tension and compression strength of braces. The vertical force is

$$Q_b = (\beta\, \omega\, R_y F_y A - \omega\, R_y F_y A)\sin(\gamma) = (\beta - 1)\, \omega R_y F_y A \sin(\gamma)$$

$$Q_b = (1.08 - 1)1.25(1.1)(40 \text{ ksi})(6.5 \text{ in}^2)\sin(37.6°) = 17 \text{ kips}$$

This unbalanced force is much smaller in BRBF than in SCBF. The corresponding horizontal force is

$$Q_h = (\beta\, \omega\, R_y F_y A + \omega\, R_y F_y A)\cos(\gamma) = (\beta + 1)\, \omega\, R_y F_y A \cos(\gamma)$$

$$Q_h = (1.08 + 1)1.25(1.1)(40 \text{ ksi})(6.5 \text{ in}^2)\cos(37.6°) = 589 \text{ kips}$$

Because BRBF do not exhibit dramatic postelastic force redistributions between braces in compression and those in tension (as do SCBF), this horizontal component can be distributed according to the elastic force distribution: 60 percent from one side and 40 percent from the other.

3. Calculation of deflections

For a given length and strength, the stiffness of a buckling-restrained brace may vary significantly between brace types. There may also be some leeway in setting brace stiffness by varying the yield length within the acceptable range for the brace type. In this design, it is assumed that the brace stiffness is based on a yield length of 50 percent of the centerline-to-centerline diagonal dimension.

Drifts must be compared to the allowable drifts, as was done for SCBF. Additionally, for designs in which significant overstrength is provided, drifts can be used to reduce the required testing range. For this design it is assumed that braces will be sized for the required force and that therefore no design overstrength is provided.

For BRBF from Appendix R of AISC 341 Table R3-1 with rigid beam-column connections, the displacement amplification factor, C_D, is 5.0. The calculated deflections are given in Table 1D-1. Note that the calculated displacements tend to be higher for BRBF than for SCBF.

Table 1D-1. Design story drifts for frame BF-2 and allowable story drifts

	Story Height	Displacement from model	Drift from model	Design Story Drift	Allowable Drift
Story Number	h_{sx} (ft)	δ_{xe} (in)	Δ_{xe} (in)	$\Delta_x = C_d\,\Delta_{xe}$ (in)	$\Delta_{Ax} = 0.020\, h_{sx}$ (in)
5	12	2.309	0.503	2.52	2.88
4	12	1.806	0.458	2.29	2.88
3	12	1.348	0.464	2.32	2.88
2	12	0.884	0.465	2.33	2.88
1	14	0.419	0.419	2.10	3.36

4. Required testing

Brace testing requirements are given in the appendix of the *Seismic Provisions*, AISC 341-05. For each brace used in a building design, there must be two applicable tests: a subassemblage test incorporating brace-end-rotation effects, and a second test which may be uniaxial. No upward extrapolation of brace strength is permitted of the subassemblage test, although unlimited downward extrapolation is allowed; limited extrapolation in either direction is allowed of the second test. The provisions require this second test to be between 50 and 150 percent of the strength of the brace in the design; thus a test can be used to qualify a brace between 67 and 200 percent of the strength of the tested brace.

For the brace sizes in the design, the following test series satisfies the requirements: a 600-kip subassemblage test, applicable to all braces in the design; a 300-kip uniaxial test, applicable to braces from 200 kips to 600 kips; and a 200-kip uniaxial test, applicable to braces from 133 kips to 400 kips. For specific testing requirements, including similarity between tested braces and those used in the construction, see Appendix T in AISC 341.

The test protocol is also described in Appendix T. For the uniaxial tests, the protocol is based on the maximum drift, which in this design corresponds to 5 times the brace yield displacement. Because the required strength governs the brace design (as opposed to drift control), the elastic displacement at the design base shear corresponds to the resistance factor, ϕ, times yield displacement. The maximum displacement is equal to the elastic displacement multiplied by the displacement amplification factor, C_D. Typically, the maximum brace displacement can be expressed in terms of the yield displacement as follows

$$\Delta_{bm} = C_D\,\Delta_{be} = C_D\,\Delta_{by}/\phi = 5.0\,\Delta_{by}/0.9 = 5.5\,\Delta_{by}$$

(to account for the inaccuracies of this method, this value is amplified by a factor of 2.0).

This is the maximum value and is given in Appendix T.

The brace test protocol for the uniaxial tests is given below; displacements are given in terms of brace yield displacement, Δ_{by}. For the purposes of computing the cumulative inelastic displacement, the actual yield displacement of the brace should be determined in the test, although an estimate of Δ_{by} can be made from the brace elastic properties. The value of the maximum brace displacement, Δ_{bm}, can be taken from the analysis, although the multiple of 5.5 is always appropriate for strength-controlled braces. Here, Δ_{by} is calculated to be 0.20 inch; the corresponding value of Δ_{bm} for the test is 1.10 inch.

Table 1D-2. Brace test protocol

Cycle	Deformation (in)	Inelastic Deformation (in)	Cumulative Inelastic Deformation (in)
2 @ Δ_{by}	0.20	0.00	0.00
2 @ $0.5\Delta_{bm}$	0.55	2.80	2.80
2 @ Δ_{bm}	1.10	7.20	10.00
2 @ $1.5\Delta_{bm}$	1.65	11.60	21.60
2 @ $2.0\Delta_{bm}$	2.20	16.00	37.60
1 @ $1.5\Delta_{bm}$	1.65	5.8	43.4

Cumulative inelastic deformation at end of protocol: 43.4 in = 217 Δ_{by}

For more information on the calculation of these quantities, see the commentary to AISC 341.

Design Example 1E
Ordinary Concentrically Braced Frame

Overview

Ordinary concentrically braced frames (OCBF) are expected to have higher strength and lower ductility than special concentrically braced frames (SCBF). Besides the lower response reduction coefficient (3.25 for OCBF as compared to 6.0 for SCBF), the main differences in the design of the two systems lie in the member strength requirements and in the ductile detailing requirements.

For one- and two-story buildings, as well as roof structures, the *AISC Seismic* requirements are waived as all elements are designed for the special seismic load combinations (Equations 16-19 and 16-20).

In the 2002 edition of the *AISC Seismic Provisions* the OCBF requirements were simple but stringent. That edition required that the special seismic load combinations be used to determine all member and connection forces. This change coincides with the *2000 NEHRP Recommended Provisions*, which limited OCBF to 35 feet in Seismic Design Categories C and D. This requirement leads to a system with an effective response reduction coefficient, R, of 2.5. However, the effective overstrength factor applied to columns and collectors is 1.0.

The 2005 edition thoroughly revised the requirements for OCBF. The provisions are expanded and more permissive.

The building considered is a two-story structure with two braced bays per side. It will be the same configuration as that used in Design Example 1A except for being two stories in height.

Outline

In this design example, AISC 341 will be used. The following design steps will be illustrated

1. Determination of design base shear

2. Determination of brace design force

3. Determination of beam design force for chevron configuration

4. Determination of connection design force

Calculations and Discussion | Code Reference

1. Determination of design base shear

The method for computing the base shear is similar to that for the special concentrically braced frame. The difference is in the value of the response-reduction coefficient R, (3.25 for OCBF). In Design Example 1A it was determined that the governing equation is for a short-period structure; with the OCBF R-coefficient, the design base shear coefficient is

$$C_s = \frac{S_{DS}}{R/I} = \frac{1.40g}{(3.25/1.0)} = .0431g$$

2. Determination of brace design force

The required strength of the first floor brace will be calculated. The forces on it are

$P_D = 8$ kips (compression)

$P_L = 6$ kips (compression)

$\rho P_E = \pm 394$ kips

The required strength of the first floor brace is

$$1.2D + 1.0E + 0.5L = 1.48D + \rho Q_E + 0.5L = 409 \text{ kips (compression)}$$

See Design Example 1A for the derivation of this formulation of the load combination to include the vertical component of the seismic load.

For this force and a length of less than 15 feet (as determined in Design Example 1A, Section 8), an HSS 9.625 × 0.500 brace has a strength of at least 419 kips (AISC Table 4-5). Its slenderness ratio is 60, less than the limit $4\sqrt{E/F_y} = 4\sqrt{29{,}000 \text{ ksi}/42 \text{ ksi}} = 105$

3. Determination of beam design force for chevron configuration

For beams in the chevron configuration as shown in Figure 1B-3, special design requirements apply.

Like the beams in SCBF, beams in V-braced OCBF need to be designed for unbalanced loading. The maximum unbalance force however is bounded by the amplified seismic load (14.3)

$$Q_b \le (R_y F_y A - 0.3F_{cr} A)\sin(\gamma)$$

$$Q_b \le (P_u - 0.3F_{cr} A)\sin\gamma$$

where

$$P_u = 1.2D + 0.5L + \Omega_o E_h + E_v$$

For the HSS 9.625 × 0.500 (from Design Example 1A)

$$R_y F_y A = 789 \text{ kips} \qquad F_{cr} = 33.6 \text{ ksi}$$

In this brace

$$P_u = 693 \text{ kip}$$

$$Q_b = [693 \text{ kips} - 0.3(33.6 \text{ ksi})(13.4 \text{ in}^2)]\sin(37.6°) = 340 \text{ kips}$$

$$Q_h = [693 \text{ kips} + 0.3(33.6 \text{ ksi})(13.4 \text{ in}^2)]\cos(37.6°) = 656 \text{ kip}$$

The beam may be designed as in Design Example 1B.

4. Determination of connection design force

The connection design force for the brace will be calculated. As with SCBF, the connection design force can be determined from the brace capacity

$$P_u = R_y F_y A_g = 1.4(42 \text{ ksi})13.4 \text{ in}^2 = 788 \text{ kips}$$

Alternatively, the special seismic load combinations may be used

$$1.2D + 1.0E_m + 0.5 = 1.48D + 2.0Q_E + 0.5L = 1.48D + 2.0Q_E + 0.5L = 803 \text{ kips}$$

where D, Q_E, and L are the same as for the brace (see Section 2 of this design example).

Assuming a ⅞-in gusset, the net-section rupture strength is

$$\phi R_n = \phi A_e F_u = \phi U A_n F_u$$

Note that the connection design force is significantly lower than the brace capacity. It is therefore possible that connection fracture will occur. Designers should be aware of the likelihood of poor performance of OCBF and should consider carefully if the system is appropriate for their project. Often SCBF offer significant advantages in member size as well as in seismic performance.

Per Design Example 1A

$L \quad = 18 \text{ in}$

$L/D = 18/9.625 = 1.9 > 1.3$

$U \quad = 1.0$

Assuming a ⅞-in-thick gusset

$A_{net} = 13.4 \text{ in}^2 - 2(7/8 \text{ in} + 1/8 \text{ in} \times 0.465 \text{ in}) = 12.5 \text{ in}^2$

$\phi R_n = 0.75(UA_{net})F_u = 0.75(1.0)(12.5 \text{ in}^2)(58 \text{ ksi}) = 544 \text{ kips} \ldots \; n.g.$

Either reinforcement should be added or a larger brace selected. See Design Example 1a for reinforcement design.

HSS 14.000 × 0.625:

$\phi R_n = 0.75[24.5 \text{ in}^2 - 2(7/8 \text{ in} + 1/8 \text{ in})\ 0.581 \text{ in}]\ 58 \text{ ksi} = 1015 \text{ kips} > 803 \text{ kips} \ldots \; o.k.$

The HSS 14.000×0.625 works without reinforcement. The reinforced HSS 9.625×0.500 will be used.

References

Aiken, I. and Kimura, I. "The Use of Buckling-Restrained Braces in the United States." *Proceedings Japan Passive Control Symposium*. Tokyo Institute of Technology, Yokohama, Japan, December, 2001.

Astaneh-Asl, A., Goel, S.C. and Hanson, R. "Cyclic Out-of-Plane Buckling of Double-Angle Gusset Plates," *Journal of Structural Engineering*, ASCE Vol. 111(5), pp. 1135–1153, 1985.

Astaneh-Asl, A. *Seismic Behavior and Design of Gusset Plates*. Structural Steel Education Council, 1998.

Bruneau, M., Uang, C.M. and Whittaker, A. *Ductile Design of Steel Structures*. MacGraw-Hill, 1998.

Cheng, J J., Kulak, G.L. and Heng-Aik Khoo. "Strength of Slotted Tubular Tension Members," *Canadian Journal of Civil Engineering*, Vol. 25, 1998, pp 982–991.

Clark, P.W., Aiken, I.D., Kasai, K., Ko, E., and Kimura, I. "Design Procedures for Buildings Incorporating Hysteretic Damping Devices," *Proceedings, 68th Annual Convention*, Structural Engineers Association of California, Santa Barbara, California, 1999.

DeWolf, J.T., and J.F. Pelliccione. "Cross-Bracing Design." *Journal of the Structural Division*, ASCE, 105(7), 1979, pp. 1379–1391.

Dowswell, Bo, "Effective Length Factors for Gusset Plate Buckling." AISC *Engineering Journal*, Second Quarter, 2006, pp 91–101.

El-Tayem, A., and Goel, S.C. "Effective Length Factor for Design of X-Bracing Systems," *Engineering Journal*, AISC, 24(4), 1986, pp. 41–45.

Federal Emergency Management Agency. *FEMA-368: NEHRP Recommended Provisions for Seismic Regulations for New Buildings and Other Structures*. Washington, D.C.: Federal Emergency Management Agency, 2000.

Goel, S.C. "Cyclic Post Buckling Behavior of Steel Bracing Members," *Stability and Ductility of Steel Structures under Cyclic Loading*, CRC Press, Boca Raton, pp. 75–104, 1992.

Hassan, O. and Goel, S.C. *Seismic Behavior and Design of Concentrically Braced Steel Structures*. Dept. of Civil Engineering, University of Michigan, 1991.

Hohbach, D., Kennedy, D.S., and Sabelli, R. "Design of Steel Braced-Frame to Foundation Connections." *Proceedings of the 71stAnnual Convention*, Structural Engineers Association of California, 2002.

Kalyanaraman, V., Sridhara, B.N. and Thairani, V. "Core Loaded Earthquake Resistant Bracing Design." *Proceedings, 2nd Intern. Conf. on Steel Construction*, San Sebastian, Spain, 1998.

Khatib, I.F., Mahin, S.A., and Pister, K.S. *Seismic Behavior of Concentrically Braced Steel Frames*. Report No. UCB/EERC-88/01. University of California, Berkeley, 1988.

Kitipornchai, S., and Finch, D.L. "Stiffness Requirements for Cross Bracing." *Journal of Structural Engineering*, ASCE, 112(12), 1986, pp. 2702–2707.

Korol, R.M. "Shear Lag in Slotted HSS Tension Members," *Canadian Journal of Civil Engineering*, Vol. 23, 1996, pp. 1350–1354.

León, R.T., and Yang, C.S. *Special Inverted-V-Braced Frames with Suspended Zipper Struts*. Georgia Institute of Technology, 2003.

López, W.A. "Design of unbonded braced frames," *Proceedings of the 70th Annual Convention*. Structural Engineers Association of California, 2001.

López, W.A. "Lessons learned from large-scale tests of unbonded braced frame subassemblages." *Proceedings of the 71st Annual Convention*, Structural Engineers Association of California, 2002.

Picard, A., and Beaulieu, D. "Design of Diagonal Cross Bracings. Part 1: Theoretical Study." *Engineering Journal*, AISC, 24(3), 1987, pp. 122–126.

Popov, E. *Mechanics of Materials*, Prentice Hall Englewood Cliffs, New Jersey, 1952.

Roeder, C. and Lehman, D. "Performance and Behavior of Gusset-plate Connections." *NASCC Conference Proceedings*, AISC, 2009.

Sabelli, R., Wilford, C., Abey, K., Pottebaum, W., and Hohbach, H. "Pushover Analysis of Inverted-V Braced Frames and its Implications for the Design of Multistory Systems." *Proceedings of the 67th Annual Convention, T7.1-20*. Structural Engineers Association of California, 1998.

Sabelli, R. *EERI/FEMA NEHRP Professional fellowship report:* Research on improving the seismic behavior of earthquake-resistant steel braced frames. Oakland, California: Earthquake Engineering Research Institute, 2001.

Sabelli, R., and D. Hohbach. "Design of Cross-Braced Frames for Predictable Buckling Behavior," *Journal of Structural Engineering, ASCE*, 1999, pp. 163–168.

Sabelli, R., Mahin, S.A., and Chang, C. Seismic demands on steel braced-frame buildings with buckling-restrained braces. Engineering Structures (in press), 2003.

Sabelli, R. Recommended provisions for buckling-restrained braced frames. *Engineering. Journal*, AISC (in press), 2004.

SEAONC. *Recommended Provisions For Buckling-restrained Braced Frames*. (Structural Engineers Association of Northern California). San Francisco, 2001.

SEAOC Seismology Committee, Ad Hoc Foundations Committee. "USD/LRFD/Limit State Approach to Foundation Design," *Proceedings of the 70th Annual Convention*. Structural Engineers Association of California, Sacramento, California, 2001.

Stoman, S.H. "Effective Length Spectra for Cross Bracings." *Journal of Structural Engineering*, ASCE, 115(12), 1989, pp. 3112–3122.

Tamboli, A. Handbook of Structural Steel Connection Design and Details, McGraw-Hill, New York, 1999.

Tang and Goel, S.C. "Brace fractures and analysis of Phase I structure," *Journal of Structural Engineering*, 115 (8), 1989, pp. 1960–1976.

Thornton, W. "On the Analysis and Design of Bracing Connections," *National Steel Conference Proceedings*. American Institute of Steel Construction, pp. 26.1–26.33 Chicago, Illinois, 1991.

Timoshenko, S.P. Theory of Elastic Stability, McGraw-Hill, New York, 1936.

Tremblay, R. "Seismic Behavior and Design of Concentrically Braced Steel Frames." *AISC Engineering Journal, Third Quarter*, 2001.

Tremblay, R., Degrange, G., and Blouin, J. Seismic rehabilitation of a four-storey building with a stiffened bracing system. *Proc. 8th Can. Conf. on Earthquake Engrg*., Canadian Association for Earthquake Engineering, Vancouver, B.C., pp. 549–554, 1999.

Tremblay, R., Inelastic Seismic Response of Steel Bracing Members. *Journal of Constructional Research*. 58. pp. 665–701, 2002.

Tremblay, R. Archambault, M.H.; Filiatrault, A. *Seismic Behaviour of Ductile Concentric Steel X-Bracings*. Earthquake Engineering: Seventh Canadian Conference, Canadian Association for Earthquake Engineering, Montreal, pp. 549–556, 1995.

Tremblay, R. "Behaviour and Design of Muti-storey Zipper Concentrically Braced Steel Frames for the Mitigation of Soft-Storey Response," in STESSA 2003, Mazzolani, F. (ed.) Balkema, 2003.

Wang, D.Q., and Boresi, A.P. "Theoretical Study of Stability Criteria for X-Bracing Systems." *Journal of Engineering Mechanics*, ASCE, 118(7), pp. 1357–1364, 1992.

Williams, George C., and Richards, Ralph M. "Analysis and design of large diagonal bracing connections." *Structural Engineering Review*, Vol. 8, No. 1 pp. 1–27, 1996.

Design Example 2
Eccentrically Braced Frame

Overview

Use of eccentrically braced frames (EBF) in steel frame buildings in high seismic regions is a fairly recent development. This system was introduced in the 1988 UBC. While the concept has been thoroughly tested in laboratories, it has seen limited testing in actual earthquakes. Many structural engineers, however, feel that it offers superior earthquake resistance compared to concentric braced frames.

Structural design of EBF frames defines the EBF links as the yielding elements in the system. Therefore, all other elements are designed for forces equal to or greater than what is required to develop link yielding. Thus beams outside links, braces, columns, foundations, collectors, and connections should be adequate to withstand forces generated during link yielding.

Typical EBF frame elevations are shown in Figure 2-1. The five-story steel frame building shown schematically in Figure 2-2 is to have eccentric braced frames for its lateral-force-resisting system. The floor and roof diaphragms consist of lightweight concrete fill over steel decking. A typical floor/roof plan for the building is shown in Figure 2-3.

Eccentrically braced frames may be configured with several geometric patterns including centrally located links as chosen in this problem or with links located adjacent to columns. When links are located adjacent to columns, a seismic SMRF connection is required at the link beam-column intersection. Several papers and many practitioners recommend that configurations using centrally located links be chosen to avoid link beam-column SMRF connections which increase the risk from brittle failure. Braces may be oriented to slope up to central link beams (inverted V braces) or down (V braces) to central link beams. Also, a two-story frame section can be designed with upper and lower braces meeting at a common link beam located between the two levels. A typical frame elevation is shown in Figure 2-4.

It is also desirable to prevent single-story yield mechanisms. Some options for this include using inverted braces at two levels with common link beams,which ensures two-story yield mechanisms, or zipper columns at either side of link beams extending from the second level to roof, which ensures multi-story mechanisms.

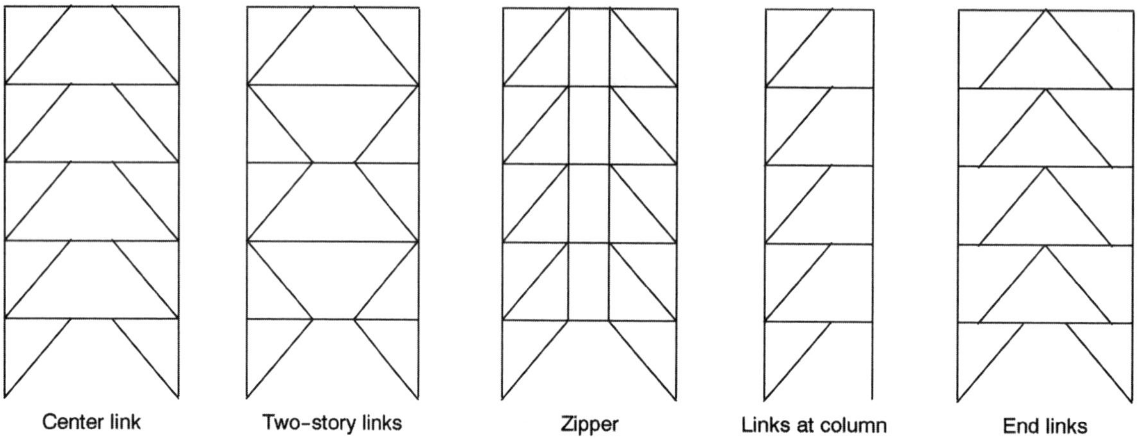

Figure 2-1. Eccentrically braced frame elevations

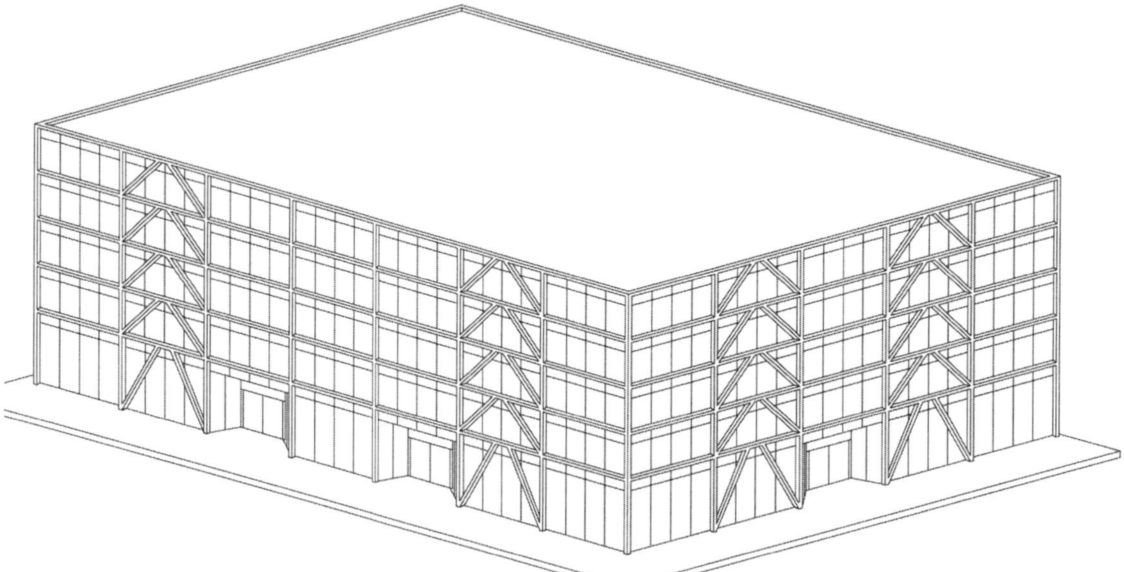

Figure 2-2. Eccentrically braced frame building.

Outline

1. **Site ground motion**
2. **Design base shear coefficient**
3. **Design base shear and vertical distribution of shear**
4. **Horizontal distribution of shear**
5. **Reliability/redundancy factor**
6. **Combined effect of horizontal and vertical earthquake induced forces**
7. **EBF member design using Load and Resistance Factor Design (LRFD)**
8. **Typical EBF details**

Given Information

Roof weights:

Roofing	6.0 psf
Insulation	3.0
Steel deck + fill	47.0
Roof framing	8.0
Ceiling	3.0
Mech/elec	2.0
Total	69.0 psf

Floor weights:

Floor covering	1.0 psf
Steel deck + fill	47.0
Framing (beams + cols.)	13.0
Ceiling	3.0
Mech/elec	2.0
Total	66.0 psf

Exterior curtain wall, steel studs, gypsum board, EIFS skin, weight: 20.0 psf

Live loads:

Roof	20.0 psf
Floor	50.0 psf
Partition walls (10 psf)	5.0 psf seismic

Seismic and site data:

Mapped spectral response accelerations (5% critical damping, Site Class B):

$S_S = 2.1g$ at 0.2-second response

$S_1 = 0.93g$ at 1.0-second response §22.0

I = 1.0 (standard occupancy) §11.5.1

Site Class D §11.4.2

Structural Materials:

Wide flange shapes and plates	ASTM A992, Grade 50 ($F_y = 50$ ksi)
Weld electrodes	E70XX
Lightweight concrete fill	$f'_c = 3000$ psi

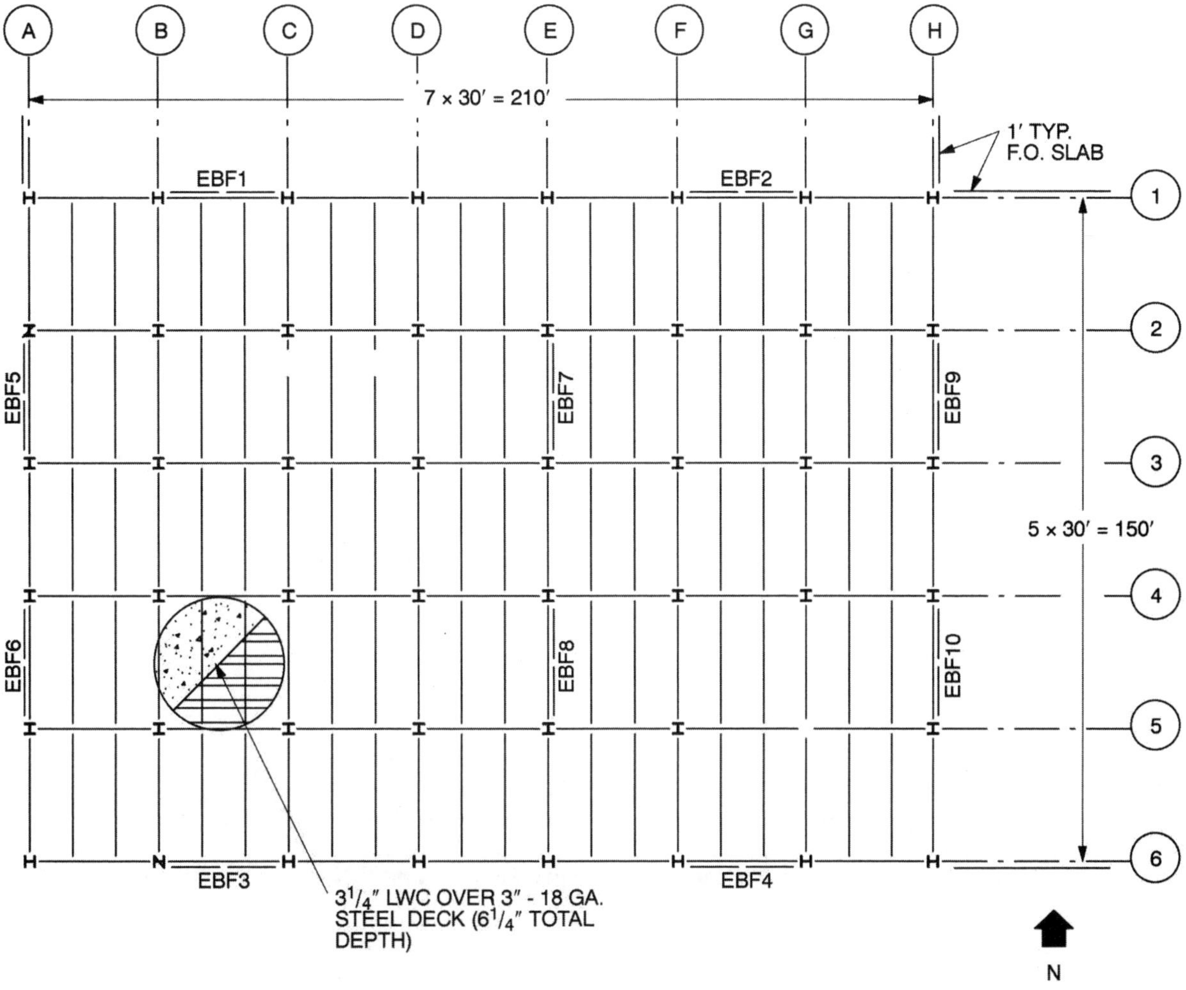

Figure 2-3. Typical floor and roof framing plan

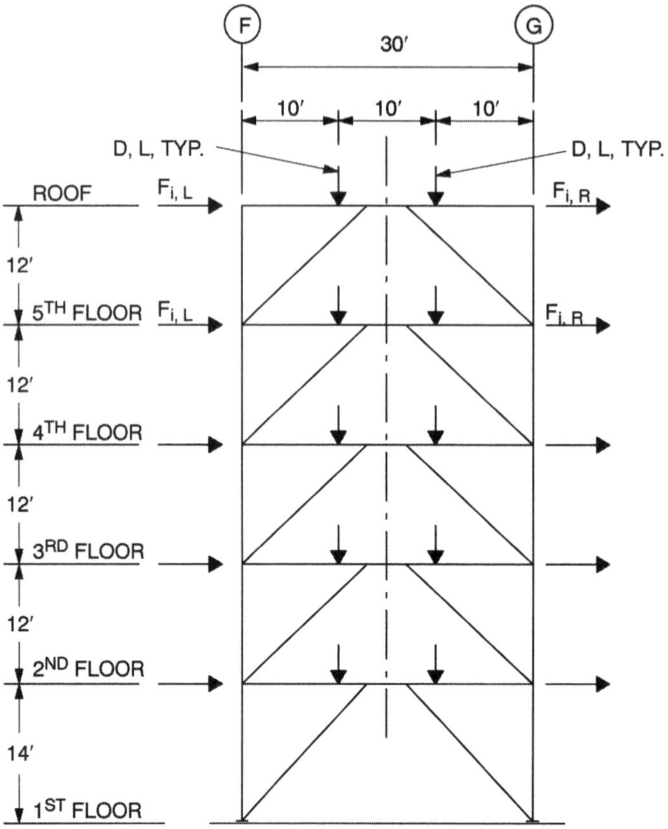

Figure 2-4. Typical frame elevation at frame EBF 4

Calculations and Discussion	Code Reference

1. Site ground motion

The maximum considered earthquake spectral response accelerations for short periods, S_{MS} and at 1.0-second period S_{M1} adjusted for site class effects, are determined by Equations 11.4-1 and 11.4-2.

$F_a = 1.0$ — T 11.4-1

$F_v = 1.5$ — T 11.4-2

$S_{MS} = F_a S_S = (1.0)(2.1g) = 2.1g$ — Eq 11.4-1

$S_{M1} = F_v S_1 = (1.5)(0.93g) = 1.4g$ — Eq 11.4-2

Five-percent damped design spectral response acceleration at short periods S_{DS} and at 1.0-second period S_{D1} adjusted for site class effects, shall be determined by Equations 11.4-3 and 11.4-4.

$S_{DS} = 2/3\ S_{MS} = (2/3)(2.1g) = 1.40g$ — Eq 11.4-3

$S_{D1} = 2/3\ S_{M1} = (2/3)(1.4g) = 0.93g$ — Eq 11.4-4

2. Design base shear coefficient §12.2-1

The coefficients for an eccentrically braced framed building using moment resisting connections at columns away from links are

Response modification coefficient, $R = 8$ T 12.2-1

System overstrength factor, $\Omega_o = 2$

Deflection amplification factor, $C_d = 4$

The approximate fundamental building period is

$$T_a = C_T h_n^x = (0.030)(62 \text{ ft})^{0.75} = 0.66 \text{ sec} \qquad \text{Eq 12.8-7}$$

Per Table 12.8-2, the values $C_t = 0.03$ and $x = 0.75$ can be found for a steel eccentrically braced frame.

The seismic base shear V in a given direction shall be determined in accordance with the following equation 12.8

$$V = C_s W \qquad \text{Eq 12.8-1}$$

The calculation of seismic response coefficient, C_s

$$C_s = \frac{S_{DS}}{\frac{R}{I}} = \frac{1.40g}{\left(\frac{8.0}{1.0}\right)} = 0.175g \qquad \text{Eq 12.8-2}$$

but need not exceed

$$C_s = \frac{S_{D1}}{\left(\frac{R}{I}\right)T} = \frac{0.93g}{\left(\frac{8.0}{1.0}\right)0.66 \text{sec}} = 0.175g \text{ for } T \le T_L \qquad \text{Eq 12.8-3}$$

or

$$C_s = \frac{S_{D1}T_L}{T^2\left(\frac{R}{I}\right)} \text{ for } T > T_L \qquad \text{Eq 12.8-4}$$

T_L = Long-period transition period determined per §11.4-5 from Figures 22-15 to 22-20. T_L will generally not have an effect on short-period structures.

The value of seismic response coefficient C_s shall not be less than:

$$C_s = 0.044 S_{DS} \ge 0.01 = (0.44)(1.40)(1.0) = 0.062 \qquad \text{Eq 12.8-5, Supplement No. 2}$$

In addition, for buildings located where S_1 is equal to or greater than 0.6g, the value of the seismic response coefficient C_s shall not be taken as less than

$$C_s = \frac{0.5S_1}{\left(\frac{R}{I}\right)} = \frac{0.5(0.93)}{\left(\frac{8}{1}\right)} = 0.058 \qquad \text{Eq 12.8-6}$$

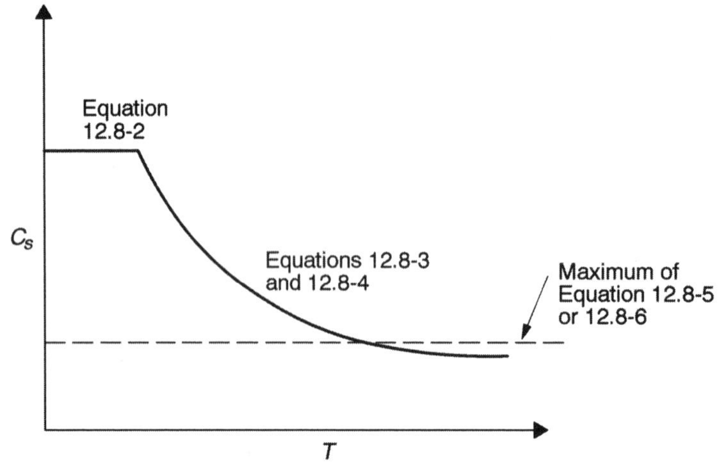

Figure 2-5. Typical response spectrum

Therefore, Equation 12.8-2 controls the base shear calculation for a short-period building and the seismic coefficient is thus

$$\underline{\underline{V = 0.175W}} \qquad \text{(Eq 12.8-1)}$$

3. Design base shear and the vertical distribution of shear

The floor area at each level is 32,224 square feet. The perimeter of the exterior curtain wall is 728 feet. The roof parapet height is 4 feet. Assume that the curtain wall weights distribute to each floor by tributary height.

The building mass calculation is shown in Table 2-1.

Table 2-1. Building mass calculation

Level	Floor Area (sf)	w_i (psf)	$W_r/_f$ (kips)	Length exterior walls (ft)	h walls (ft)	w_i walls (psf)	W walls (kips)	W_i (kips)
Roof	32,224	74	2385	728	10	20	146	2530
5	32,224	76	2449	728	12	20	175	2624
4	32,224	76	2449	728	12	20	175	2624
3	32,224	76	2449	728	12	20	175	2624
2	32,224	76	2449	728	13	20	189	2638
Totals	161,120		12,181				860	13,040

3a. Design base shear — 12.8

Using the design base shear coefficient from Part 1, the base shear for the east-west direction is

$$V = 0.175W = 0.175(13{,}040 \text{ kips}) = \underline{\underline{2282}} \text{ kips}$$

3b. Vertical distribution of shear — 12.8-3

The total lateral force (i.e., design base shear) is distributed over the height of the building in accordance with 12.8-3. The following equations apply.

$$F_x = C_{vx}V \qquad \text{Eq 12.8-11}$$

$$C_{vx} = \frac{w_x h_x^k}{\sum_{i=1}^{n} w_i h_i^k} \qquad \text{Eq 12.8-12}$$

k is a distribution exponent related to building period as follows.

$k = 1$ for buildings with a period of 0.5 second or less

$k = 2$ for buildings with a period of 2.5 seconds or greater

for $T = 0.66$ sec, $k = 1.08$

Interpolate for buildings with period between 0.5 second and 2.5 seconds.

Using the building mass tabulated in Table 2-1 above, the vertical distribution of shear is determined as shown in Table 2-2 below.

Table 2-2. Vertical distribution of shear

Level	w_x (kips)	w (kips)	h_x (ft)	h (ft)	$w_x h_x^k$ (kip-ft)	C_{vx} (%)	F_x (kips)	ΣV_i (kips)
R	2530		62		218,240	33	749	749
		2,530		12				
5	2624		50		179,394	27	615	1364
		5,154		12				
4	2624		38		133,379	20	458	1822
		7,778		12				
3	2624		26		88,530	13	304	2125
		10,401		12				
2	2638		14		45,619	7	157	2282
		13,040		14				
Totals	13,040				665,162	100	2282	

4. Horizontal distribution of shear §12.8-4

The center of mass of the building is located at the center of the floor plan. This location will be identified as coordinates ($x = 0$, $y = 0$). X is oriented the long direction (east-west). Y is oriented transverse (north-south).

The center of rigidity can be calculated per Table 2-3, assuming that each EBF frame has a stiffness or rigidity equal to one. The center of rigidity is located at ($x = 5$, $y = 0$). For analysis in the east west direction there is no actual eccentricity ($\bar{y} = 0$). Accidental eccentricity (5%) is always added into the analysis.

Table 2-3. Calculation of location of center of rigidity

FRAME	x to cgm	y to cgm	R	xR	$\bar{x} = \Sigma xR/\Sigma R$	yR	$\bar{y} = \Sigma yR/\Sigma R$
Long.							
1		−75	1			−75	
2		−75	1			−75	
3		75	1			75	
4		75	1			75	
Sum			4			0	0
Trans.							
5	−105		1	−105			
6	−105		1	−105			
7	15		1	15			
8	15		1	15			
9	105		1	105			
10	105		1	105			
Sum			6	30	5		
Totals							

The centers of mass and rigidity coincide in the east-west direction; however, §12.8.4.2 requires an additional accidental torsional eccentricity e equal to 5 percent of the building dimension perpendicular to the direction of force regardless of the relative location of the centers of mass and rigidity.

$$e_{ew} = (0.05)(150) = 7.5 \text{ ft (east-west)}$$

$$e_{ns} = (0.05)(210) = 10.5 \text{ ft (north-south)}$$

The calculation of building torsion can be performed as shown in Table 2-4:

Table 2-4. Calculation of building torsion in orthogonal directions

Base Shear (k)	Actual Eccentricity N-S x (ft)	Actual Eccentricity E-W y (ft)	Accidental Eccentricity N-S x (ft)	Accidental Eccentricity E-W y (ft)	Sum Eccentricity N-S x (ft)	Sum Eccentricity E-W y (ft)	Torsion N-S Y (kft)	Torsion E-W X (kft)
1	5	0	10.5	7.5	15.5	7.5	15.5	7.5

Assume that all frames have the same rigidity because all are similar EBFs. This assumption can be refined in a subsequent analysis, after members have been sized and an elastic deflection analysis has been completed. Many designers estimate the torsional contribution for a symmetric building by adding 5–10 percent to the element forces. However, in this example the numerical application of the code provisions will be shown.

Assume $R_1 = R_2 = \ldots R_{14} = 1.0$, where R_i is the rigidity of each EBF frame.

The calculation of direct shear plus torsion for a given frame is based on the following formula.

$$V_i = R_i\left(\frac{V}{\Sigma R}\right) + R_i\left(\frac{Vec}{\Sigma R_{xy}c^2}\right) \pm \left(\frac{Ve_{acc}c}{\Sigma R_{xy}c^2}\right)$$

Table 2-5 gives the distribution of direct shear and torsional shear components as percentages of shear force (based on geometry).

Table 2-5. Calculation of direct shear plus torsion as percentage of story shear

Frame ID	X (ft)	Y (ft)	R_i	XR_i	YR_i	X^2R_i	Y^2R_i	$J = \Sigma Rd^2$	V_i/V_x (%)	T_x(c/J) (%)	ΣV_x (%)	V_i/V_y (%)	T_y(c/J) (%)	ΣV_y (%)
Long.														
1		75	1		–75		5625		25	–0.84	25.00		–1.74	
2		75	1		–75		5625		25	–0.84	25.00		–1.74	
3		75	1		75		5625		25	0.84	25.84		1.74	
4		75	1		75		5625		25	0.84	25.84		1.74	
Trans.														
5	–110		1	–110		12100				–1.23		16.67	–2.55	16.67
6	–110		1	–110		12100				–1.23		16.67	–2.55	16.67
7	10		1	10		100				0.11		16.67	0.23	16.90
8	10		1	10		100				0.11		16.67	0.23	16.90
9	100		1	100		10000				1.12		16.67	2.32	18.98
10	100		1	100		10000				1.12		16.67	2.32	18.98
Totals								66900	100	0.00			0.00	

1. X and Y are distances from the center of mass (i.e., the center of the building) to frames in the X and Y directions, respectively.
2. V_x and V_y are direct shears on frames in the X and Y directions, respectively.
3. T_x and T_y are shear forces on frames which resist torsional moments on the building. These shear forces are either in the X or Y directions and can be additive or subtractive with direct shear forces.

Based on the direct and torsional shear values tabulated in Table 2-3, and on the vertical distribution of shear tabulated in Table 2-2, the story forces to be used for design of the typical eccentric braced frame (EBF4) are as follows.

Table 2-6. Story shear forces for design of frame EBF4

Frame ID	Level	Story V_x (kips)	T_x (kips)	$V_{i,4}$ (kips)	$T_{i,4}$ (c/J) (kips)	Story $F_{x,4}$ (kips)
4	5	749	5615	187	47	234
4	4	1364	10231	341	86	427
4	3	1822	13663	455	115	570
4	2	2125	15941	531	134	665
4	1	2282	17115	570	144	714

5. Redundancy factor §12.3.4

The redundancy factor ρ must be determined. The factor was added to the 1997 UBC code to penalize non-redundant systems. It now varies from a minimum of 1.0 to a maximum of 1.3, and needs to be determined for each principal direction.

For structures assigned to Seismic Design Category D, E, or F, ρ shall be 1.3 unless one of the following two conditions is met, whereby ρ is permitted to be taken as 1.0.

a. Each story resisting more than 35 percent of the base shear in the direction of interest shall comply with Table 12.3-3, or

b. Structures that are regular in plan at all levels, provided that the seismic force resisting systems consist of at least two bays of seismic force-resisting perimeter framing on each side of the structure in each orthogonal direction at each story resisting more than 35 percent of the base shear.

This building meets conditions in which the perimeter frames resist 100 percent or 66 percent of the seismic base shear. Therefore ρ = 1.0 in both directions.

6. Combined effect of horizontal and vertical earthquake induced forces §12.4.2

$$E = E_h + E_v \quad \text{Eq 12.4-1}$$

$$E = E_h - E_v \quad \text{Eq 12.4-2}$$

The horizontal seismic load effect E_h shall be determined as follows

$$E_n = \rho Q_E \quad \text{Eq 12.4-3}$$

The vertical seismic load effect E_v shall be determined as follows

$$E_v = 0.2S_{DS}D \quad \text{Eq 12.4-4}$$

Therefore

$$E = \rho Q_E + 0.2S_{DS}D$$
$$= 1.0(Q_E) + 0.2(1.40g)D$$
$$= Q_E + 0.28D$$

7. EBF member design using (LRFD)

7a. Seismic forces for initial member design

Seismic forces on a typical EBF, in this case EBF4 on line 6, will be determined. The forces, E, applied to EBF4 are calculated first by determining the seismic load along line 6. The unit shear load along line 6, V_{i6}, is thus V_{i6}/210 feet. Frame EBF4 has a tributary collector length of 210 feet/2 = 105 feet, and tributary lengths on the west side of the frame of 60 feet and on the east side of the frame of 45 feet. The frame forces are thus $F_{4iL} = V_{i6}$ (60 feet) and $F_{4iR} = V_{i6}$ (45 feet). The compression force in the link is equal to half the story shear tributary to the frame minus the frame force at the right side $(F_{4iL} + F_{4iR})/2 - F_{4iR}$. Table 2-7 summarizes the forces at each level of frame EBF4.

Table 2-7. Axial forces through shear links on frame EBF4

Level	Frame $\rho F_{x,4}$ (kips)	Line 6 $\rho v_{i,4}$ (klf)	ρF_{x4iL} (west) (kips)	ρF_{x4iR} (east) (kips)	$\rho C = \rho T$ (link) (kips)
R	234	2.23	133.9	100.5	16.7
5	427	1.83	110.1	82.6	13.8
4	570	1.36	81.9	61.4	10.2
3	665	0.91	54.3	40.7	6.8
2	714	0.47	28.0	21.0	3.5

7b. Link length

The inelastic behavior of a link is influenced by its length e. The shorter the link length, the greater the influence of shear forces on the inelastic performance. Shear yielding tends to occur uniformly along the link length. Shear yielding of short links is very ductile with an inelastic capacity in excess of that predicted by calculations.

The following is a summary of link behavior as a function of the link length e. M_p is the flexural strength of the link and V_p is the shear strength.

$1.3\frac{M_p}{V_p} \le e \le 1.6\frac{M_p}{V_p}$ assures shear behavior and is the recommended upper limit for shear links. For links less than $\frac{1.0M_p}{V_p}$, the link may not yield as expected.

$e \ge 1.6\frac{M_p}{V_p}$ elastic behavior is controlled by shear behavior. However, the region is a transition between shear governed behavior and bending governed behavior.

$e = 2.0\dfrac{M_p}{V_p}$ link behavior is theoretically balanced between shear and flexural yielding.

$e \geq 2.6\dfrac{M_p}{V_p}$ elastic deformation is controlled by flexural yielding.

The shorter the link length, the stiffer the EBF frame will be; however, the link rotation will be greater. AISC 341 sets limits on link plastic rotation of 0.080 radians due to δ_x deflections. For most designs, link lengths of 1.3 to $1.6M_p/V_p$ work well.

7c. Preliminary EBF frame member sizes

Preliminary sizes of the EBF frame beams are determined by calculating the required shear area (dt_w) due to the story forces and frame geometry. The combinations of factored loads are given in load combinations 1 to 7 in §2.3.2.

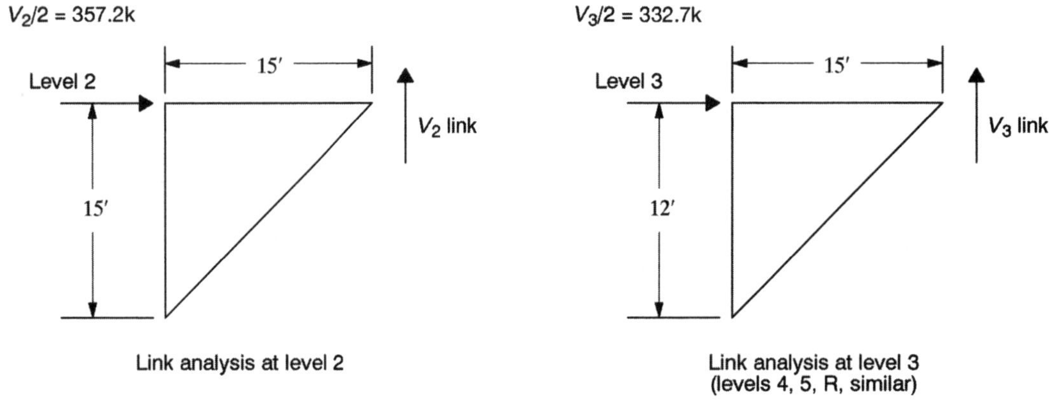

Figure 2-6. Preliminary link analysis

For initial sizing, shear forces in the links may be approximated as

$$V_{i,\text{link}} = \frac{V_i(h)}{l} = \frac{V_i/2(h)}{l/2}$$

$$V_{2,\text{link}} = \left(\frac{357.2\text{ kips}(15\text{ ft})}{15\text{ ft}}\right) = 357.2\text{ kips}$$

$$V_{3,\text{link}} = \left(\frac{332.7\text{ kips}(12\text{ ft})}{15\text{ ft}}\right) = 266.2\text{ kips}$$

$$V_{4,\text{link}} = \left(\frac{285.2\text{ kips}(12\text{ ft})}{15\text{ ft}}\right) = 228.1\text{ kips}$$

$$V_{5,\text{link}} = \left(\frac{213.5\text{ kips}(12\text{ ft})}{15\text{ ft}}\right) = 170.8\text{ kips}$$

$$V_{R,\text{link}} = \left(\frac{117.2\text{ kips}(12\text{ ft})}{15\text{ ft}}\right) = 93.8\text{ kips}$$

The values for, $t_w(d - 2t_f)$, V_p and M_p are calculated as

$$\text{minimum } t_w(d - 2t_f) = \frac{V_{i,\text{link}}}{0.90(0.60F_y)}$$

$$V_p = 0.6F_y t_w(d - 2t_f)$$

$$M_p = Z_x F_y$$

Preliminary beam sizes are determined as shown in Table 2-8.

Table 2-8. Preliminary link analysis and sizing for frame EBF4

Level	Story H (ft)	$V_i/2$ (kips)	$F_i/2$ (kips)	$V_{i,\text{link}}$ Req. Link Shear	min. t_w $(d - 2t_f)$ (in^2)	Link Beam Size	d (in)	t_w $(d - 2t_f)$ (in)	Z_x (in^3)	M_p (k-in)	V_p (kips)	e (in)	eV_p/M_p	Ω
R	12	117	117	94	3.47	W12×45	12	3.65	65	3235	110	38	1.29	1.17
5	12	96	214	171	6.33	W16×77	17	6.83	150	7500	205	48	1.31	1.20
4	12	72	285	228	8.45	W18×97	19	9.01	211	10550	270	50	1.28	1.19
3	12	48	333	266	9.86	W18×106	19	9.94	230	11500	298	50	1.30	1.12
2	15	24	357	357	13.23	W18×158	20	13.64	356	17800	409	60	1.38	1.15

The most efficient link sections usually

- optimize the required shear area, i.e., minimize $t_w(d - 2t_f)$
- are the deepest sections possible while complying with the compact web criteria, i.e., maximize $t_w(d - 2t_f)$
- have compact flanges and sufficient flexural strength to ensure shear yield of the section under ultimate load
- satisfy the deflection and link rotation limitations

The recommended link length is $e_{\max} = 1.3\frac{M_p}{V_p}$ (Engelhardt and Popov, 1989).

A computer model has been created for EBF4 and the results of the computer analysis, including forces and displacements are determined. The computer model is analyzed with moment resisting connections which more closely estimate the real behavior of the frame with end moments much less than M_p.

For the first story, the EBF member design will be based on use of a W18×158 link beam at Level 2.

LRFD design provisions for EBF frames are contained in Section 15 of the AISC document, *Seismic Provisions for Structural Steel Buildings*, including Supplement No. 1 published in 2005. This document is known as AISC 341 or ANSI/AISC 341-05.

7d. Link shear strength

The link shear strength, V_n, can be found from the minimum values of V_p or $2M_p/e$. The values for V_p are calculated as follows

$$\phi_v V_n \geq V_{i,\text{link}}$$

where V_n = nominal shear strength of the link, equal to the lesser of V_p or $2M_p/e$.

$$V_p = 0.60F_y t_w(d - 2t_f) \qquad \text{AISC 341, §15.2b}$$

$$M_p = Z_x F_y$$

Preliminary beam sizes are determined as shown in Table 2-9. Note that seismic forces for LRFD procedures use both E_h and E_v. The E_v seismic force is additive to dead load D and is included in the load combination 5.

$$1.2D + 0.5L + 1.0E$$

$$E = \rho Q_E + 0.2S_{DS}D$$

$$E_h = \rho Q_E$$

$$E_v = 0.2S_{DS}D = 0.2(01.40g)D = 0.28D$$

Substituting for E_h and E_v, in load combination 5

$$1.2D + 0.5L + 1.0(\rho Q_E + 0.2S_{DS}D)$$

$$= 1.2D + 0.5L + 1.0(1.0Q_E + 0.28D)$$

$$= 1.48D + 0.5L + Q_E$$

Table 2-9. Preliminary link analysis and sizing

Design seismic forces at EBF Frame

LEVEL	Story Forces	Frame Forces, E_h			Frame Forces, E_h		V_i	F_i	$V_i/2$
		Left	Right	C, T link	F_{il}	F_{ir}			
R	234	133.9	100.5	16.7	133.9	100.5	234.4	234.4	117.2
5	193	110.1	82.6	13.8	110.1	82.6	427.1	192.7	213.5
4	143	81.9	61.4	10.2	81.9	61.4	570.3	143.2	285.2
3	95	54.3	40.7	6.8	54.3	40.7	665.4	95.1	332.7
2	49	28.0	21.0	3.5	28.0	21.0	714.4	49.0	357.2

Preliminary link beam sizes and properties

LEVEL	Story Height	$V_i/2$	$F_i/2$	V_{li}	$(d - 2t_f)t_w$ Req'd	Size	d	t_w	t_f
R	12	117.2	117.2	93.8	3.47	W12×45	12.06	0.34	0.58
5	12	96.3	213.5	170.8	6.33	W16×77	16.52	0.46	0.76
4	12	71.6	285.2	228.1	8.45	W18×97	18.59	0.54	0.87
3	12	47.5	332.7	266.2	9.86	W18×106	18.73	0.59	0.94
2	15	24.5	357.2	357.2	13.23	W18×158	19.72	0.81	1.44

Preliminary link beam analysis

LEVEL	$(d - 2t_f)t_w$	Results	Z_x	M_p	ϕV_p	1.3 M_p/V_p	1.6 M_p/V_p	Link e	Ratio M_p/V_p	Ω	CDR
R	3.65	OK	64.70	3235	99	42.6	52.5	38	1.29	1.17	1.05
5	6.83	OK	150.00	7500	184	52.9	65.1	48	1.31	1.20	1.08
4	9.01	OK	211.00	10550	243	56.3	69.4	50	1.28	1.19	1.07
3	9.94	OK	230.00	11500	268	55.7	68.5	50	1.30	1.12	1.01
2	13.64	OK	356.00	17800	368	62.8	77.3	60	1.38	1.15	1.03

For the first (ground level) story, the EBF link beam design will be based on use of a W18×158 link beam at Level 2. Note that §6.1 of AISC 341 limits the yield strength of the link beam to $F_y = 50$ ksi.

7e. Link rotation angle

The frame displacement at the second level δ_{xe2} was determined from a separate computer analysis (not shown) using the design base shear (without ρ, in this case ρ = 1.0).

$$\delta_{xe2} = 0.34 \text{ in}$$

The corresponding inelastic displacement, $\delta\rho_{xe2}$ may be estimated from a static analysis by the following formula

$$\delta_{x2} = \frac{\delta_{xe2}(C_d)}{I} = \frac{(0.34 \text{ in})(4.0)}{1.0} = 1.36 \text{ in} \qquad \text{Eq 12.8-15}$$

The link rotation is computed as a function of the frame story drift and frame geometry. For a frame of story height H, bay width L, link length e, and dimensions $a = (L - e)/2$, the link rotation may be calculated by the following formula. Link rotations θ must be limited to 0.080 radians for links of length $1.6M_p/V_p$ or less per AISC 341, §15.2c

$$\theta = \frac{\delta_{x2}}{H}\left(1 + \frac{2a}{e}\right) = \frac{1.36 \text{ in}}{180}\left(1 + \frac{2(150 \text{ in})}{60}\right) = 0.045 \text{ radians} \le 0.080 \ \ldots \ o.k.$$

Comments:
The above formula makes the assumption that all deformation occurs within the link rotation at a particular level. It has been observed that there is significant contribution to deformations from column elongation and shortening. A more accurate analysis of link rotation can be made looking at joint displacements and calculating rotations based on relative joint displacements. Another simple method is to perform an analysis using very stiff column section properties in the model and force all deformations into the link beam for purposes of evaluating the link rotations.

The term C_d is used to calculate the link rotation angle, which is limited to 0.08 radians for $e = 1.6M_p/V_p$ or less.

7f. Link shear strength — AISC 341, §15.2b

The nominal shear strength of the link V_n is equal to the lesser of V_p or $2M_p/e$. Solving for the design shear strength $\phi_v V_n$

$$\phi_v V_n \geq V_{i,\text{link}} \text{ at any given level}$$

$$\phi_v V_n = 0.9(0.60)F_y t_w(d - 2t_f) = 0.9(0.6)(50 \text{ ksi})(0.810 \text{ in})(19.7 \text{ in} - 2(1.440 \text{ in}) = 368 \text{ kips}$$

$$\phi 2M_p/e = 0.9(2)M_p/e = 0.9(2)F_y Z_x/e = 0.9(2)(50 \text{ ksi})(356 \text{ in}^3)/60 \text{ in} = 534 \text{ kips}$$

$$\phi_v V_n = 368 \text{ kips}$$

$$V_n = \frac{298 \text{ k}}{0.9} = 409 \text{ kips}$$

The design overstrength factor for this link beam Ω is calculated as follows

$$\Omega = \frac{V_n}{V_{i,\text{link}}} = \frac{409 \text{ kips}}{357 \text{ kips}} = 1.15$$

The minimum link design shear overstrength ratio is controlled by the ϕ factor. Thus, the minimum Ω is $\Omega_{min} = 1.0/\phi = 1.0/0.9 = 1.11$. The significance of the overstrength ratio is that the link will not yield until seismic forces overcome the link yield point. The overstrength factor Ω is a relationship between code seismic forces and design overstrength forces that will likely yield the link. Note that the Ω factor does not include the R_y factor for expected yield stress of the steel.

The link beam in this example has been sized for strength and stiffness. In beams above the level under discussion, it was found necessary to add cover plates for the beams outside the links (for increased beam capacity outside the link). The author was balancing the design between ratios of M_p/V_p of approximately 1.3 and the requirement for cover plates outside the link. Thus, the author chose to use cover plates to meet strength requirements for EBF beams outside the link. The trade-off would have been to lessen the ratio of M_p/V_p (by shortening the link lengths) and not to use cover plates. The author believed the performance of the link was more important that the cover plate requirement, and thus was not able to size beams to meet requirements outside the link without beam cover plates for this configuration of EBF frame.

7g. Beam compact flange

Check the W18×158 beam to assure that the flanges are compact to prevent flange buckling.

$$\frac{b}{t} = \frac{b_f}{2t_f} = \frac{11.306 \text{ in}}{2(1.440 \text{ in})} = 3.92 \le 0.30\sqrt{E/F_y} = 0.30\sqrt{29{,}000/50} = 7.225 \;\therefore\; o.k.$$

AISC 341, §15.2a, §8.2b, T I-8-1

7h. Link length

The length of the link will determine whether the link yields in shear or in bending deformations. To assure the desired shear yielding behavior (see discussion in Part 5b), the link beams have been limited to lengths less than $1.6M_p/V_p$. From part 7f, $V_p = V_n$ and M_p are calculated

$$V_p = V_n = 409 \text{ kips}$$

$$M_p = Z_xF_y = (356 \text{ in}^3)(50 \text{ ksi}) = 17{,}900 \text{ kip-in}$$

Check that the $1.6M_p/V_p$ criteria are not exceeded

$$\frac{eV_p}{M_p} = \frac{(60 \text{ in})(409 \text{ kips})}{17{,}900 \text{ k-in}} = 1.37 \approx 1.4 \;\therefore\; o.k.$$

Second floor link length of 60 inches is o.k.

7i. Verification of link shear strength

The strength of the link is used to establish the minimum strength required of elements outside the link. The link shear strength, V_p, was determined using the web area $(d - 2t_f)t_w$ of the beam. When the beam has reached flexural capacity, shear in the link may be less than the shear strength of the section. If this is the case, the flexural capacity of the section will limit the shear capacity of the link. AISC 341, §15.2b requires that the shear strength of the section be the minimum of shear yielding strength or shear required for plastic moment yielding behavior.

$$V_p = 409 \text{ kips}$$

$$\frac{2M_p}{e} = \frac{2(50 \text{ ksi})(356 \text{ in}^3)}{60 \text{ in}} = 593 \text{ kips}$$

The controlling nominal shear capacity, V_n, is the minimum of V_p or $2M_p/e$. From Part 7f, $V_p = 409$ kips. By selecting the W18×158 section as the link beam, the controlling mode of yielding is shear yielding in the link and flexural yielding will not develop.

7j. Required beam brace spacing

The limiting unbraced length for full plastic bending capacity, L_p, is determined as follows. Lateral beam braces for the top and bottom flanges at the ends of the link beams are still required.

$$L_p = 1.76 r_y \sqrt{\frac{E}{F_{yf}}} = 1.76(2.74\text{ in})\sqrt{\frac{29{,}000}{50}} = 116.1\text{ in} = 9\text{ ft }8\text{ in}$$ AISC 360, F2

Therefore the beam bracing at 10 feet, 0 inches is not quite adequate. A recalculation using the combined r_y with cover plates would prove that the beam bracing is adequate. (Note: the composite steel deck and lightweight concrete fill is not considered effective in bracing the bottom flange.)

7k. Beam and link axial loads

The summation of story forces down to level 3, V_3 in Table 2-4, (the sum of level shears from the roof to level 3) is 665.4 kips. The frame forces in level 2 at the left connection and right connection are $F_{2L} = 28.0$ kips and $F_{2R} = 21.0$ kips.

If the required axial strength of the link P_u is equal to or less than $0.15P_y$, the effect of axial force on the link design shear strength need not be considered.

Therefore, the axial force in the link is

$$C_{\text{link}} = T_{\text{link}} = \frac{(28.0\text{ kips} - 21.0\text{ kips})}{2} = 3.5\text{ kips}$$

The maximum axial stress in the link must be checked for the requirements of §15.2b of AISC 341

$$f_a = \frac{\Omega(3.5\text{ kips})}{A_g} = \frac{1.15(3.5\text{ kips})}{46.3\text{ in}^2} = 0.09\text{ ksi} \le 0.15F_y = 7.5\text{ ksi}$$

Therefore the effect of axial force on the link design shear strength need not be considered.

7l. Beam compact web — AISC 341, §8.2b

The section should be checked using Table I-8-1, of AISC 341. The web should be compact along the full length of the beam. AISC 341 does not allow the use of doubler plates for a link beam.

For a W18×158 beam

$$A = 46.3\text{ in}^2$$

$$\frac{h}{t_w} = \frac{d - 2t_f}{t_w} = \frac{19.72\text{ in} - 2(1.440\text{ in})}{0.810\text{ in}} = 20.8$$

Maximum axial force in beam outside the link

$$P_{2L} = \Omega\left(\frac{V_3}{2} + F_{2L}\right) = 1.15\left(\frac{665.4 \text{ kips}}{2} + 28.0 \text{ kips}\right) = 414.8 \text{ kips}$$

$$\frac{P_u}{\phi_b P_y} = \frac{414.8 \text{ kips}}{0.90(50 \text{ ksi})(46.3 \text{ in}^2)} = 0.20 \geq 0.125 \qquad \text{AISC 341, T I-8-1}$$

For $P_u/\phi_b P_y \geq 0.125$, allowable d/t_w to prevent local buckling is determined from the equation below.

$$\left(\frac{h}{t_w}\right) = 1.12\sqrt{\frac{E}{F_y}}\left(2.33 - \frac{P_u}{\phi_b P_y}\right) = 1.12\sqrt{\frac{29{,}000}{50}}(2.33 - 0.20) = 57.5$$

$\therefore\ h/t_w = 20.8 \leq 57.5$ . . . *o.k.* AISC 341, §15.2b

7m. Combined link loads

The combined bending plus axial strength of the link must be checked and compared with the yield stress. In the link, axial and bending stresses are resisted entirely by flanges.

$$P_u = 3.5 \text{ kips}(\Omega) = 3.5 \text{ kips } (1.15) = 4.0 \text{ kips}$$

$$\frac{P_u}{P_y} = \frac{4.0 \text{ kips}}{(50 \text{ ksi})(46.3 \text{ in}^2)} = 0.002 \leq 0.15 \qquad \text{AISC 341, §15.2b}$$

$$M_u = V_p\frac{e}{2} = 409 \text{ kips}\frac{(60 \text{ in})}{2} = 12{,}270 \text{ kip-in}$$

$$A_f = B_f t_f = (11.3 \text{ in})(1.440 \text{ in}) = 16.27 \text{ in}^2$$

$$Z_f = (d - t_f)(b_f t_f) = (19.72 \text{ in} - 1.440 \text{ in})(16.27 \text{ in}^2) = 297.4 \text{ in}^3$$

$$\frac{P_u}{2A_f} + \frac{M_u}{Z_f} = \frac{4.0 \text{ kips}}{2(16.27 \text{ in}^2)} + \frac{12{,}270 \text{ kip-in}}{297.4 \text{ in}^3} = 41.4 \text{ ksi} \leq 50 \text{ ksi}$$

$\therefore$ link combined axial plus bending capacity is OK.

7n. Beam analysis (outside of link) AISC 341, §15.6b

It has been determined (by author) that link beams have difficulty resisting the link beam moments increased by 1.1 and R_y when using a lower bound strength not including R_y. AISC 341 allows the LRFD design strength to be increased by increasing P_n, based on R_yF_y this can be roughly approximated as R_yP_n for moderate slenderness.

This solution has the beam outside the link plus the braces resisting the link beam moment. The analysis in this example includes the use of flange cover plates to increase the bending capacity of the beam outside the link.

It should be noted that the two-story link configuration reduces the need for the beams outside the links to resist large moments and axial forces. The lateral frame shear is able to transfer down the braces without transferring through the beams in the two-story configuration.

The beam outside the link (in the example) is required to resist 110 percent of the bending and axial forces corresponding to the link beam yield, using its nominal strength R_yV_n. The combined beam bending plus axial interaction equations are referenced from AISC 360-05, Section H. Axial load analysis is referenced from AISC 360-05, Section E, and bending analysis is referenced from AISC 360-05, Section F.

Forces are from the hand evaluation of EBF frame behavior and from the computer model (not shown):

Axial force in beam outside link

$$P_E = 361 \text{ kips}$$

From computer model, the load combination 5, including $E_v = 0.28D$, is

$$1.2D + 0.28D + 0.5L + 1.0\rho Q_E\text{:}$$

$$1.48D + 0.5L;\ P_{D+L} = 21 \text{ kips}$$

From EBF4 frame analysis

$$M_E = 12{,}270 \text{ kip-in}$$

$$P_u = 1.1\Omega R_y P_E + \mathrm{P}_{D+L}$$

$$P_u = (1.1)(1.15)(1.1)(361 \text{ kips}) + (21 \text{ kips}) = 531 \text{ kips}$$

From computer analysis, load combination 5

$$1.2D + 0.28D + 0.5L + 1.0\rho Q_E;\ M_{D+L} = 619 \text{ kip-in}$$

$$M_u = \frac{1.1R_yV_pe}{2} + M_{D+L} = \frac{1.1(1.1)(409 \text{ kips})(60 \text{ in})}{2} + (619 \text{ kip-in}) = 15{,}465 \text{ kip-in}$$

Beam section properties:

Combined section properties are given in Table 2-12, the reader should understand how to convert typical beam section properties to those with cover plates:

The beams at Level 2 through the roof all require cover plates and thus have transformed section properties for use in the following equations.

For W18×158 beam with ¼ × 6-in cover plates top and bottom (combined section properties):

$A = 49.0 \text{ in}^2$; $Z_x = 386 \text{ in}^3$; $Z_f = 327 \text{ in}^3$; $I_x = 3{,}359 \text{ in}^4$; $S_x = 332 \text{ in}^3$; $r_y = 2.69$ in; $I_y = 356 \text{ in}^4$; $J = 25.4 \text{ in}^4$

Beam slenderness parameters:

$$\frac{kl}{r_y} = \frac{(1.0)(120\text{ in})}{2.69\text{ in}} = 44.6$$

$$F_e = \frac{\pi^2 E}{\left(\frac{kl}{ry}\right)^2} = \frac{(3.1416)(29{,}000\text{ ksi})}{(44.6)^2} = 143{,}890\text{ psi}$$ AISC 360, §E3 (E3-4)

When

$$\frac{kl}{r} \le 4.71\sqrt{\frac{E}{F_y}} = 4.71\left(\frac{29{,}000\text{ ksi}}{50\text{ ksi}}\right)^{1/2} = 113.4$$

$$F_{cr} = (0.658^{Fy/Fe})F_y = 0.658^{50{,}000\text{ psi}/143{,}890\text{ psi}}(50\text{ ksi}) = 43.2\text{ ksi}$$

AISC 360, §E3 (E3-2)

$$\phi_c = 0.90$$

Nominal axial strength is calculated as follows

$$P_n = A_g F_{cr} = (49\text{ in}^2)(43.2\text{ ksi}) = 2116\text{ kips}$$ AISC 360, §E3 (E3-1)

$$R_y P_n = R_y P_n = 1.1(2116\text{ kips}) = 2328\text{ kips}$$ AISC 341, §15.6b

Bending capacity calculations are calculated

$$\phi_b = 0.90$$ AISC 360, §F1

$M_n = M_p$ for a limit state if flexural yielding. AISC 360, §F2 (F2-1)

$$M_p = Z_x F_y = (386\text{ in}^3)(50\text{ ksi}) = 19{,}300\text{ kip-in}$$

Check lateral torsional buckling stability and allowable strength

$$M_n = C_b\left[M_p - (M_p - 0.75F_y S_x)\left(\frac{L_b - L_p}{L_r - L_p}\right)\right] \le M_p$$ AISC 360, §F2.2 (F2-2)

$$C_b = 1.0$$

Unbraced length

$$L_b = 120\text{ in}$$

Limiting laterally unbraced length for full plastic yielding

$$L_p = 1.76 r_y \sqrt{\frac{E}{F_{yf}}} = 1.76(2.69)\sqrt{\frac{29{,}000}{50}} = 114.02$$ AISC 360 §F2.2 (F2-5)

Limiting laterally unbraced length for inelastic lateral torsional buckling

$$L_r = 513.6\text{ in}$$ AISC 360, Table 3-2

Limiting buckling moment

$$0.75F_yS_x = 0.75(50{,}000 \text{ psi})(332 \text{ in}') = 12{,}450 \text{ kip-in}$$

$$M_n = C_b\left[M_p - (M_p - 0.75F_yS_x)\left(\frac{L_b - L_p}{L_r - L_p}\right)\right]$$

$$= 1.0\left[19{,}300 - (19{,}300 - 12{,}450)\left(\frac{120 \text{ in} - 114 \text{ in}}{514 - 114}\right)\right]$$

$$= 19{,}197 \text{ kip-in} \le M_p = 19{,}300 \text{ kip-in}$$

$$\therefore\ M_n = 19{,}197 \text{ kip-in}$$

$$R_yM_n = 1.1(19{,}197 \text{ kip-in}) = 21{,}117 \text{ kip-in}$$

Comparison of lateral torsional buckling moment with plastic yield moment indicates that plastic yield moment is the controlling yield behavior. AISC-LRFD Section H, combined axial plus bending interaction equations are as follows

For the case

$$\frac{P_r}{P_c} = \frac{P_u}{\phi_c R_y P_n} \ge 0.2$$ AISC 360, Hl-la

$$\frac{P_u}{\phi_c R_y P_n} + \frac{8}{9}\frac{M_{rx}}{\phi_b R_y M_{cx}} \le 1.0$$

For the case

$$\frac{P_r}{P_c} = \frac{P_u}{\phi_c R_y P_n} < 0.2$$ AISC 360, H1-1b

$$\frac{P_u}{\phi_c R_y P_n} + \frac{M_{rx}}{\phi_b R_y M_{cx}} \le 1.0$$

Thus, for this example

$$\frac{P_r}{P_c} = \frac{P_u}{\phi_c R_y P_n} = \frac{531 \text{ kips}}{0.9(2328 \text{ kips})} = 0.25 \ge 0.2$$

$$\frac{P_u}{\phi_c R_y P_n} + \frac{8}{9}\frac{M_{rx}}{\phi_b R_y M_{cx}} = \frac{531 \text{ kips}}{(0.9)(2328 \text{ kips})} + \left(\frac{8}{9}\right)\frac{15{,}465 \text{ kip-in}}{0.90(21{,}117 \text{ kip-in})} = 0.98 \le 1.0 \quad \ldots\ o.k.$$

Therefore, the W18×158 beam outside the link is okay. The EBF beams require cover plates and thus utilize combined section properties in the above equations.

7o. Beam stiffeners — AISC 341, §15.3

There are two types of stiffeners required in links: link stiffeners at ends at brace connections and intermediate stiffeners. These are shown in Figure 2-12.

Link end stiffeners

Full depth web stiffeners are required on both sides of the link beam at the brace connections. The stiffeners are used to prevent web buckling and to ensure ductile shear yielding of the web.

The stiffeners shall have a combined width not less than $b_f - 2t_w$ and a thickness not less than $0.75t_w$ or ⅜ inch, whichever is larger. For the W18×158 beam

$$b_f - 2t_w = 11.3 \text{ in} - 2(0.81 \text{ in}) = 9.68 \text{ in.} \quad \text{Use } 2(5.0 \text{ in})$$

The minimum thickness of the stiffeners is

$$0.75t_w = 0.75(0.81 \text{ in}) = 0.61 \text{ in.} \quad \text{Use ⅝-inch stiffeners}$$

Therefore, use 5 × ⅝-inch stiffeners each side of beam (total 4).

Intermediate stiffeners

AISC 341, §15.3a requires intermediate full depth web stiffeners (Figure 2-12) where link lengths are $1.6M_p/V_p$ or less.

Where link lengths are $1.6M_p/V_p$ or less, the spacing shall not exceed $30t_w - d_w/5$ for link rotation of 0.08 radians and $52t_w - d_w/5$ for link rotations of 0.02 radians. Linear interpolation may be used between link rotations of 0.02 and 0.08 radians. Thus

$$30t_w - \frac{d}{5} = 30(0.81 \text{ in}) - \left(\frac{19.72 \text{ in}}{5}\right) = 20.4 \text{ in}$$ AISC 341, §15.3a

$$52t_w - \frac{d}{5} = 52(0.81 \text{ in}) - \left(\frac{19.72 \text{ in}}{5}\right) = 38.2 \text{ in}$$ AISC 341, §15.3a

The link rotation is 0.045 radians for the beam, interpolation must be used to determine the maximum spacing of intermediate stiffeners. This is shown below.

$$\left(\frac{0.080 \text{ rad} - 0.045 \text{ rad}}{0.080 \text{ rad} - 0.020 \text{ rad}}\right)(38.2 \text{ in} - 20.4 \text{ in}) + 20.4 \text{ in} = 30.8 \text{ in}$$

The link length is 60 inches. Use three equal spacings of 20 inches.

The link beam is a W18, stiffener depth is 19.72 in – 2(0.81 in) = 18.1 in. Under §15.3e, AISC 341, intermediate stiffeners of depth greater than 25 inches are required to be placed on both sides of the beam. One-sided stiffeners are allowed for link depths less than 25 inches. The width must not be less than

$$b_f/2 - t_w = 11.30 \text{ in}/2 - 0.81 \text{ in} = 4.84 \text{ in}$$

Therefore use 5 × ⅝-inch stiffeners on one side of the W18 beam.

Web stiffener welds

The web stiffener welds are required to develop a stiffener force of

$$A_{st}F_y = (4.75\text{ in})(0.50\text{ in})(50\text{ ksi}) = 118\text{ kips} \qquad \text{AISC 341, §15.3}$$

The minimum size of fillet weld, per AISC-LRFD Table J2.4, is ¼ inch to the link web and 5⁄16 inch to the link flange. Using E70XX electrodes and ¼-inch fillet welds each side, the weld capacity is $0.6F_{EXX}$, $\phi = 0.75$. The required weld length on the beam web is

$$l_{\text{required}} = \frac{118\text{ kips}}{0.75(0.60)(70\text{ ksi})2(0.25\text{ in})(0.707)} = 10.6\text{ in}$$

Therefore, use ¼-inch fillet welds, both sides of the stiffener, at web. Avoid welding the web stiffeners in the beam "*k*" region.

The web stiffener welds at flanges are required to develop a stiffener force of

$$\frac{A_{st}F_y}{4} = \frac{(5.0\text{ in})(0.625\text{ in})(50\text{ ksi})}{4} = 39.0\text{ kips} \qquad \text{AISC 341, §15.3c}$$

The required weld length on the beam flanges is

$$l_{\text{required}} = \frac{39.0\text{ kips}}{0.75(0.60)(70\text{ ksi})2(0.3125\text{ in})(0.707)} = 2.81\text{ in}$$

Note: use 5⁄16-inch weld at both sides of the stiffener at flanges.

7p. Tabulated link beam design

In the following Table 2-10 are tabular calculations that show the results from procedures from Parts 7a through 7o applied to all beams in the frame EBF4. The link beam design for all levels is as shown below in tabular form following the equations given above (each row/level is a continuation of the table above).

Table 2-10. Tabular design of link beams at all levels

Link Beam Section Properties

LEVEL	Link Beam	e	a	h	F_y	A	b_f	t_f	d	t_w	A_{web} $t_w(d-2t_f)$	Size
R	W18×97	38	161	144	50	28.50	11.15	0.87	18.59	0.54	9.01	OK
5	W18×97	48	156	144	50	28.50	11.15	0.87	18.59	0.54	9.01	OK
4	W18×97	50	155	144	50	28.50	11.15	0.87	18.59	0.54	9.01	OK
3	W18×106	50	155	144	50	31.10	11.20	0.94	18.73	0.59	9.94	OK
2	W18×158	60	150	168	50	46.30	11.30	1.44	19.72	0.81	13.64	OK

Beam Section Properties

LEVEL	Link Beam	A_f	Z_f	Z_x	I_x	S_x	r_x	I_y	S_y	r_y	J	C_w
R	W18×97	19.4	171.8	211.0	1750.0	188.0	7.84	201.0	36.07	2.66	5.86	15800
5	W18×97	19.4	171.8	211.0	1750.0	188.0	7.84	201.0	36.07	2.66	5.86	15800
4	W18×97	19.4	171.8	211.0	1750.0	188.0	7.84	201.0	36.07	2.66	5.86	15800
3	W18×106	21.1	187.3	230.0	1910.0	204.0	7.84	220.0	39.29	2.66	7.48	17400
2	W18×158	32.5	297.5	356.0	3060.0	310.0	8.13	347.0	61.42	2.74	25.40	28900

Combined Section Properties (Beams plus Cover Plates)

LEVEL	Link Beam	Plate b	t	A_t	Z_x	Z_f	I_x	S_x	r_y	I_y	J	C_w	X1	X2
R	W18×97	0.0	0.000	29	211	172	1750	188	2.66	201	5.9	15800	2748	0.00259
5	W18×97	6.0	0.375	33	254	214	2155	223	2.55	215	5.9	15800	2499	0.00340
4	W18×97	6.0	0.375	33	254	214	2155	223	2.55	215	5.9	15800	2499	0.00340
3	W18×106	6.0	0.375	36	273	230	2321	238	2.56	234	7.5	17400	2742	0.00241
2	W18×158	6.0	0.250	49	386	327	3359	332	2.69	356	25.4	28900	4264	0.00044

Beam Compact Flange/Web

LEVEL	V_p	ϕ	M_p	eV_p/M_p coeff.	$b_f/2t_f$	Compact Flange $0.3(E/F_y)^2$	Compact Flange	$1.1R_y$ M_s/V_s	P_u	P_y	P_u ϕP_n	h/t_w	Compact Web λp	Compact Web
R	243.4	2.88	10550	0.97	6.41	7.22	OK	52.4	133.94	1425.00	0.11	31.5	63.0	OK
5	243.4	1.58	10550	1.23	6.41	7.22	OK	52.4	227.29	1425.00	0.19	31.5	57.9	OK
4	243.4	1.19	10550	1.28	6.41	7.22	OK	52.4	295.39	1425.00	0.24	31.5	56.3	OK
3	268.4	1.12	11500	1.30	5.96	7.22	OK	51.8	339.49	1555.00	0.26	28.6	56.0	OK
2	368.3	1.15	17800	1.38	3.92	7.22	OK	58.5	360.69	2315.00	0.18	20.8	58.0	OK

Combined Link Stresses/Unsupported Length

LEVEL	Shear above	F_i	P_{max} Link Beam	P_{link}	M_{link}	Comb Link Loads	Allowable Link Stress	Link Stresses	V_{pa}	M_{pa}	$2\phi M_{pa}/2$	V_u link	Value M_p/V_p	L_u max ASD
R	0	133.9	133.9	16.7	5138.4	30.3	50	OK	242.3	11279	534.3	242.3	0.88	112.6
5	234.4	110.1	227.3	15.5	6490.6	38.2	50	OK	240.3	10463	392.4	240.3	1.11	108.1
4	427.1	81.9	295.4	13.7	6761.1	39.7	50	OK	238.1	9868	355.3	238.1	1.15	108.1
3	570.3	54.3	339.5	12.0	7456.1	40.1	50	OK	261.9	10607	381.9	261.9	1.17	108.6
2	665.4	28.0	360.7	10.3	12276.4	41.4	50	OK	363.8	17731	531.9	363.8	1.24	113.9

Beam Outside Link, Design Forces

LEVEL	P	M $V_{se}/2$	Ω	Beam factor	P_{comp} $1.48D + .5L$	M_{comp} $1.48D + .5L$	LF	R_y	P_{bu} design	M_{bu} design
R	134	5138	2.88	1.10	1.4	264.0	1.10	1.1	472	6481
5	227	6491	1.58	1.10	14.9	381.6	1.10	1.1	464	8235
4	295	6761	1.19	1.10	20.9	420.0	1.10	1.1	454	8601
3	339	7456	1.12	1.10	21.7	488.4	1.10	1.1	490	9510
2	361	12276	1.15	1.10	21.0	613.0	1.10	1.1	529	15467

Axial Compression Parameters

LEVEL	SECTION	L_u (ft)	kl/r	Limit kl/r	F_e	F_{cr} (ksi)	ϕ_c	P_n (k)	R_yP_n (k)
R	W18×97	10	45.2	113	140	43.07	0.90	1227	1350
5	W18×97	10	47.1	113	129	42.52	0.90	1403	1544
4	W18×97	10	47.1	113	129	42.52	0.90	1403	1544
3	W18×106	10	46.9	113	130	42.58	0.90	1516	1668
2	W18×158	10	44.7	113	144	43.22	0.90	2131	2344

Flexural Strength Parameters and Combined Axial plus Bending Results (LTB = Lateral Torsional Buckling mode)

LEVEL	ϕ_b (ksi)	$M_n = M_p$ Yielding (kin)	C_b LTB (k)	L_b (in)	L_p (in)	L_r	$.7F_yS_x$ (kin)	M_n LTB (kin)	M_n	R_yM_n	$P_u/\phi P_n$	AISC LRFD H1-1a	AISC LRFD H1-1b
R	0.9	10550.0	1.0	120	113	364	6590	10432.7	10433	11476	0.41	0.97	NA
5	0.9	12683.6	1.0	120	108	364	7799	12455.4	12455	13701	0.35	0.95	NA
4	0.9	12683.6	1.0	120	108	364	7799	12455.4	12455	13701	0.35	0.97	NA
3	0.9	13649.3	1.0	120	109	382	8339	13426.7	13427	14769	0.35	0.98	NA
2	0.9	19297.8	1.0	120	114	514	11629	19180.7	19181	21099	0.25	0.97	NA

Link Rotations

Level	Δ_s	Story Drift Δ_s	Story Drift Δ_m	H	a	e	rot
R	1.47	0.29	1.15	144	161	38	0.0758
5	1.18	0.30	1.20	144	156	48	0.0623
4	0.88	0.31	1.23	144	155	50	0.0616
3	0.58	0.28	1.13	144	155	50	0.0566
2	0.29	0.29	1.16	180	150	60	0.0387

7q. AISC seismic brace design — AISC 341, §15.6

The braces are required to be designed for $1.25R_yV_p$ times the yielding link strength plus 1.25 times gravity loads combinations.

$$P_E = 1.25\Omega R_y P_{\text{computer}} \text{ due to } E_h \text{ loads}$$

$$M_E = 1.25R_yV_pe/2$$

Using strength design procedures outlined in AISC LRFD Section H, obtaining forces from a computer analysis, and showing calculations in tabular form (Table 2-11), the design forces for braces (P and M) are calculated. Column shear forces are not a controlling factor and are not shown for brevity.

Table 2-11. Tabular design of braces for EBF4 at all levels

Brace Section Properties

LEVEL	SECTION	A (in^2)	Z_x (in^3)	S_x (in^3)	L (ft)	r_x (in)	r_y (in)	kl/r_y	F_y (ksi)
5	W12×136	40	214	186	18	5.57	3.16	69.3	50
4	W12×136	40	214	186	18	5.57	3.16	69.3	50
3	W12×152	45	243	209	18	5.66	3.19	68.7	50
2	W12×152	45	243	209	18	5.66	3.19	68.0	50
1	W12×210	62	348	292	19	5.88	3.28	69.9	50

Brace Design Loads

LEVEL	SECTION	P	M	Ω	Beam factor	P_{comp} $1.465D + .5L$	M_{comp} $1.465D + .5L$	LF	R_y
R	W12×152	124	936	2.88	1.25	19.0	132.0	1.25	1.3
5	W12×152	232	1800	1.58	1.25	31.0	168.0	1.25	1.3
4	W12×152	310	2472	1.19	1.25	31.0	180.0	1.25	1.3
3	W12×152	368	2832	1.12	1.25	31.0	192.0	1.25	1.3
2	W12×210	451	3936	1.15	1.25	30.0	168.0	1.25	1.3

Brace Axial Design Parameters

LEVEL	SECTION	L_u (ft)	kl/r	λ_c	F_{cr} (ksi)	ϕ_c	P_n (k)
R	W12×152	18	69.3	0.916	35.18	0.9	1403.7
5	W12×152	18	69.3	0.916	35.18	0.9	1403.7
4	W12×152	18	68.7	0.908	35.40	0.9	1582.4
3	W12×152	18	68.0	0.899	35.66	0.9	1593.8
2	W12×210	19	69.9	0.924	34.97	0.9	2161.2

Brace Bending Design Parameters

LEVEL	ϕ_b (ksi)	$M_n = M_p$ Yielding (kin)	C_b LTB (k)	L_b (in)	L_p (in)	L_r	M_r (kin)	M_n LTB (kin)	M_n
R	0.9	10700.0	1.0	219	134	736	7440	10239.7	10240
5	0.9	10700.0	1.0	219	134	736	7440	10239.7	10240
4	0.9	12150.0	1.0	219	135	743	8360	11627.3	11627
3	0.9	12150.0	1.0	217	135	743	8360	11641.5	11641
2	0.9	17400.0	1.0	229	139	1008	11680	16806.7	16807

Brace Combined Axial plus Bending Results

LEVEL	SECTION	$P_u/\phi P_n$	AISC LRFD H1-1a	AISC LRFD H1-1b
5	W12×136	0.51	0.95	NA
4	W12×136	0.53	1.00	NA
3	W12×152	0.47	0.90	NA
2	W12×152	0.52	0.98	NA
1	W12×210	0.48	0.92	NA

7r. AISC-Seismic column design — AISC 341, §15.8

Shown in Table 2-12 is the design of the columns for frame EBF4 for the requirements of AISC 341. The columns are required to resist an axial force corresponding to $1.1R_yV_n$ which is the shear strength of the links, to assure that the yielding mechanism is within the link beams. Design forces (P and M) are calculated as $1.1\Omega R_y$ times seismic forces plus 1.1 times factored gravity load combinations. Column shear forces are not a controlling factor and are not shown for brevity.

Table 2-12. Tabular design of columns for EBF4 at all levels

Column Section Properties

LEVEL	SECTION	A (in^2)	Z_x (in^3)	S_x (in^3)	L (ft)	r_x (in)	r_y (in)	kl/r_y	F_y (ksi)
5	W12×58	17	86	78	12	5.29	2.51	87.3	50
4	W12×58	17	86	78	12	5.29	2.51	87.3	50
3	W12×58	17	86	78	12	5.29	2.51	87.3	50
2	W12×136	40	214	186	12	5.57	3.16	68.6	50
1	W12×136	40	214	186	15	5.57	3.16	72.6	50

Column Design Loads

LEVEL	SECTION	P	M	Ω	Col factor	P_{comp} $1.465D + .5L$	M_{comp} $1.465D + .5L$	Col factor	R_y
R	W12×58	0	252	2.88	1.10	4.0	144.0	1.10	1.3
5	W12×58	64	396	1.58	1.10	23.0	144.0	1.10	1.3
4	W12×58	197	384	1.19	1.10	51.0	132.0	1.10	1.3
3	W12×136	378	576	1.12	1.10	78.0	96.0	1.10	1.3
2	W12x136	579	840	1.15	1.10	110.0	72.0	1.10	1.3

Column Axial Design Parameters

LEVEL	SECTION	L_u (ft)	kl/r	λ_c	F_{cr} (ksi)	ϕ_c	P_n (k)
R	W12×58	12	57.4	0.759	39.30	0.85	668.0
5	W12×58	12	57.4	0.759	39.30	0.85	668.0
4	W12×58	12	57.4	0.759	39.30	0.85	668.0
3	W12×136	12	45.6	0.603	42.95	0.85	1713.7
2	W12×136	15	57.0	0.753	39.43	0.85	1573.3

Column Bending Design Parameters

LEVEL	ϕ_b (ksi)	$M_n = M_p$ Yielding (kin)	C_b LTB (k)	L_b (in)	L_p (in)	L_r	M_r (kin)	M_n LTB (kin)	M_n
R	0.9	4320.0	1.0	144	106	323	3120	4112.1	4112
5	0.9	4320.0	1.0	144	106	323	3120	4112.1	4112
4	0.9	4320.0	1.0	144	106	323	3120	4112.1	4112
3	0.9	10700.0	1.0	144	134	666	7440	10638.7	10639
2	0.9	10700.0	1.0	180	134	666	7440	10418.3	10418

Column Combined Axial plus Bending Results

LEVEL	SECTION	$P_u/\phi P_n$	AISC LRFD H1-1a	AISC LRFD H1-1b
5	W12×58	0.01	NA	0.33
4	W12×58	0.30	0.55	NA
3	W12×58	0.69	0.88	NA
2	W12×136	0.47	0.57	NA
1	W12×136	0.80	0.94	NA

7s. Foundation design considerations

In EBF design, special consideration should be given to the foundation design. The basis for design of the EBF is that the yielding occurs in the EBF links. Thus all other elements should have the strength to develop the link beam yielding strengths.

The code does not require the foundation design to be capable of developing the link beam strengths. However, if a minimum code foundation design is performed, the foundation will generally not develop the EBF link beam strengths, and yielding will occur in the foundation. This is not consistent with the design philosophy for EBF frames.

The SEAOC Blue Book recommends that the foundation be designed to develop the strength of the EBF frame. The intention is to have adequate foundation strength and stability to assure the development of link beam yield mechanisms to achieve the energy dissipation anticipated in the eccentric braced frames. A static push-over of an EBF frame can give a good indication of the foundation adequacy.

7t. Final frame member sizes

Table 2-13. Final frame member sizes for EBF4

Level	Beams	Links	Beam Cover Plates		Columns	Braces
			width (in)	thickness (in)		
R	W18×97	38	0	0.000		
5	W18×97	48	6	0.375	W12×58	W12×136
4	W18×97	50	6	0.375	W12×58	W12×136
3	W18×106	50	6	0.375	W12×58	W12×152
2	W18×158	60	6	0.250	W12×136	W12×152
1					W12×136	W12×210

8. Typical EBF details

Figures 2-7 through 2-14 are examples of typical EBF connection details. These details are not all-inclusive and other options not shown are possible. If EBF link beams are provided with welded connections to columns as used in this example, an increase in R values from 7 to 8 is achieved. Collector beams may be welded or not depending on a designer's particular design. These details are shown for both wide-flange and tube section braces.

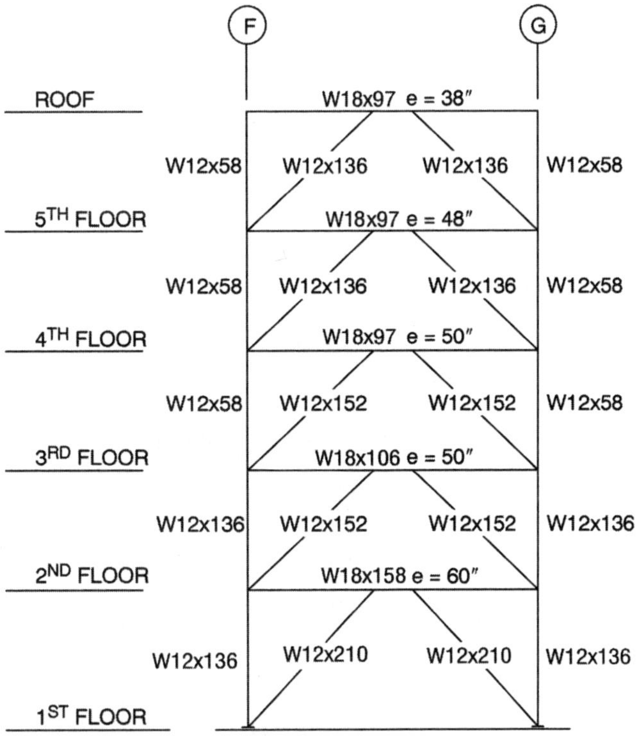

Figure 2-7. EBF4 Frame member sizes

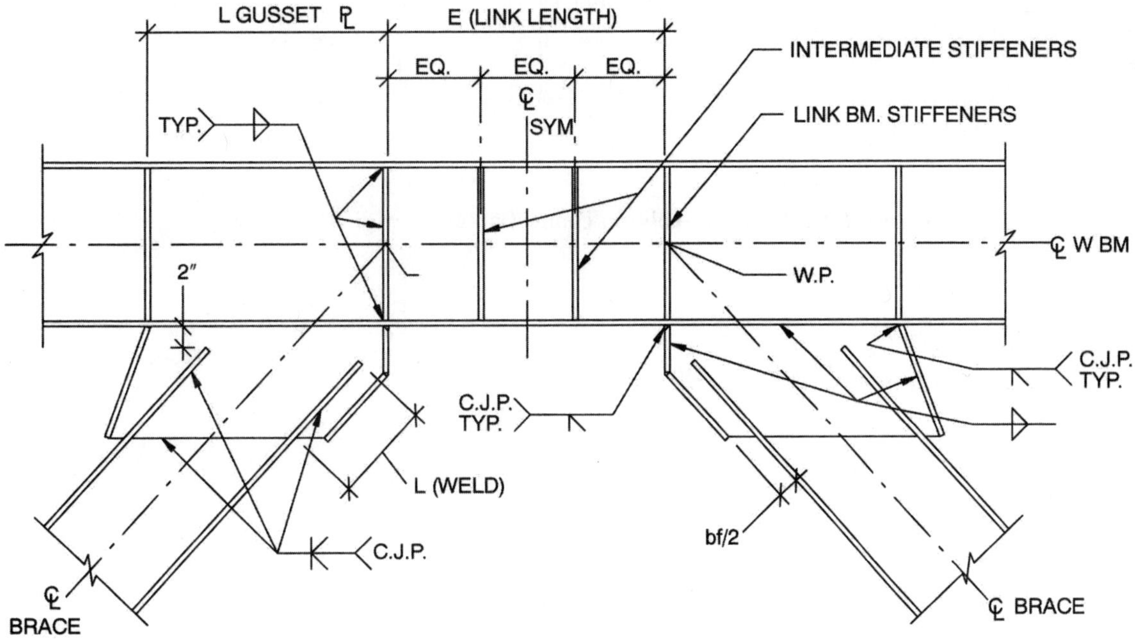

Figure 2-8. EBF brace-beam connection at link using wide flange brace

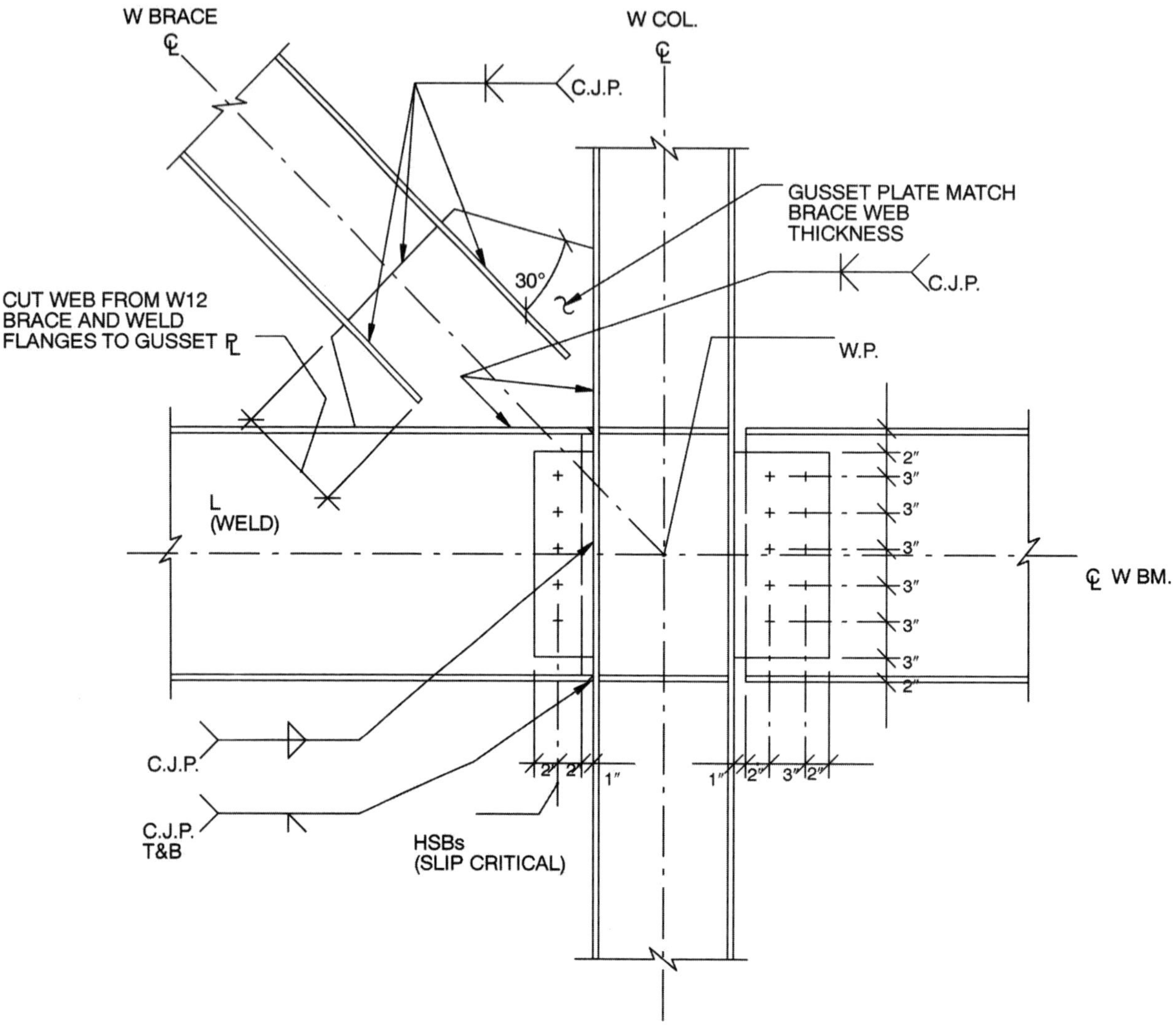

Figure 2-9. EBF brace-column connection using wide flange brace

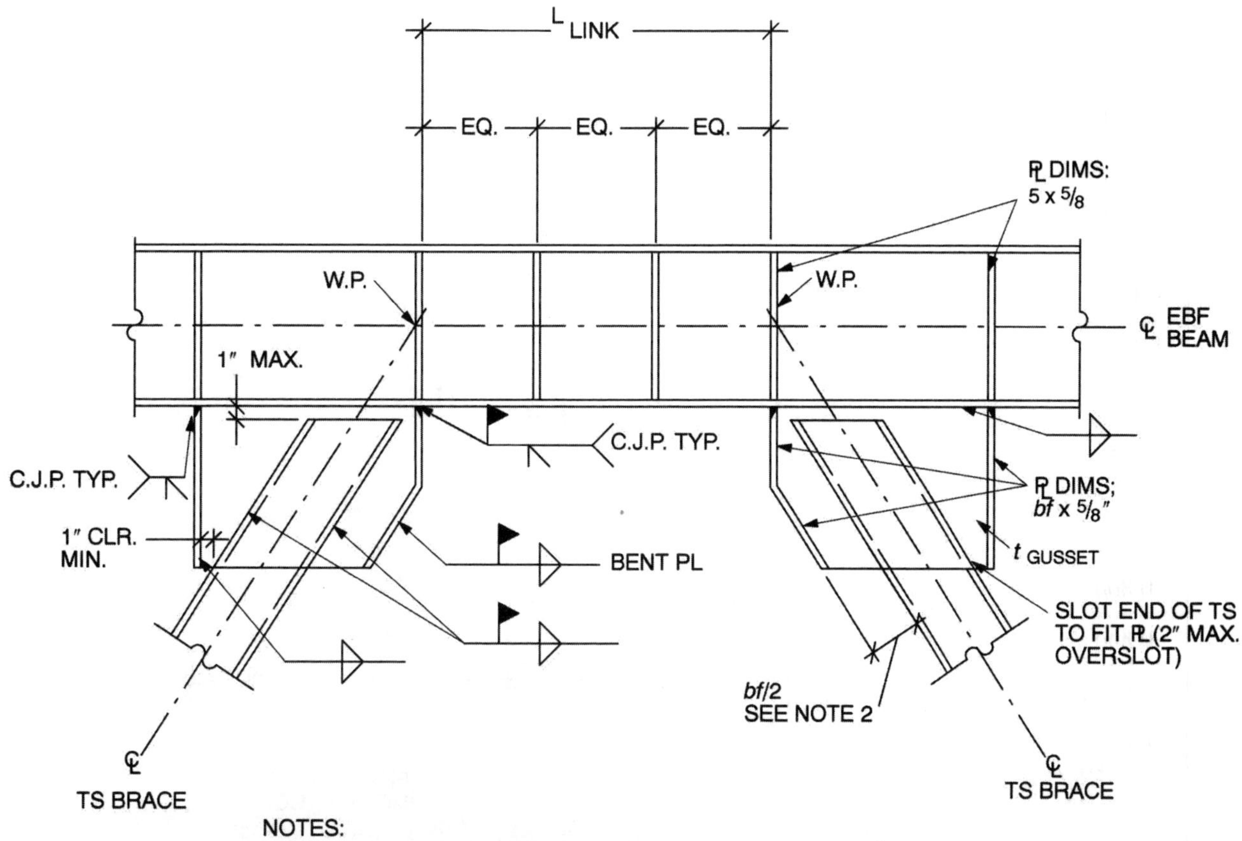

NOTES:

1. b_f = FLANGE WIDTH OF BEAM
   t_w = WEB THICKNESS OF BEAM
   t_f = FLANGE THICKNESS OF BEAM

Figure 2-10. EBF beam/brace connection at link using TS brace

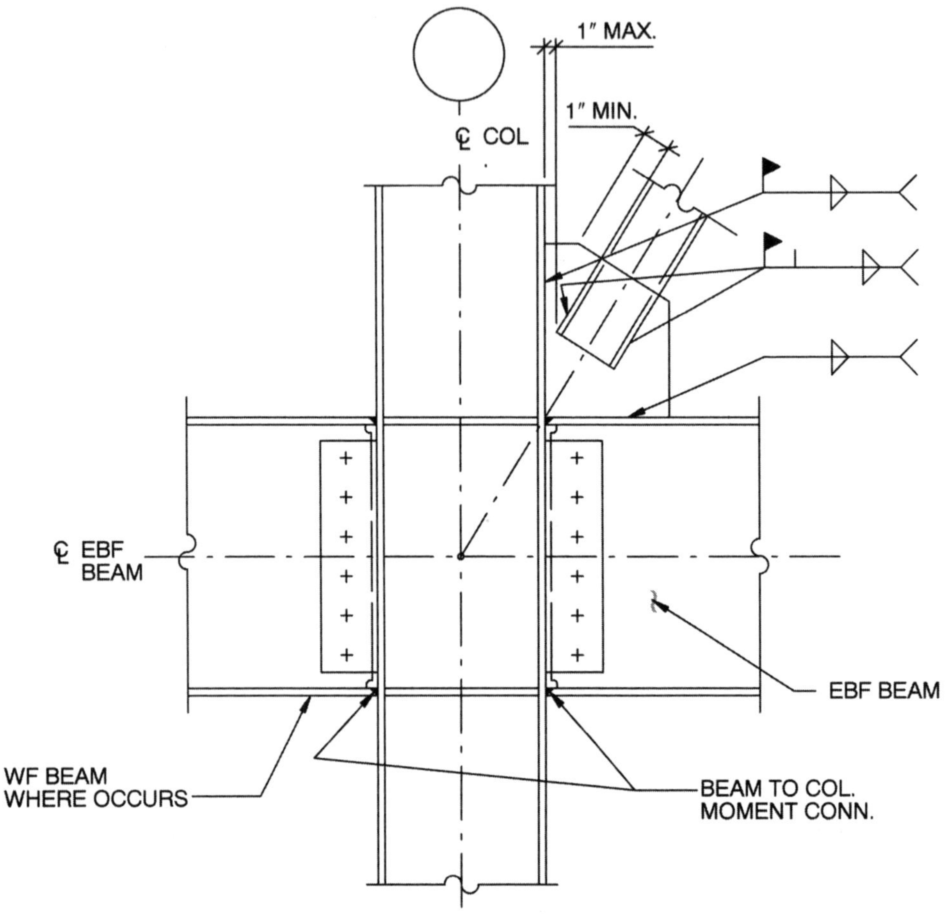

Figure 2-11. Brace/beam connection with TS brace

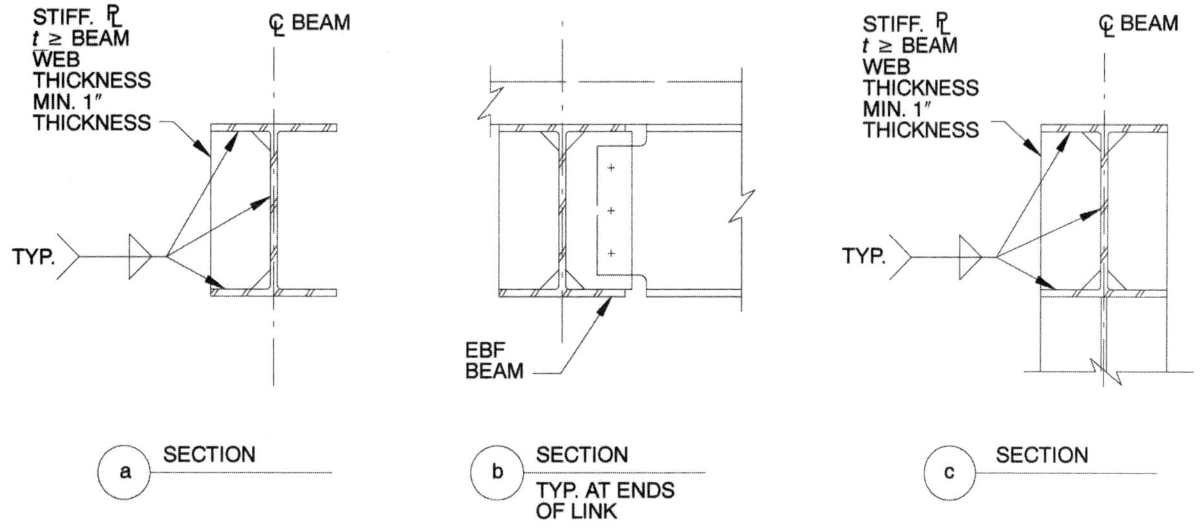

Figure 2-12. EBF stiffeners at links

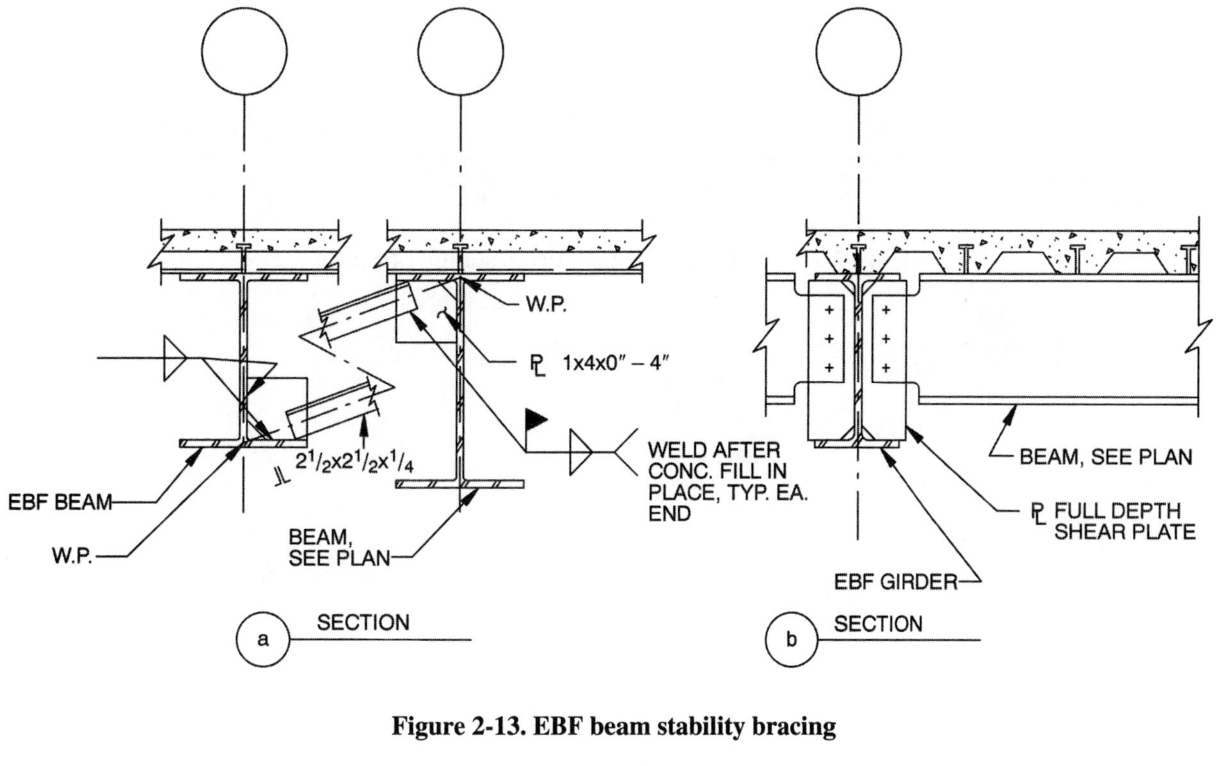

Figure 2-13. EBF beam stability bracing

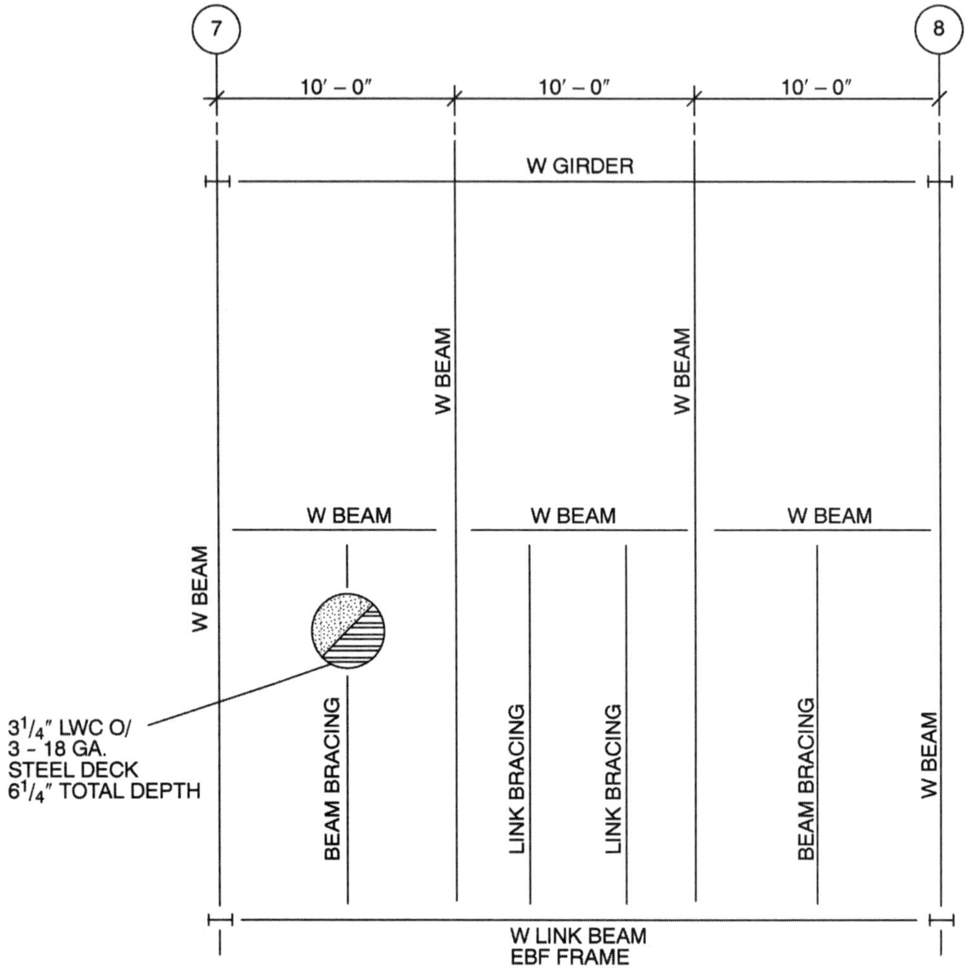

Figure 2-14. Partial plan of EBF beam stability bracing

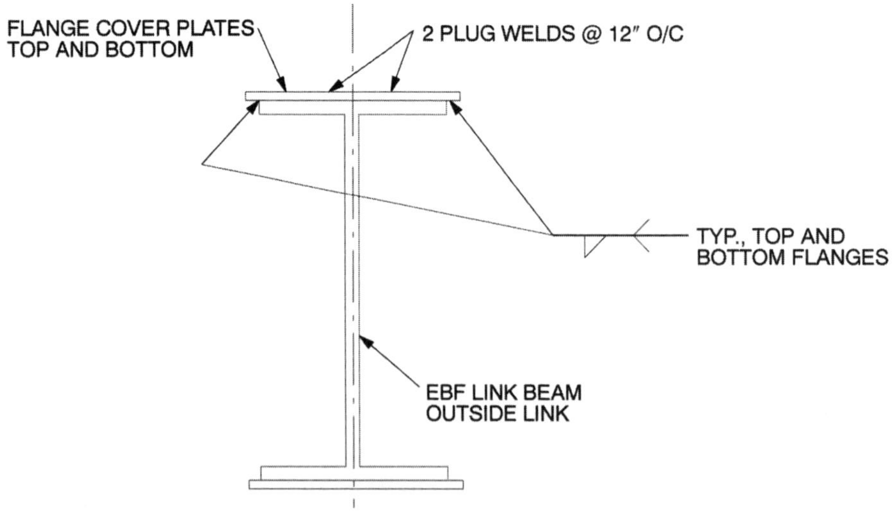

Figure 2-15. Link beam cover plates (beam outside the link)

Commentary

EBF frames are considered a highly reliable seismic system because of their ability to yield with a known behavior at controllable locations and to demonstrate very good hysteretic behavior during cyclical loading. The possibility exists of discrete post earthquake repairs in local areas if yielding of a frame occurs during an earthquake. Significant deformation can occur in the floor beams. The construction of these frames is not difficult, and the cost is only slightly greater than the cost of special concentric braced frame systems.

The author found that by designing an EBF link beam that meets all of the most desirable attributes of EBF design, that the beam outside the link might require cover plates to achieve the required strength. The designer will struggle with optimization of the link design and the requirement for cover plates outside the link. The author believes that optimization of the link is the most important element in the system and feels that if cover plates are required outside the link, that is a cost worth paying.

References

AISC-341-05, American Institute of Steel Construction, AISC, *Seismic Provisions for Structural Steel Buildings*, March 2005, and *Supplement No. 1*, November 2005, Chicago, Illinois.

AISC-360-05, American Institute of Steel Construction, AISC, *Manual of Steel Construction*, Thirteenth Edition, Chicago, Illinois, 2005.

Becker and Ishler, "*Seismic Design Practice for Eccentrically Braced Frames, Based on the 1994 UBC*," Steel Tips, Structural Steel Educational Council, Moraga, California, December 1996.

Engelhardt and Popov, "*On Design of Eccentrically Braced Frames*," *Earthquake Spectra*, Vol. 5, No. 3, Earthquake Engineering Research Institute, Oakland, California, 1989.

IBC, "*International Building Code*," 2009 Edition, International Code Council, Washington, D.C.

Kasai and Popov, "*General Behavior of WF Steel Shear Link Beams*," Journal of Structural Engineering, Vol. 112, No. 2, American Society of Civil Engineers, Reston, Virginia, February 1986.

Popov, Kasai, and Engelhardt, "*Advances in Design of Eccentrically Braced Frames*," Earthquake Spectra, Vol. 3, No. 1, Earthquake Engineering Research Institute, Oakland, California, 1987.

Popov, Engelhardt, and Ricles, "*Eccentrically Braced Frames: US Practice*," Engineering Journal, AISC, 2nd Quarter, American Society of Civil Engineers, Reston, Virginia, 1989.

Design Example 3
Special Moment Frame

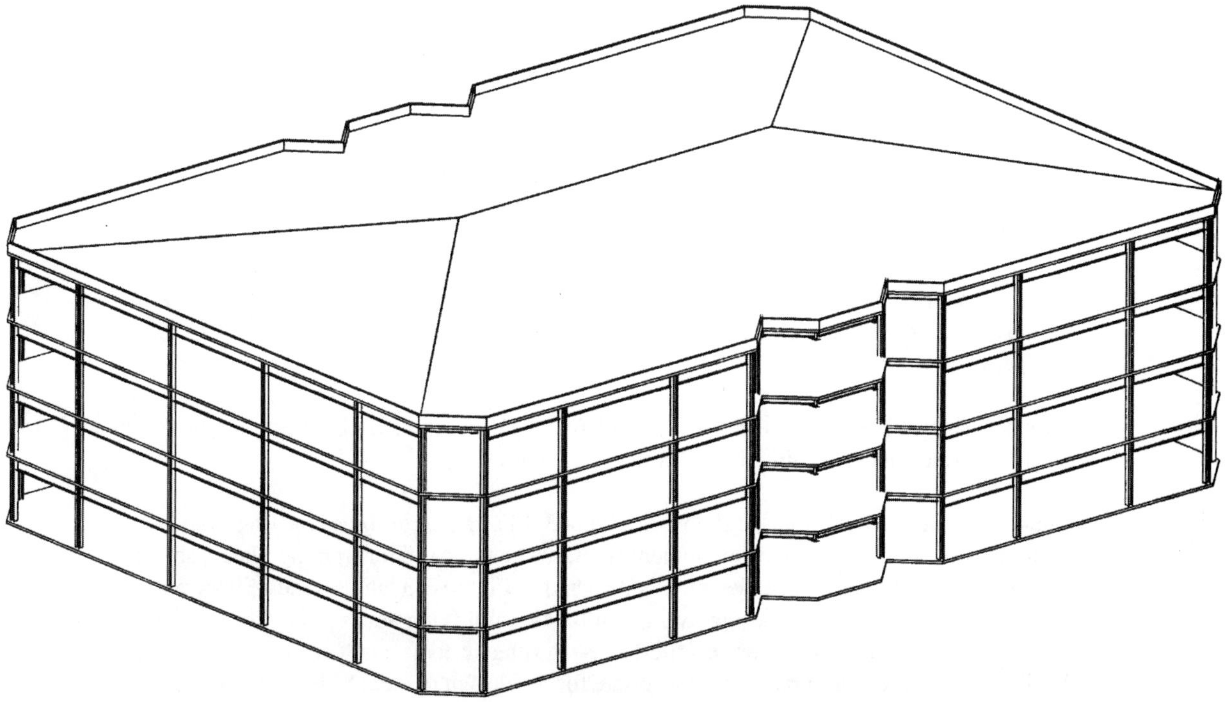

Figure 3-1. Four-story steel office building with steel SMF

Foreword

This design example illustrates the use of the 2009 IBC and ASCE/SEI 7-05 provisions for design of a steel special moment frame (SMF). During the course of the development of this Volume 3, the American Institute of Steel Construction published new editions of three Standards. These standards are

1. The use of wood diaphragms as part of the lateral force resisting system. ANSI/AISC 341-05, Seismic Provisions for Structural Steel Buildings, Including Supplement No. 1 (AISC 341),

2. ANSI/AISC 358-05, Prequalified Connections for Special and Intermediate Steel Moment Frames for Seismic Applications (AISC 358),

3. ANSI/AISC 360-05, Specification for Structural Steel Buildings (AISC 360).

These standards comprise the regulations that govern the design of structural steel special moment frames. Many of the provisions in AISC 358 and AISC 341 are rooted in the seminal SMF publication FEMA-350 *Recommended Seismic Criteria for New Steel Moment-Frame Buildings*.

Overview

In 2005, AISC published a standard (AISC 358) that provides design, detailing, fabrication and quality criteria for prequalifed SMF connections. This standard comprises documentation meeting the requirements of ASIC 341, Appendix P. The following is an excerpt from the AISC 358 introduction.

> *This Standard specifies design, detailing, fabrication and quality criteria for connections that are prequalified in accordance with the AISC Seismic Provisions for Structural Steel Buildings (herein referred to as the AISC Seismic Provisions) for use with special moment frames (SMF) and intermediate moment frames (IMF). The connections contained in this Standard are prequalified to meet the requirements in the AISC Seismic Provisions only when designed and constructed in accordance with the requirements of this Standard. Nothing in this Standard shall preclude the use of connection types contained herein outside the indicated limitations, or the use of other connection types, when satisfactory evidence of qualification in accordance with Appendix S of the AISC Seismic Provisions is presented to the authority having jurisdiction.*

This design example follows the provisions of AISC 358 for the RBS connection type for the steel SMF (the only prequalified SMF connection type).

The 4-story steel office structure shown in Figure 3-1 has a lateral force resisting system comprising structural steel special moment frames. The typical floor framing plan is shown in Figure 3-2. A typical frame elevation is depicted in Figure 3-3, and a general floor plan is shown in Figure 3-4. This design example utilizes simplifying assumptions for ease of calculation, consistency between seismic-force-resisting systems, and computational efficiency. While the example is accurate and appropriate for the design of steel SMF structures, different methodologies for analysis, connection design, and inelastic behavior can be utilized, including the use of proprietary SMF connection design; this example does not explore every possible option.

Outline

This example will illustrate the following parts of the design process.

1. Earthquake loads general; scope.

2. Seismic ground motion values. (2009 IBC)

3. Building weights and mass distribution.

4. Seismic design requirements for building structures. (ASCE/SEI 7-05)

5. SMF member design. (ANSI/AISC 341, 358, 360)

6. SMF beam-column connection design. (ANSI/AISC 358, 341, 360)

Given Information

Roof weights (dead load):

Roofing	5.0 psf
Insulation	2.0
Concrete fill on metal deck	43.0
Ceiling	3.0
Mechanical/electrical	3.0
Steel framing	6.0
Total	62.0 psf

Live load (Table 1607.1): 20.0 psf

Floor weights (dead load):

Floor covering	1.0 psf
Concrete fill on metal deck	43.0
Ceiling	3.0
Mechanical/electrical	3.0
Steel framing	9.0
Partitions (seismic DL)	10.0
Total	69.0 psf

Live load (Table 1607.1): 80.0 psf

Exterior wall system weight;steel studs,gypsum board, fascia panels 10.0 psf

Structural materials

Wide flange shapes	ASTM A992 (F_y = 50 ksi)
Plates	ASTM A572, Grade 50
Weld electrodes	E70X-XX

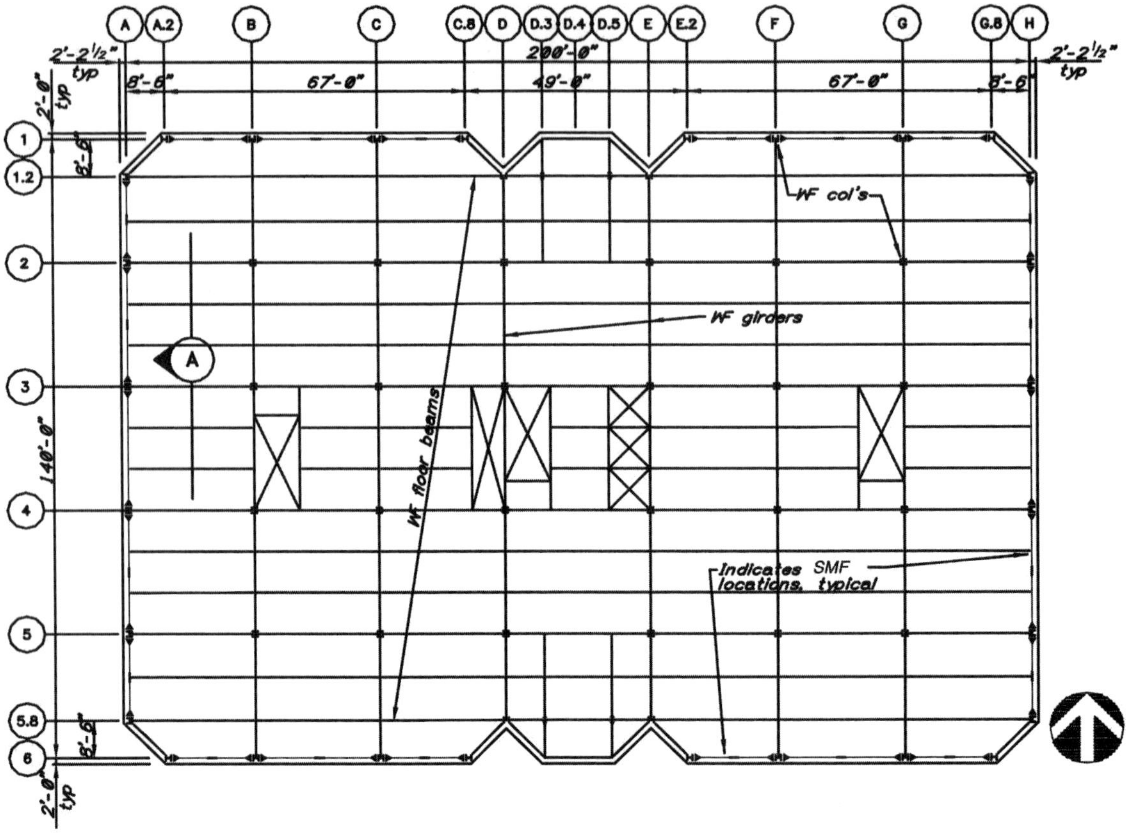

Figure 3-2. Typical floor framing plan

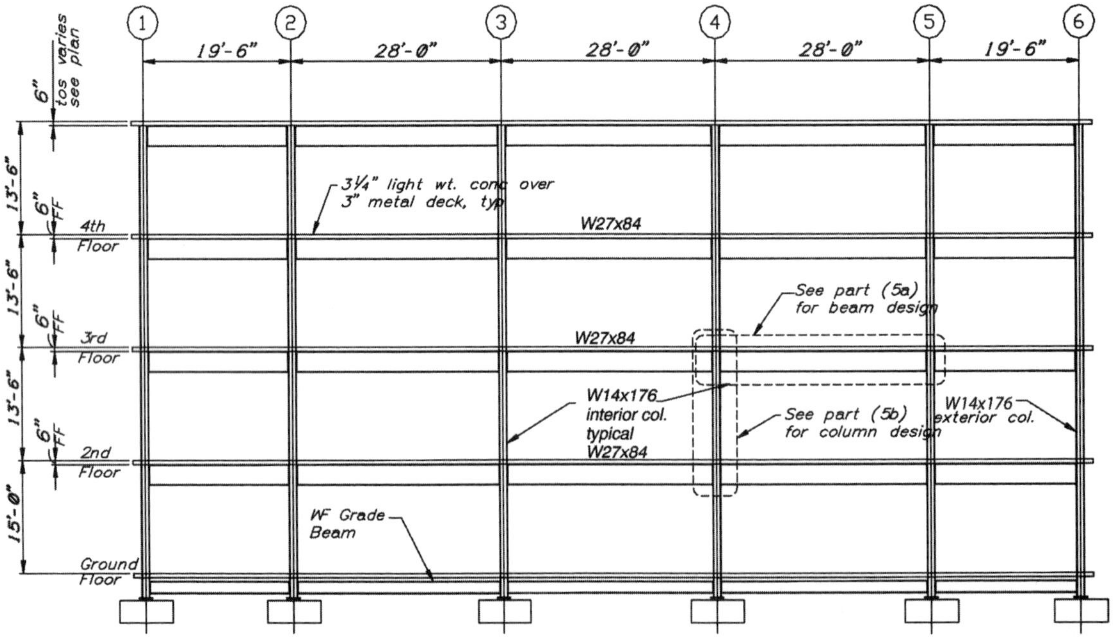

Figure 3-3. Frame elevation at Line A

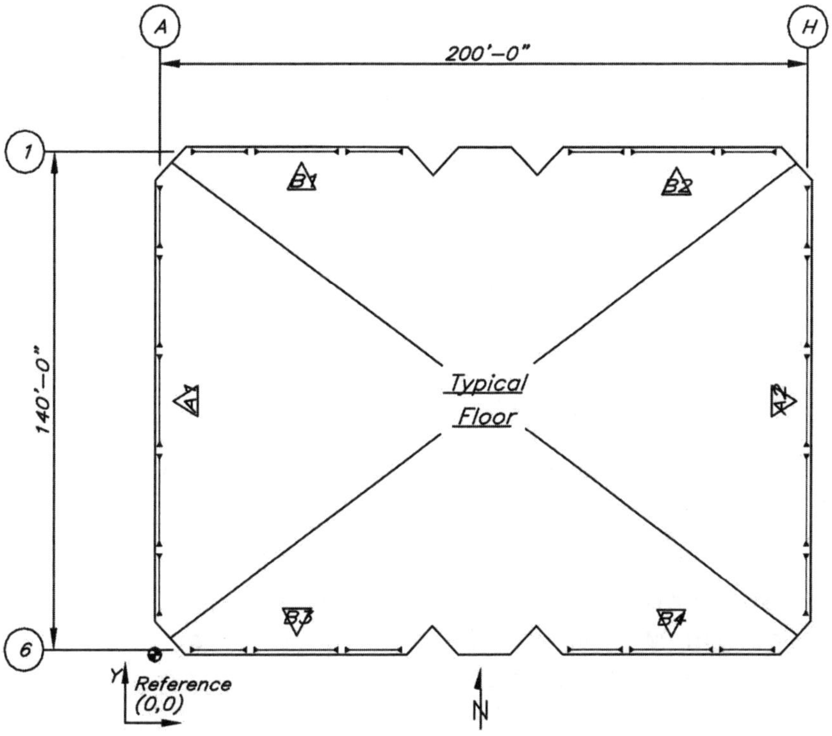

Figure 3-4. Typical floorplan

Calculations and Discussion	Code Reference

1. Scope §1613.1

Note: As promulgated in the 2009 IBC, every structure, and portion thereof, including nonstructural components that are permanently attached to structures and their supports and attachments, shall be designed and constructed to resist the effects of earthquake motions in accordance with ASCE/SEI 7-05, excluding Section 14 and Appendix 11A. The Seismic Design Category for a structure is permitted to be determined in accordance with Section 1613 of the 2009 IBC or ASCE/SEI 7-05.

Section 2 of this example comprises seismic design requirements associated with the 2009 IBC Section 1613 and pertinent design parameters and equations from ASCE/SEI 7-05. Both the 2009 IBC and ASCE/SEI 7-05 require the use of pertinent AISC standards (listed above). Please note that not every aspect of this design example follows all best practices for system design, especially in the selection of member sizes related to expected inelastic behavior; representative sizes are depicted for the articulation of methods and procedures required for the adequate and accurate design of a steel SMF per accepted national standards.

2. Seismic ground motion values §1613.5

2a. Mapped acceleration parameters §1613.5.1

The maximum considered earthquake (MCE) spectral response acceleration (5-percent damped, site class B) are taken from maps presented in Figures 1613.5(1) through 1613.5(14). For a site near Fairfield, California, the spectral accelerations are

$S_s = 1.5g$ @ $T = 0.2$ sec F 1613.5(3)

$S_1 = 0.6g$ @ $T = 1.0$ sec F 1613.5(4)

2b. Site class definitions §1613.5.2

Use default "stiff soil profile" in absence of geotechnical investigation

Site Class D T 1613.5.2

2c. Site coefficients and adjusted maximum considered earthquake spectral response acceleration parameters §1613.5.3

Site coefficients F_a and F_v for site class D, with the mapped spectral accelerations noted above, are given in Tables 1613.5.3 (1&2), as

$F_a = 1.0$ T 1613.5.3(1)

$F_v = 1.5$ T 1613.5.3(2)

The MCE spectral response acceleration, adjusted for site class effects, is determined by Equations 16-37 and 16-38

$S_{MS} = F_a S_s = 1.0(1.5) = 1.5g$ (short period) Eq 16-36

$S_{M1} = F_v S_1 = 1.5(0.6) = 0.9g$ (1-second period) Eq 16-37

2d. Design spectral response acceleration parameters §1613.5.4

IBC Equations 16-39 and 16-40 determine the design spectral response accelerations at short periods (S_{DS}) and at 1-second period (S_{D1})

$S_{DS} = (2/3)\ S_{MS} = (2/3)1.5g = 1.0g$ Eq 16-38

$S_{D1} = (2/3)\ S_{M1} = (2/3)\ 0.9g = 0.6g$ Eq 16-39

2e. Determination of seismic design category §1613.5.6

Seismic design categories are dependent on occupancy category and the severity of the design earthquake ground motion at the site. For this example office building, the occupancy category is most closely aligned with Category II (Table 1604.5). Seismic design category is then determined from Table 1613.5.6(1) and Table 1613.5.6(2) using values of S_{DS} and S_{D1}, respectively.

Short period response: $S_{DS} = 1.0g$ => Category D T 1613.5.6(1)

1-second period response: $S_{D1} = 0.6g$ => Category D T 1613.5.6(2)

As the mapped spectral acceleration $S_1 = 0.6g$ is less than $0.75g$, assignment of the structure to Category E is not required. §1613.5.6

Importance factor for Occupancy Category II; $I = 1.0$ ASCE T 11.5-1

3. Building weights and mass distribution

The following information is required for the rest of the design example.

Calculate the building weight and center of gravity at each level. Distribute the exterior curtain wall to each level by tributary height.

Table 3-1. Building mass properties

Roof Level Mass Properties:
Roof: $w_{DL} = 62.0$ psf; Exterior Walls: $w_{wall} = 10$ psf; Wall Area = (6.5 + 3.75)(696 ft) = 7308 ft²

Mark[1]	w_{DL} (psf)	Area (sf)	W_i (kips)	X_{cg} (ft)	Y_{cg} (ft)	$W(X_{cg})$	$W(Y_{cg})$
Roof	62.0	29,090	1804	100	70	180,358	126,251
Walls	10.0	7,308	75	100	70	7,482	5,237
Totals			1877			187,666	131,367
	$X_{cg} = 187{,}666/1877 = 100$ ft ; $Y_{cg} = 131{,}367/1877 = 70$ ft						

4th, 3rd & 2nd Level Mass Properties
Floor: $w_{DL} = 69.0$ psf; Exterior Walls: $w_{wall} = 10$ psf; Wall Area = (13.5)(696 ft) = 9396 ft²

Mark[1]	w_{DL} (psf)	Area (sf)	W_i (kips)	X_{cg} (ft)	Y_{cg} (ft)	$W(X_{cg})$	$W(Y_{cg})$
Floor	69.0	29,090	2007	100	70	200,721	140,504
Walls	10.0	9,396	94	100	70	9,396	6,577
Totals			2101			210,117	147,081
	$X_{cg} = 210{,}117/2101 = 100$ ft ; $Y_{cg} = 147{,}081/2101 = 70$ ft						

Table 3-2. Mass properties summary

Level	W (kips)	X_{cg} (ft)	Y_{cg} (ft)	M	MMI
Roof	1877	100	70	4.9	24,127
4th	2101	100	70	5.4	27,006
3rd	2101	100	70	5.4	27,006
2nd	2101	100	70	5.4	27,006
Total	8180			21.1	

Note: Mass (M) and mass moment of inertia (MMI) are used in analysis for determination of fundamental period, T. They are computed as

a) $M = (W/386.4)$ (kip-sec^2/in)

b) $MMI = M/A\ (I_x + I_y)$(kip-sec^2-in)

4. Seismic design requirements for building structures — Chapter 12

4a. Structural system selection; selection and limitations — §12.2.1

The building framing comprises structural steel framing with lateral resistance provided by "special steel moment frames" (System type C.1, Table 12.2-1):

$R = 8$ response modification factor

$\Omega_o = 3$ system overstrength factor

$C_d = 5.5$ deflection amplification factor

No height limit is specified, and system detailing requirements are referenced in the 2009 IBC and ASCE/SEI 7-05, §12.2.5.5.

4b. Special moment frames in structures assigned to Seismic Design Categories D through F — §12.2.5.5

The frame is continuous to the foundation for this example, so the specific requirements of §§12.3.3.2 and 12.3.3.4 do not need to be checked.

4c. Diaphragm flexibility, configuration, irregularities, and redundancy; rigid diaphragm condition — §12.3.1.2

The diaphragms consist of concrete filled metal deck with span-to-depth ratios less than 3. There are no horizontal irregularities as shown in §12.3.2. Therefore, the diaphragms will be considered rigid in the design/analysis.

4d. Irregular and regular classification — §12.3.2

Check the horizontal and vertical configurations. The structure does not contain any irregularities as defined in Tables 12.3-1 and 12.3-2.

4e. Redundancy factor, ρ, for Seismic Design Categories D through F §12.3.4.2

Note: For structures assigned to Seismic Design Category D, E, or F, ρ shall equal 1.3 unless one of the following two conditions is met, whereby p is permitted to be taken as 1.0: (we will examine condition 12.3.4.2.b)

b. Structures that are regular in plan at all levels, provided that the seismic force-resisting systems consist of at least two bays of seismic force-resisting perimeter framing on each side of the structure in each orthogonal direction at each story resisting more than 35 percent of the base shear.

For this design example, the moment frames are uniform at all levels and are assumed to resist approximately the same relative lateral force at each story. By observation, the design example meets the requirements of §12.3.4.2.b, thereby permitting the use of ρ = 1.0.

4f. Seismic load effect §12.4.2

To determine the seismic load for use in load combination #5 of IBC 2009 §1605.2, the value E is determined as

$$E = E_h + E_v \qquad \text{Eq 12.4-1}$$

and

$$E = E_h - E_v \qquad \text{Eq 12.4-2}$$

E = seismic load effect,

E_h = effect of horizontal seismic forces as defined in §12.4.2.1

E_v = effect of vertical seismic forces as defined in §12.4.2.2

4g. Horizontal seismic load effect §12.4.2.1

E_h shall be determined as

$$E_h = \rho Q_E \qquad \text{Eq 12.4-3}$$

Q_E = effects of horizontal seismic forces from V or F_p. Where required in §§12.5.3 and 12.5.4, such effects shall result from application of horizontal forces simultaneously in two directions at right angles to each other.

ρ = redundancy factor, as defined in §12.3.4

4h. Vertical seismic load effect §12.4.2.2

E_v shall be determined as

$$E_v = 0.2 S_{DS} D$$ Eq 12.4-4

S_{DS} = design spectral response acceleration parameter at short periods obtained from Table 1613.5.4

D = effect of dead load

The exceptions indicate that E_v is permitted to be taken as zero for Equations 12.4-1, 12.4-2, 12.4-5, and 12.4-6 where S_{DS} is equal to or less than 0.125, or when using Equation 12.4-2 to determine demands on the soil-structure interface of foundations.

4i. Seismic load effect including overstrength factor §12.4.3

For use in load combination 5 of §2.3.2 (§12.4.2.3 Strength Design), E shall be taken equal to E_m as determined in accordance with Equation 12.4-5

$$E_m = E_{mh} + E_v$$ Eq 12.4-5

For use in load combination 7 of §2.3.2 (§12.4.2.3 Strength Design), E shall be taken equal to E_m as determined in accordance with Equation 12.4-6

$$E_m = E_{mh} - E_v$$ Eq 12.4-6

E_m = seismic load effect including overstrength factor

E_{mh} = effect of horizontal seismic forces including structural overstrength as defined in §12.4.3.1

E_v = vertical seismic load effect as defined in §12.4.2.2

4j. Horizontal seismic load effect with overstrength factor §12.4.3.1

E_{mh} shall be determined as

$$E_{mh} = \Omega_o Q_E$$ Eq 12.4-7

Q_E = effects of horizontal seismic forces from V or F_p as specified in §12.8.1 and §13.3.1, respectively. Where required in §§12.5.3 and 12.5.4, such effects shall result from application of horizontal forces simultaneously in two directions at right angles to each other.

Ω_o = overstrength factor

EXCEPTION: The value of E_{mh} need not exceed the maximum force that can develop in the element as determined by a rational, plastic mechanism analysis or nonlinear response analysis utilizing realistic expected values of material strengths.

4k. Load combinations with overstrength factor §12.4.3.2

The following load combinations are presented for structures not subject to flood or atmospheric ice loads when determining seismic load combinations as directed in IBC 2009 §1605.1.

Basic Combinations for Strength Design with Overstrength Factor (reference §2.3.2 and §2.2)

$$(1.2 + 0.2S_{DS})D + \Omega_o Q_E + L + 0.2S$$ §12.4.3.2, Equation 5

$$(0.9 - 0.2S_{DS})D + \Omega_o Q_E + 1.6H$$ §12.4.3.2, Equation 7

1. The load factor on L is permitted to be taken as 0.5 in combination 5 for all occupancies in which L_o in Table 1607.1 is less than or equal to 100 psf except for garages or public assembly areas.

2. The load factor on H is zero in combination 7 if the structural action due to H counteracts the structural action due to E. Also, resistance due to lateral earth pressure for structural action related to other forces, shall not be included in H, but shall be included in the design resistance.

Basic Combinations for Allowable Stress Design are similar but not presented here. Allowable Stress Design Load Combinations will not be used in this example for consistency in design between the element design and connection design. AISC 358 specifically requires LRFD design of the SMF connection (in this example, RBS is used for the connection design).

4l. Direction of loading §12.5

Seismic Design Categories D through F §12.5.4

This example building does not have a shared column or intersecting wall, so the use of the procedures outlined in §12.5.3a will satisfy the requirements related to the effects of loading direction. Section 12.5.3a indicates that an equivalent lateral force analysis procedure (§12.8) with loading applied independently in any two orthogonal directions and the most critical load effect due to direction of application of seismic forces on the structure considering 100 percent of the forces for one direction plus 30 percent of the forces for the perpendicular direction. The combination requiring the maximum component strength shall be used and is reported in Part 5.

4m. Analysis procedure selection §12.6

Dependant on the seismic design category, structural characteristics and occupancy category, analytical procedures are permitted per Table 12.6-1.

For Seismic Design Category D, regular structures with $T < 3.5T_S$ are permitted to be analyzed with any of the procedures identified in Table 12.6-1 (T in part 4p)

$$3.5T_s = 3.5S_{D1}/S_{DS} = 3.5(0.6/1.0) = 2.1 \text{ seconds} > T$$ §11.4.5 & T12.6-1

4n. Effective seismic weight §12.7.2

This example assumes that the subject building does not have express storage areas, permanent equipment stored within the useable space, or snow loads. Therefore, the example uses a 10 psf load for partitions in the "given" portion of the example problem statement. This meets the requirement of §12.7.2 item 2 for inclusion into the effective seismic weight used in the analysis model.

4o. Structural modeling §12.7.3

Because the building is regular, a 3-D model is not required. However, a three-dimensional analysis is required to determine force distributions both horizontally and vertically. This can be accomplished using a number of different techniques, but is easily determined through the use of a linear elastic three-dimension mathematical computer model. The following steps outline the equivalent lateral force procedure of §12.8.

4p. Period determination §12.8.2

The fundamental period, T, shall not exceed the product of the coefficient for upper limit on calculated period, C_u, from Table 12.8-1 and the approximate fundamental period, T_a, determined from Equation 12.8-7.

C_u (at $S_{D1} = 0.6$) = 1.4 T12.8-1

Approximate fundamental period §12.8.2.1

$T_a = C_t(h_n)^x$ Eq 12.8-7

$C_t = 0.028$ for Steel Moment Resisting Frames T 12.8-2

$h_n = 55.5$ feet (Figure 3.3)

$x = 0.8$ T 12.8-2

$T_a = 0.028(55.5)^{0.8} = 0.696$ sec

Using a computer model with assumed member sizes and estimated building weights, the period is determined from structural dynamics

North-South (y): $T_y = 1.30$ sec

East-West (x): $T_x = 1.16$ sec

The upper limit on calculated periods above shall not exceed C_uT_a

maximum value: C_uT_a §12.8.2

$\therefore T = T_x = T_y = (0.696\text{ sec})(1.4) = 0.974$ sec

4q. Calculation of seismic response coefficient §12.8.1.1

Where C_s is the seismic response coefficient (per §12.8.1.1)

$$C_s = \frac{S_{DS}}{(R/I)} = \frac{1.0g}{(8/1.0)} = 0.125g$$ Eq 12.8-2

S_{DS} = the design spectral response acceleration parameter in the short period range as determined from IBC §1613.5.4

R = the response modification factor in Table 12.2-1

I = the occupancy importance factor determined in accordance with §11.5.1

The value of C_s need not exceed

$$C_s = \frac{S_{D1}}{T(R/I)} \text{ for } T \le T_L = \frac{0.6g}{0.974\left(\frac{8}{1.0}\right)} = 0.077g$$ Eq 12.8-3

T_L = 8 seconds §11.4.5 & F 22-16

And C_s shall not be less than

$$C_s = 0.044SDS_I = 0.044(1.4g)(1.0) = 0.0616g$$ Eq 12.8-5, Supplement No. 2

For structures with $S_1 \ge 0.6g$, C_s shall not be less than

$$C_s = \frac{(0.5)S_1}{(R/I)} = \frac{0.5(0.6g)}{(8/1.0)} = 0.038g$$ Eq 12.8-6

S_{D1} = the design spectral response acceleration parameter at a period of 1.0 second, as determined from IBC §1315.5.4

T = the fundamental period of the structure(s) determined in §12.8.2

T_L = long-period transition period(s) determined in §11.4.5

S_1 = the mapped maximum considered earthquake spectral response acceleration parameter determined in accordance with IBC §1613.5.1

4r. Seismic base shear §12.8.1

The design seismic base shear, V, for a given direction is

$$V = C_s W$$ Eq 12.8-1

Equation 12.8-3 governs seismic base shear

$$V = C_s W = 0.077W = 0.077(8180 \text{ kips}) = 629.9 \text{ kips, say } 630 \text{ kips}$$ Eq 12.8-1

4s. Vertical distribution of seismic forces §12.8.3

Vertical distribution of force to each level, F_x, is applied as

$$F_x = C_{vx} V \qquad \text{Eq 12.8-11}$$

and

$$C_{vx} = \frac{w_x h_x^k}{\sum_{i=1} w_i h_i^k} \qquad \text{Eq 12.8-12}$$

where

C_{vx} = vertical distribution factor,

V = total design lateral force or shear at the base of the structure

w_i and w_x = the portion of the total effective seismic weight of the structure (W) located or assigned to level i or x

h_i and h_x = the height from the base to level i or x

k = an exponent related to the structure period as follows:
for structures having a period of 0.5 second or less, $k = 1$
for structures having a period of 2.5 seconds or more, $k = 2$
for structures with periods between 0.5 second and 2.5 seconds, determine k using linear interpolation between 1 and 2

For this structure

$$T = 0.974 \text{ sec}$$

$\therefore$ Using linear interpolation

$$k = \frac{2-1}{2.5-0.5} \times (0.974 - 0.5) + 1 = 1.237$$

The lateral force, F_x, is applied at each level as

$$F_x = \frac{w_x h_x^k}{\sum_{i-1} w_i h_i^k}(V) = \left(\frac{w_x h_x^k}{\sum_{i-1} w_i h_i^k}\right)(630) \qquad \text{Eq 12.8-12}$$

Table 3-3. Vertical distribution of shear

Level	w_x (kips)	h_x (ft)	$w_x h_x^k$ (kip-ft)	$\frac{w_x h_x^k}{\Sigma w h_x^k}$	F_x (kips)	ΣV (kips)
Roof	1877	55.5	269,870	0.399	251.4	—
4th	2101	42.0	213,986	0.316	199.4	251.4
3rd	2101	28.5	132,455	0.196	123.4	450.8
2nd	2101	15.0	59,876	0.089	55.8	574.2
Total	8180		676,187	1.000	630.0	630.0

4t. Horizontal distribution of forces §12.8.4

The design story shear in any story is the sum of lateral forces at the floor levels above

$$V_x = \sum_{i=1}^{n} F_i \qquad \text{Eq 12.8-13}$$

Unless shown to be flexible as defined in §12.3.1.3, structures with concrete fill over metal deck floor systems are typically assumed to have rigid diaphragms. For rigid diaphragms, seismic forces are distributed to the moment frames according to their relative stiffness. For structures with rigid diaphragms, the inherent torsion must be considered resulting from the eccentricity between the locations of the center of mass and the center of rigidity (§12.8.4.1). In addition to the inherent rigidity, an accidental torsion must be applied equal to that caused by displacing the center of mass 5 percent of the building dimension perpendicular to the direction of the applied lateral force (§12.8.4.2).

For the structural computer model of this design example, this can be achieved by combining the direct seismic force applied at the center of mass at each level with a torsional moment at each level that results in the greatest effect (for consideration of accidental torsion per §12.8.4.2).

North-South

$$M_t = 0.05(204 \text{ ft})F_x = (10.2)F_x$$

East-West

$$M_t = 0.05(144 \text{ ft})F_x = (7.2)F_x$$

Table 3-4. Horizontal distribution of shear

Level	F_x (kips)	*N/S* M_t (kip-ft)	*E/W* M_t (kip-ft)
Roof	251.4	2564	1810
4th	199.0	2034	1436
3rd	123.4	1259	888
2nd	55.8	569	402

Note: M_t = horizontal torsional moment

Using the direct seismic forces and torsional moments noted above, the force distribution to the frames is generated by computer analysis. For the computer model, member sizes are initially proportioned within the constraints specified by AISC 358 for SMF reduced beam section (RBS) connections, as discussed in Parts 5 and 6 below.

From the preliminary computer analysis, the shear force at the ground level is determined for each frame column. As shown in Figure 3-4, there are a total of six rigid frames: A1, A2, B1, B2, B3, and B4. Frames A1 and A2 are identical. Frames B1, B2, B3, and B4 are also identical. Recognizing that the building is symmetric, the frame forces are the same for Frames A1 and A2, as well as for Frames B1 through B4. Frame forces at the base of each frame type, A1 and B1 are summarized in Tables 3-5 and 3-6.

Table 3-5. North-South direction, frame type A1

Column Shears (kips)	Line A/1.2 (kips)	Line A/2 (kips)	Line A/3 (kips)	Line A/4 (kips)	Line A/5 (kips)	Line A/5.8 (kips)	Total (kips)
Direct Seismic	35.9	65.3	62.2	62.2	65.3	35.9	326.8
Torsion Force	2.2	4.0	3.9	3.9	4.0	2.2	20.2
Direct + Torsion	38.1	69.3	66.1	66.1	69.3	38.1	347.0

Table 3-6. East-West direction, frame type B1

Column Shears (kips)	Line 1/A.2 (kips)	Line 1/B (kips)	Line 1/C (kips)	Line 1/C.8 (kips)	Total (kips)
Direct Seismic	29.4	52.3	52.3	29.4	163.4
Torsion Force	1.2	2.0	2.0	1.2	6.4
Direct + Torsion	30.6	54.3	54.3	30.6	169.8

As a check on the computer output, compare the total column shears with the direct seismic base shear of 630 kips.

North-South:

$$\Sigma F_{\text{typeA}} = 2(326.8) = 653.6 > 630 \text{ kips} \ \ldots \ o.k.$$

East-West:

$$\Sigma F_{\text{typeB}} = 4(163.4) = 653.6 > 630 \text{ kips} \ \ldots \ o.k.$$

The summation of the column shears is about 3.7 percent greater than the design base shear input to the computer model. This is mostly due to the inclusion of P-Δ effects in the computer analysis performed for this example. Per §12.8.7, P-Δ effects are to be considered when the stability coefficient, θ, is greater than 0.10. Although included in the computer analysis (to be conservative), P-Δ effects are not required to be considered in this design. This will be explained further in the following steps.

4u. Story drift determination §12.8.6

The design story drift, Δ, is the story displacement at the center of mass of one floor level relative to the displacement of the center of mass at the floor level below. It is derived from a linear-elastic static analysis using the design seismic forces calculated above. Calculated lateral displacements, δ_{xe}, are increased by the deflection amplification factor, C_d, to include both elastic and estimated inelastic displacements resulting from the design basis ground motion. With $C_d = 5.5$ for a steel SMF, the design story deflection is

$$\delta_x = \frac{C_d \delta_{xe}}{I_e} = (5.5)\delta_{xe}$$ Eq 12.8-15

When calculating lateral deflections for drift analysis, §12.8.6.2 removes the upper limit on the building period, $C_u T_a$, specified in §12.8.2 in determining seismic design forces.

Without the upper limit, C_u, on building period T, the design base shear per Equations 12.8-3 and 16-36 is

North-South: $T_y = 1.30$ sec

$$C_s = \frac{S_{D1}}{T(R/I)} \text{ for } T \le T_L = \frac{0.6g}{1.30\left(\frac{8}{1.0}\right)} = 0.0577g$$ Eq 12.8-3

$$V_{\text{n/s}} = C_S W = 0.058(8180) = 472 \text{ kips}$$ Eq 12.8-1

East-West: $T_x = 1.16$ sec

$$C_s = \frac{S_{D1}}{T(R/I)} \text{ for } T \le T_L = \frac{0.6g}{1.16\left(\frac{8}{1.0}\right)} = 0.0647g$$ Eq 12.8-3

$$V_{\text{e/w}} = C_S W = 0.065(8180) = 529 \text{ kips}$$ Eq 12.8-1

Using these modified design base shears, the force distribution to each level is adjusted and applied to the computer model. The structure displacements, drift ratios and stability coefficients are derived as shown below in Table 3-7.

The following calculation is used to develop the stability coefficient

$$\theta = \frac{P_x \Delta}{V_x h_{sx} C_d} \qquad \text{Eq 12.8-16}$$

where

P_x = the total vertical design load at and above Level x; where computing P_x, no individual load factor need exceed 1.0

Δ = the design story drift as defined in §12.8.6 occuring simultaneously with V_x

V_x = the seismic shear force acting between Levels x and x-1

h_{sx} = the story height below Level x

C_d = the deflection amplification factor in table 12.2-1

The stability coefficient, θ, shall not exceed θ_{max} as determined from

$$\theta_{max} = \frac{0.5}{\beta C_d} = \frac{0.5}{(1.0)(5.5)} = 0.091 \le 0.25; \quad \beta = 1.0 \text{ (conservative)} \qquad \text{Eq 12.8-17}$$

When the stability coefficient, θ, exceeds 0.10, but less than θ_{max}, the incremental factor related to P-Δ effects on displacements and member forces shall be determined by rational analysis or displacements and member forces can be multiplied by $1.0/(1 - \theta)$. If θ is greater than θ_{max}, the structure is potentially unstable and shall be redesigned.

Table 3-7. Interstory displacements and drifts

North-South Displacements and Drifts					
Story	Height h (in)	$\delta_x = \delta_{xe} C_d$ (in)	Δ Drift (in)	Drift Ratio (Δ/h)	Stability Coefficient θ
4th	162	(2.22)5.5 = 12.21	(12.21 – 10.51) = 1.70	0.010	0.019
3rd	162	(1.91)5.5 = 10.51	(10.51 – 7.70) = 2.81	0.017	0.037
2nd	162	(1.40)5.5 = 7.70	(7.70 – 4.13) = 3.57	0.022	0.057
1st	180	(0.75)5.5 = 4.13	(4.13 – 0.0) = 4.13	0.023	0.072
East-West Displacements and Drifts					
Story	Height h (in)	$\delta_x = \delta_{xe} C_d$ (in)	Δ Drift (in)	Drift Ratio (Δ/h)	Stability Coefficient θ
4th	162	(1.90)5.5 = 10.46	(10.46 – 9.18) = 1.28	0.008	0.013
3rd	162	(1.67)5.5 = 9.18	(9.18 – 6.61) = 2.57	0.016	0.030
2nd	162	(1.20)5.5 = 6.61	(6.61 – 3.51) = 3.10	0.019	0.044
1st	180	(0.64)5.5 = 3.51	(3.51 – 0.0) = 3.51	0.020	0.055

Notes: Interstory drift ratio = Δ/story height (h)

4v. Story drift limit; moment frames in structures assigned to Seismic Design Categories D through F §12.12.1.1

Per §12.12.1 and §12.12.1.1, allowable story drift limits are given in Table 12.12-1. For steel special moment frame structures 4 stories or less with interior walls, partitions, ceilings, and exterior wall systems that have been designed to accommodate the story drifts within Occupancy Category II, (this example structure) the allowable story drift is 0.025 h_{sx}. Use the design story drift (Δ) ratios calculated in Table 3-7 and compare to the allowable drift ratio (0.025).

A review of drift ratios tabulated in Table 3-7 shows that all interstory drift ratios are less than 0.025, using seismic forces corresponding to the actual period, T, in base shear Equation 12.8-1. Stability coefficients are all less than 0. 10, so P-Δ effects need not be considered. However, AISC 358 indicates that the global drift consideration of reduced beam section (RBS) SMF design must consider the reduced stiffness associated with the reduced beam section. The actual reduction in stiffness could be calculated and included in the mathematical model to determine "accurate" story drifts based on increased beam flexibility. However, a general value of system stiffness can be used to simplify the design and is allowed to be taken as 1.1 times the elastic story drift for a 50 percent flange reduction per AISC 358, §5.8. Using this approach for the north-south displacements at the 1st story, the critical drift ratio is

$$(\Delta/h) = (0.023) \times (1.1) = 0.025 \le 0.025 \quad o.k.$$

and

$$\theta \times 1.1 = 0.079 < 0.091 \quad o.k.$$

Most published reference documents recommend a mathematical model using center-line to center-line dimensions of framing members, while also allowing for realistic assumptions for the stiffness of panel zones or modification of effective span length for beams and columns to more accurately model frame stiffness. Many different analyses using different computer software, beam-column joint models and other parametric considerations indicate that a centerline analysis (assuming no panel zone stiffness) represents a reasonable assumption for the majority of steel special moment frame designs. This assumption allows for balanced connection behavior that does not rely on an overly rigid panel zone to bear out the system analysis. This simplified analysis assumption (center-to-center modeling of frame elements) with a 10% increase on expected elastic story drift is generally acceptable for the design of steel SMF in the static linear elastic analysis/design regime.

5. SMF member design

In this part of the design example, representative SMF beam and column members of Frame A1 are designed in accordance with ASCE/SEI 7-05. SMF structures assigned to Seismic Design Category D are to conform to the requirements of the 2005 AISC Seismic Provisions for Structural Steel Buildings (AISC 341). This example problem uses the RBS SMF connection type. The RBS is a connection that has been extensively tested as part of the SAC Project, the state-of-the-art research that comprises FEMA 350, California healthcare facility projects, University of California/NSF funded research projects, and various other experimental testing programs. The RBS connection could be utilized through compliance vehicles presented by Appendix S or Appendix P of AISC 341. This example will use Appendix P of AISC 341, which integrates/references AISC 358.

AISC 358, *Prequalified Connections for Special and Intermediate Steel Moment Frames for Seismic Applications,* is the applicable national standard that explicitly complies with the requirements of AISC 341 Appendix P (*Prequalification of Beam-Column and Link-to-Column Connections*). The connection design for this example problem will follow the AISC 358 standard for compliance with AISC 341, ASCE/SEI 7-05 and the IBC 2009. Member and connection strengths are calculated in accordance to the methods outlined in AISC 358 and AISC 341 (typically strength based).

Steel moment frame designs are typically drift controlled. Frame members are chosen to provide sufficient stiffness to meet drift limits imposed by the model building code, in this case IBC 2009, which relies upon limits promulgated in ASCE/SEI 7-05. The selected members are then checked for SMF member design requirements per AISC 341. The selection of beam-column combinations must also conform to the connection limitations set forth in AISC 358 and AISC 360 as appropriate. After confirming that drift limitations are reasonably satisfied through selection of beam and column members to determine acceptable system performance, the W27×84 beam at the 3rd floor, Line A and the W14×176 column that comprises an interior connection were chosen to illustrate this design example. While refinement of column sizing could be utilized to "tune" the structure to ensure progressive mechanism formation throughout the structure, that effort is beyond the scope of this example.

5a. Determine member material specifications and strength properties

AISC 341 §6 indicates the material specification and strength properties for steel members used in seismic force resisting systems. Where the term "expected strength" is used, the value is to be taken as

$$F_{ye} = (R_y)(F_y)$$

The "lower bound" strength is the minimum specified yield strength, F_y.

For this design example, material strengths are taken as

W27×84 beam, Shape Group 2, A992

$$F_y = 50 \text{ ksi}$$

$$F_u = 65 \text{ ksi}$$

W14×176 column, Shape Group 4, A992

$$F_y = 50 \text{ ksi}$$

$$F_u = 65 \text{ ksi}$$

Per AISC 341, Table I-6-1, for ASTM A992

$$R_y = 1.1$$

5b. Check frame for *P*-Δ instability

AISC 341 (Commentary C3) identifies that a moment frame should be checked for *P*-Δ geometric nonlinearities considering the effects of large lateral drifts. The largest drift is at the first story, so we will check that condition using the following methodology as recommended in the AISC 341 Commentary.

$$\Psi_i = \frac{P_i R \Delta_i}{V_{yi} H} \qquad \text{AISC 341, C3-1}$$

P_i = from Table 3-7 plus 25% of the live load

$P_{LLRoof} = (29{,}090\ \text{ft}^2)(0.25)(20\ \text{psf}) = 145.5\ \text{kips}$

$P_{LLFloor} = (29{,}090\ \text{ft}^2)(0.25)(80\ \text{psf}) = 581.8\ \text{kips}$

$R = 8$

Δ_t = as indicated in Table 3-7

V_{yi} = as indicated in (C3-2) below as long as all columns meet the strong-column-weak-beam criterion (assumed true, confirm later)

H = Story height from Table 3-7

$$V_{yi} = \frac{2\sum_{j=1}^{m} M_{pGj}}{H} \qquad \text{AISC 341, C3-2}$$

$M_{pG} = ZF_y = (244\ \text{in.}^3)(50\ \text{ksi}) = 12{,}200\ \text{kip-in} = 1017\ \text{kip-ft}$

There are two frames with 3 identical girders each in the N-S direction

$$V_{yi} = \frac{2[2(3)(1017\ \text{k-ft})]}{13.5\ \text{ft}} = 904 \qquad \text{AISC 341, C3-2}$$

$$\Psi_{2,3,4} = \frac{P_i R \Delta_i}{V_{yi} H} = \frac{[(8180 + 145.5 + 581.8)k/2](8)(0.75\ \text{in})}{(904k)(15\ \text{ft} \cdot 12\ \text{in/ft})} = 0.183 \qquad \text{AISC 341, C3-1}$$

As indicated in the AISC 341 Commentary, a value of Ψ_i less than 0.3 indicates that the frame is considered stable for that story (and in this case, the "worst case" for a given frame line assuming that all frames are identical).

5c. Design typical beam at 3rd floor

The third-floor beam in Frame A1 is selected to illustrate SMF beam design. This W27×84 beam is shown in Figure 3-5.

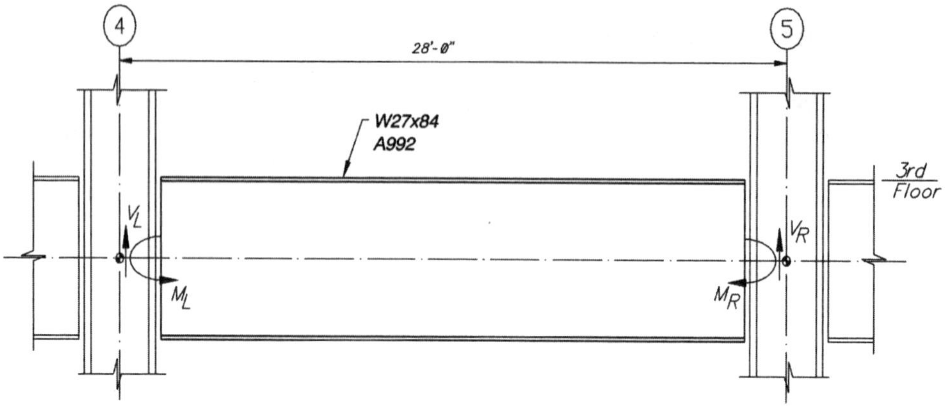

Figure 3-5. Typical beam at third floor of Frame A1

From a review of the computer output prepared in support of this design example, the moments and shears at the right end of the beam are greatest. The unfactored moments and shears at the face of the column at Line 5 are

$$M_{DL} = 1042 \text{ kip-in}$$
$$M_{LL} = 924 \text{ kip-in}$$
$$M_{seis} = \pm 3083 \text{ kip-in}$$
$$V_{DL} = 16.4 \text{ kips}$$
$$V_{LL} = 13.3 \text{ kips}$$
$$V_{seis} = \pm 19.0 \text{ kips}$$

Note that seismic forces identified above include both vertical and horizontal components E_v and E_h. The vertical component, E_v, is added to the dead load in Equation 12.4-1, and subtracted from the dead load in Equation 12.4-2

$$(1.2 + 0.2S_{DS})D + \rho Q_E + 0.5L + 0.2S$$ §12.4.2.3

$$E = \rho Q_E \pm 0.2S_{DS}(D)$$ Eqs 12.4-1, 12.4-3, 12.4-4

$$E_h = \rho Q_E = 1.0Q_E$$

$$E_v = 0.2S_{DS}D = 0.2(1.0g)D = 0.2D$$

Using the Basic Combinations of §2.3.2

$$1.2D + 1.6L: \quad M_{D+L} = 1.2(1042) + 1.6(924) = 2729 \text{ kip-in}$$

$$V_{D+L} = 1.2(16.4) + 1.6(13.3) = 41.0 \text{ kips}$$

Using the Seismic Load Combinations, Basic Combinations for Strength Design of §12.4.2.3

$$(1.2 + 0.2S_{DS})D + \rho Q_E + 0.5L + 0.2S$$

$$M_{D+L+E} = [1.2 + 0.2(1.0)](1042) + 1.0(3083) + 0.5(924) + 0 = 5004 \text{ in-kip}$$

$$V_{D+L+E} = [1.2 + 0.2(1.0)](16.4) + 1.0(19) + 0.5(13.3) + 0 = 48.6 \text{ kips}$$

$$(0.9 - 0.2S_{DS})D - \rho Q_E - 1.6H$$

$$M_{D-E} = [0.9 - 0.2(1.0)](1042) - 1.0(3083) - 0 = -2354 \text{ in-kip}$$

$$V_{D-E} = [0.9 - 0.2(1.0)](16.4) - 1.0(19) - 0 = -7.5 \text{ in-kips}$$

Prior to evaluating demand/capacity ratios for member strength, SMF beams must be checked for stability and proportions per AISC 358, §5.3, which references AISC 341, §9.4a, requiring that beam elements comply with §8.2b, referencing Table I-8-1 for limiting width-thickness ratios for elements subjected to compression forces.

Check the flange width-thickness ratio

$$\frac{b}{t} = \frac{b_f}{2t_f} \le 0.3\sqrt{\frac{E}{F_y}} = 0.3\sqrt{\frac{29{,}000 \text{ ksi}}{50 \text{ ksi}}} = 7.22$$ AISC 341, T I-8-1

For W27×84: $\frac{b_f}{2t_f} = 7.81 > 7.22 \;\ldots\; n.g.$

However, per AISC 358, §5.3.1(6) for RBS connections, b_f may be measured at the end of the center 2/3 of the reduced W27×84 beam section, provided that gravity loads do not shift the plastic hinge location a significant distance from the center of the reduced section. As we assume that we will reduce the beam flange width by 50 percent at the center of the RBS section, we use the following calculation with the assumed dimensions (reference Part 6a, Figure 3-8)

$$y = b - r + \sqrt{r^2 - x^2} = 2.25 - 28 + \sqrt{28^2 - 7.33^2} = 1.27 \text{ in}$$

(Properties of the Circle, AISC *Seismic Design Manual* Table 17-26)

$$\frac{b_f - y}{2t_f} = \frac{10.0 - 1.27}{2(0.64)} = 6.82 < 7.22 \;\ldots\; o.k.$$

Check the web width-thickness ratio

$$\frac{h}{t_w} \le 2.45\sqrt{\frac{E}{F_y}} = 2.45\sqrt{\frac{29{,}000 \text{ ksi}}{50 \text{ ksi}}} = 59.0$$ AISC 341, T I-8-1

$$\frac{h}{t_w} = 52.7 < 59.0 \;\ldots\; o.k.$$

Check beam depth, weight and span-to-depth ratio limits per AISC 358, §5.3

Maximum beam depth → W36 > W27 . . . *o.k.*

Maximum beam weight → 300 lb/ft > 84 lb/ft . . . *o.k.*

Maximum flange thickness → 1.75 in > 0.64 in . . . *o.k.*

$$\text{Minimum span-to-depth ratio} \rightarrow 7 < \frac{12(28)}{27} = 12.4 \; . . . \; o.k.$$

Check beam lateral bracing requirements per AISC 358, §5.3.1(7), which references AISC 341, §9.8

Maximum brace spacing, L_b

$$L_b = \frac{0.086 r_y E}{F_y} = \frac{0.086(2.07 \text{ in})(29{,}000 \text{ ksi})}{50 \text{ ksi}} = 103.25 \text{ in} = 8.60 \text{ ft}$$

Place minimum bracing at quarter points: $L = 28/4 = 7.0$ ft on center

$$P_{br} = 0.06 \frac{M_r}{h_o}$$

$M_r = R_y Z F_y$, where $R_y = 1.1$ AISC 341, T I-6-1

$$= 1.1(244 \text{ in}^3)(50 \text{ ksi}) = 13{,}420 \text{ in-k} = 1118 \text{ ft-k}$$

$$P_{br} = 0.01 \frac{13{,}420 \text{ in-k} \times 1.0}{26.06 \text{ in}} = 5.15 \text{ kips}$$

The length of the brace is assumed to be measured from the centerline of the W27×84 to the centerline of the adjacent gravity beam. Assuming a 12-ft, 6-in beam spacing, the length of the brace is

$$L = \sqrt{[12.5 \text{ ft}(12 \text{ in/ft})]^2 + (26.7)^2} = 152.4 \text{ in} = 12.7 \text{ ft}$$

Based on the *Steel Construction Manual*, 13th edition, we will examine a *L*5×5×5/16 as an eccentrically loaded single angle ($A_g = 3.65$ in^2).

$$\phi R_n = 22.0 \text{ kips} > R_u = 5.15 \text{ kips}$$

AISC 341 identifies that a minimum stiffness is also required to provide adequate lateral bracing. This type of brace is considered a nodal brace, providing a rigid brace support, the required brace stiffness is

$$\beta_{br} = \frac{10 M_r C_d}{\phi L_b h_o}$$ AISC 360, A-6-8

where

$$\phi = 0.75$$

$$M_r = R_y Z F_y = 1.1(244 \text{ in}^3)(50 \text{ ksi}) = 13{,}420 \text{ in-k} = 1118 \text{ ft-k}$$

$$C_d = 2.0$$

$$L_b = 7.0 \text{ ft} = 84 \text{ in}$$

$$h_o = d_b - t_f = 26.7 \text{ in} - 0.64 \text{ in} = 26.06 \text{ in}$$

$$\beta_{br} = \frac{10(13{,}420 \text{ in-k})(2.0)}{0.75(84 \text{ in})(26.06 \text{ in})} = 163.5 \text{ k/in}$$

The brace stiffness can be calculated as

$$k = \frac{A_g E}{L}\cos^2(\theta)$$

$$\theta = \tan^{-1}\left(\frac{26.7 \text{ in}}{12.5 \text{ ft}(12 \text{ in/ft})}\right) = 10.09°$$

$$k = \frac{3.65 \text{ in}^2(29{,}000 \text{ ksi})}{152.4 \text{ in}}\cos(10.09°) = 684 \text{ k/in}$$

$$k > \beta_{br} \;\ldots\; o.k.$$

L5×5×5/16 kickers provided at 7 feet on center to brace the beam bottom flange to the top flange of an adjacent steel beam meet lateral bracing requirements.

The W27×84 beam meets stability and proportion criteria; next, check the design flexural strength (LRFD) per AISC 360.

From AISC 360 (Chapter F.2) for W27×84

$$L_p = 1.76 r_y \sqrt{\frac{E}{F_y}} = 7.3 \text{ ft} > 7.0 \text{ ft} \;\; \therefore \; M_n = F_y Z_x$$ AISC 360, F2-5

Design Flexural Strength: $\phi_b M_n = 0.9(50)(244) = 10{,}980$ kip-in

Demand-capacity ratio: $D/C = \frac{5004}{10{,}980} = 0.46 < 1.0 \;\ldots\; o.k.$

Check nominal shear capacity

For W27×84: $\frac{d}{t_w} = 58.0 < 2.45\sqrt{\frac{29{,}000}{50}} = 59.0$ AISC 341, T I-8-1

$\therefore V_n \;= 0.6 F_y A_w C_v$ AISC 360, G2-1

$C_v \;= 1.0$ AISC 360, G2-3

$\phi_v V_n = 0.9(0.6)(50)(0.46)(26.7)(1.0) = 331.6$ kips AISC 360, §G2.1

$D/C = \frac{48.6}{331.6} = 0.15 < 1.0 \;...\; o.k.$

$\therefore$ Use W27×84 beam

Note: The W27×84 beam is larger than required by strength considerations (V_u as calculated from prescriptive load calculations). However, given the constraints of meeting frame drift limits, it is a reasonable choice for this design. Limitations related to beam-column connection design and reduced beam section parameters guided the size of beam selected. It is prudent to confirm that V_u developed in Part 6h can be accommodated by this beam size.

5d. Design typical column at 2nd story

The column to be designed is the second-story column of Frame A1 as shown in Figure 3-6. The maximum strong axis moments occur at the bottom of the column, and are taken at the top flange of the second-floor beam. For brevity, the example will look at critical conditions affecting design, omitting the many different inconsequential demand values that are determined for the various analyses.

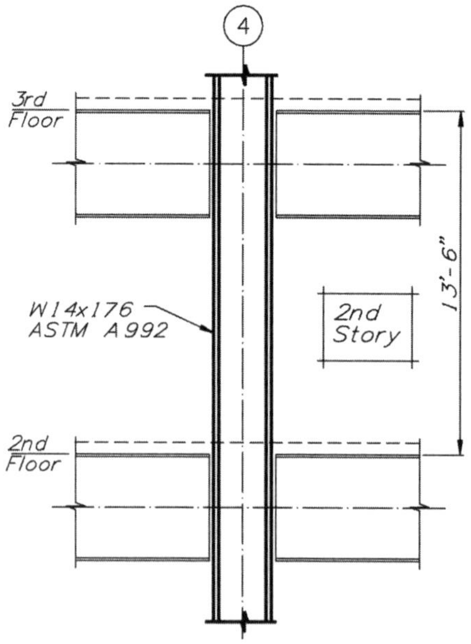

Figure 3-6. Typical second story column at Frame A1

For the second-story column at Line 4, the maximum unfactored column forces generated by the frame computer analysis are

$$M_{DL} = 236 \text{ kip-in}$$
$$M_{LL} = 201 \text{ kip-in}$$
$$M_{seis} = \pm 3970 \text{ kip-in}$$
$$V_{DL} = 3.1 \text{ kips}$$
$$V_{LL} = 2.7 \text{ kips}$$
$$V_{seis} = \pm 56.8 \text{ kips}$$
$$P_{DL} = 113 \text{ kips}$$
$$P_{LL} = 75 \text{ kips}$$
$$P_{seis} = \pm 28 \text{ kips}$$

Note that seismic forces identified above include both vertical and horizontal components E_v and E_h. The vertical component, E_v, is added to the dead load in Equation 12.4-1 and subtracted from the dead load in Equation 12.4-2

$$(1.2 + 0.2S_{DS})D + \rho Q_E + 0.5L + 0.2S$$ §12.4.2.3

$$E = \rho Q_E = 0.2S_{DS}(D)$$ Eqs 12.4-1,12.4-3,12.4-4

$$E_h = \rho Q_E = 1.0Q_E$$
$$E_v = 0.2S_{DS}D = 0.2(1.0g)D = 0.2D$$

Using the Basic Combinations of Section 2.3.2

$1.2D + 1.6L$: $P_{D+L} = 1.2(113) + 1.6(75) = 255.6 \text{ kips}$

$$M_{D+L} = 1.2(236) + 1.6(201) = 604.8 \text{ kip-in}$$
$$V_{D+L} = 1.2(3.1) + 1.6(2.7) = 8.0 \text{ kips}$$

Using the Seismic Load Combinations, Basic Combinations for Strength Design of Section 12.4.2.3

$(1.2 + 0.2S_{DS})D + \rho Q_E + 0.5L + 0.2S$:

$$P_{D+L+E} = [1.2 + 0.2(1.0)](113) + 1.0(28) + 0.5(75) + 0 = 223.7 \text{ kips}$$
$$M_{D+L+E} = [1.2 + 0.2(1.0)](236) + 1.0(3{,}970) + 0.5(201) + 0 = 4401 \text{ in-kip}$$
$$V_{D+L+E} = [1.2 + 0.2(1.0)](3.1) + 1.0(56.8) + 0.5(2.7) + 0 = 62.5 \text{ kips}$$

$(0.9 - 0.2S_{DS})D - \rho Q_E - 1.6H$:

$$P_{D-E} = [0.9 - 0.2(1.0)](113) - 1.0(28) - 0 = 51.1 \text{ kips}$$
$$M_{D-E} = [0.9 - 0.2(1.0)](236) - 1.0(3970) - 0 = -3805 \text{ in-kip}$$
$$V_{D-E} = [0.9 - 0.2(1.0)](3.1) - 1.0(56.8) - 0 = -54.6 \text{ kips}$$

Try W14×176, ASTM A992 column.

Check the flange width-thickness ratio

$$\frac{b}{t} = \frac{b_f}{2t_f} \le 0.3\sqrt{\frac{E}{F_y}} = 0.3\sqrt{\frac{29{,}000 \text{ ksi}}{50 \text{ ksi}}} = 7.22 \qquad \text{AISC 341, T I-8-1, footnote b}$$

For W14×176: $\frac{b_f}{2t_f} = 5.97 < 7.22 \; . . . \; o.k.$

Check the web width-thickness ratio

$$C_a = \frac{P_u}{\phi_b P_y} = \frac{P_u}{0.9F_y A_g} = \frac{256 \text{ kips}}{0.9(50 \text{ ksi})(51.8 \text{ in}^2)} = 0.110$$

$$\frac{h}{t_w} \le 3.14\sqrt{\frac{E_s}{F_y}}(1-1.54C_a) = 3.14\sqrt{\frac{29{,}000 \text{ ksi}}{50 \text{ ksi}}}(1-1.54(0.110) = 62.81$$

$$\frac{h}{t_w} = 13.7 < 62.81 \; . . . \; o.k. \qquad \text{AISC 341, T I-8-1}$$

Check column depth, weight and span-to-depth ratio limits per AISC 358 §5.3:

Maximum column depth → W36 > W14 . . . *o.k.*

Maximum column weight → unlimited . . . *o.k.*

Maximum flange thickness → compact per AISC 341 . . . *o.k.*

Unbraced column height (taken from top of framing at bottom to mid-depth of beam at top). Estimate; final design will need to reevaluate this condition.

$L_p = 14.2$ ft, $L_r = 73.2$ ft AISC 360, T 3-2

$h = 13.5 - (2.5/2) = 12.25$ ft

$L_b = 12.25 \text{ ft} < L_p$

The effective length factor, *K*, for the X-X axis is determined using the alignment charts for uninhibited side-sway found in AISC 360 Commentary

$$G = \frac{\Sigma(I_c/L_c)}{\Sigma(I_g/L_g)}$$

Two lateral frame beams and two lateral frame columns contribute to the lateral stiffness at the top of the column under examination

$$G_{top} = \frac{2\left(\frac{2140 \text{ in}^4}{12.25 \text{ ft}}\right)}{2\left(\frac{2850 \text{ in}^4}{28.0 \text{ ft}}\right)} = 1.68$$

Two lateral frame beams and two lateral frame columns contribute to the lateral stiffness at the bottom of the column under examination

$$G_{\text{bot}} = \frac{\left(\frac{2140 \text{ in}^4}{12.5 \text{ ft}}\right) + \left(\frac{2140 \text{ in}^4}{15.0 \text{ ft}}\right)}{2\left(\frac{2850 \text{ in}^4}{28.0 \text{ ft}}\right)} = 1.54$$

From AISC 360 Commentary Figure C-C2.4, K_x is determined to be 1.5.

From AISC 360 Commentary and Table C-C2.2c, we will assume a value of 1.2 for K_y.

Find design stress $\phi_c F_{cr} (\phi_c = 0.85)$

$$\frac{K_x L_x}{r_x} = \frac{1.5(12.25)(12)}{6.43} = 34.3$$

$$\frac{K_y L_y}{r_y} = \frac{1.2(12.25)(12)}{4.02} = 43.9$$

For $\frac{KL}{r} \leq 4.71\sqrt{\frac{E}{F_y}} = 43.9 < 113.4$

$$F_e = \frac{\pi^2 E}{\left(\frac{KL}{r}\right)^2} = \frac{\pi^2 (29{,}000)}{(43.9)^2} = 148.5 \text{ ksi}$$ AISC 360, E3-4

$$F_{cr} = \left[0.658^{\frac{F_y}{F_e}}\right] F_y = 43.43 \text{ ksi}$$ AISC 360, E3-2

The nominal design axial capacity is

$$\phi P_n = \phi_c F_{cr} A_g = (0.9)(43.43)(51.8) = 2025 \text{ kips}$$ AISC 360, E3-1

Determine the flexural strength of the W14×176, Grade 50

$$L_p = 14.2 > 12.5$$

$$\therefore M_{nx} = M_{px} = F_y Z_x = (50)(320) = 16{,}000 \text{ kip-in}$$

$$M_{ny} = M_{py} = F_y Z_y = (50)(163) = 8150 \text{ kip-in}$$

$$M_{cx} = \phi M_{nx} = (0.9)(50)(320) = 14{,}400 \text{ kip-in}$$

$$M_{cy} = \phi M_{ny} = (0.9)(50)(163) = 7335 \text{ kip-in}$$

Considering second order effects

$$B_1 = \frac{C_m}{1 - \alpha P_r / P_{e1}} \geq 1$$ AISC 360, C2-2

$$P_r = P_{nt} + P_{lt}$$

$$\alpha = 1.0$$ AISC 360, C2-3

$$C_m = 0.6 - 0.4(M_1/M_2)$$ AISC 360, C2-4

$$C_{mx} = 0.6 - 0.4(3796 \text{ in-k}/5006 \text{ in-k}) = 0.297$$

$$C_m = 0.6 - 0.4(605 \text{ in-k}/605 \text{ in-k}) = 0.200$$

$$P_{e1x} = \frac{\pi^2 EI_x}{(K_x L_x)^2} = \frac{\pi^2 (29{,}000 \text{ ksi})(2140 \text{ in}^4)}{[1.2(12.5 \text{ ft})(12 \text{ in/ft})]^2} = 18{,}905 \text{ kips}$$

$$P_{e1y} = \frac{\pi^2 EI_y}{(K_y L_y)^2} = \frac{\pi^2 (29{,}000 \text{ ksi})(838 \text{ in}^4)}{[1.2(12.5 \text{ ft})(12 \text{ in/ft})]^2} = 7403 \text{ kips}$$

Therefore,

$$B_{1x} = \frac{0.297}{1 - [(1.0 \cdot 255.6) + (1.02 \cdot 51.1)] / 18{,}905 \text{ k}} \geq 1.0$$

$$= 0.238 < 1.0$$

$$B_{1x} = 1.0$$

$$B_{1y} = \frac{0.200}{1 - 255.6 \text{ kips}/7403 \text{ kips}} \geq 1$$

$$= 0.207 < 1.0$$

$$B_{1y} = 1.0$$

$$B_2 = \frac{1}{1 - \left(\frac{\alpha \Sigma P_{nt}}{\Sigma P_{e2}}\right)} \geq 1.0$$ AISC 360, C2-3

$$P_{e2} = \frac{\pi^2 EI_x}{(K_x L_x)^2} = \frac{\pi^2 (29{,}000 \text{ ksi})(2140 \text{ in}^4)}{[1.5(12.5 \text{ ft})(12 \text{ in/ft})]^2} = 12{,}099 \text{ kips}$$

Because there are three identical frames with four contributing columns

$$\Sigma P_{nt} = 4(255.6) = 1022.4 \text{ kips}$$

$$\Sigma P_{e2} = 4(12{,}099) = 48{,}396 \text{ kips}$$

$$B_2 = \frac{1}{1 - \left(\frac{1022.4}{48{,}396}\right)} = 1.02 \geq 1.0$$

$$M_r = B_2 M_{nt} + B_2 M_{lt}$$ AISC 360, C2-1a

$$M_{ux} = 1.02(605 \text{ in-k} + 4{,}401 \text{ in-k}) = 5106 \text{ in-k}$$

$$M_{uy} = 1.02(605 \text{ in-k} + 0 \text{ in-k}) = 617 \text{ in-k}$$

Check combined stresses for the critical load combination

$$\frac{P_u}{P_n} = \frac{224}{2025} = 0.11 < 1.0 \,\&\, 0.2 \qquad \text{so} \ldots$$

$$\frac{P_u}{2\phi_c P_n} + \frac{M_{ux}}{\phi_b M_{nx}} = \frac{256}{3442.5} + \frac{5106}{14{,}400} = 0.42 < 1.0 \ \ldots \ o.k. \qquad \text{AISC 360, H1-1b}$$

Check column shear strength

$$\frac{h}{t_w} \le 2.24\sqrt{\frac{E}{F_{yw}}} = 2.24\sqrt{\frac{29{,}000 \text{ ksi}}{50 \text{ ksi}}} = 54.0$$

Since $\frac{h}{t_w} = 13.7$

$$V_n = 0.6F_yA_wC_v \qquad \text{AISC 360, G2-1}$$

$$C_v = 1.0$$

$$\phi_vV_n = 1.0(0.6)(50)(12.6)(1.0) = 378 \text{ kips} \qquad \text{AISC 360, G2-2}$$

$$V_u = 62.5 \text{ kips} < \phi_vV_n = 378 \text{ kips} \ \ldots \ o.k.$$

AISC 341, §4.1 stipulates certain special seismic load combinations, which are to be applied to the design of frame columns where referred to in the specification using ASCE 7 §12.4.3.2

$$1.2D + 0.5L + 0.2S + \Omega_oQ_E$$
$$0.9D - \Omega_oQ_E$$

The intent of these supplemental load combinations is to ensure that the columns have adequate axial strength to preclude buckling when subjected to the maximum seismic force developed in the structure as it yields in an earthquake. Per AISC 341, §8.2, columns are to be evaluated for these load combinations if $P_u/\phi P_n > 0.4$. For this design example, $P_u/\phi P_n = 0.122 < 0.4$, and the above load combinations need not be considered for this column design.

Column flange bracing must also be considered; and is addressed in Part 6 below.

$\therefore$ Use W14×176 column

Note: The W14×176 is not necessarily an optimum size based on strength and local or global stability. The size was chosen to meet drift requirements with consideration toward connection design limitations and costs associated with the beam-column connection. A more refined optimization could result from a number of detailed iterations, but this size is adequate for illustration of the design example and economical in terms of the construction cost.

6. SMF beam-column connection design

The RBS connection is a pre-qualified connection type per AISC 358. This design example follows the procedure outlined in AISC 358 with reference to AISC 341 and AISC 360. The basis for the prequalification of the RBS connection is widely attributed to initial RBS connection tests conducted at the University of Texas Ferguson Laboratory (Engelhardt et al., 1996). Extensive additional testing of RBS joint combinations performed as part of Phase 2 of the SAC program and numerous privately funded and publicly funded research projects confirmed the viability of the connection using myriad beam and column sizes. Further information on the research supporting the validity of pre-qualification parameters may be found in FEMA-355D *State of the Art Report on Connection Performance* and AISC 358.

The following calculations comprise a design methodology that is inherently iterative and requires some experience to gain proficiency. After considering code drift limits and evaluating several combinations for strong column-weak beam and panel zone strength criteria, the combination of a W27×84 beam and W14×176 column was selected for use in this design example. These frame members were shown to have adequate strength to resist the factored load combinations (Part 5), and this combination of beam and column sizes in the computer analysis results in overall frame drifts within the code limits (Part 4). The W14×176 column was chosen to elucidate details associated with doubler plates; with the understanding that an increase in column size could obviate the need for doubler plates and possibly reduce the cost of the frame. When given the option, steel fabricators have elected to use heavier columns in lieu of doubler plates for economy. As shown in Figure 3-3, the W14×176 columns are to be full-height, one length. Full-height columns without splices were found to be the least-cost option. Column splices in SMFs must comply with AISC 341, §8.4. The reduced beam section (RBS) connection developed in this design example is shown in Figures 3-7 and 3-12.

6a. Establish plastic hinge configuration and location

AISC 358, 5.8, Step 1

The fundamental design intent espoused in FEMA-350, supported by extensive successful research and experimental testing, is expressed in the AISC 358 design procedure for RBS. This intent to move the plastic hinge away from the weld between the beam flange and column flange is accomplished by providing a defined reduced beam section that locates the plastic hinge region at a pre-determined distance away from the column flange (Figure 3-7).

AISC 358, §5.8 identifies specific limitations for the dimensions associated with the radius cut for the reduced beam section.

$0.5b_{bf} \leq a \leq 0.75b_{bf}$ AISC 358, 5.8-1

$0.65d \leq b \leq 0.85d$ AISC 358, 5.8-2

$0.1b_{bf} \leq c \leq 0.25b_{bf}$ AISC 358, 5.8-3

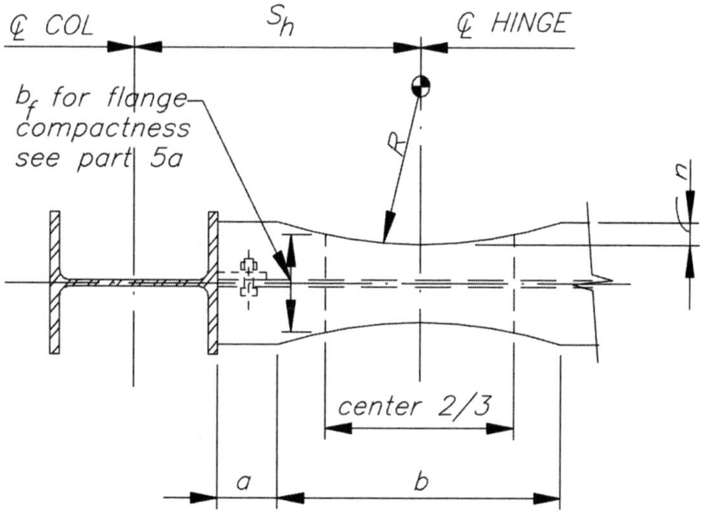

Figure 3-7. RBS geometry

W27×84

$$0.5b_f = 0.5(9.96) = 4.98 \text{ in}$$

$$0.75b_f = 0.75(9.96) = 7.47 \text{ in}$$

$$\therefore \text{Use } a = 6.0 \text{ in}$$

and

$$0.65d = 0.65(26.7) = 17.36 \text{ in}$$

$$0.85d = 0.85(26.7) = 22.70 \text{ in}$$

$$\therefore \text{Try } b = 22.0 \text{ in}$$

The initial estimate for the depth of the cut, c, should be made such that 40 to 50 percent of the flange is removed. This should limit the projection of moments at the face of the column to within 90 to 100 percent of the plastic capacity of the full beam section.

With a 45-percent reduction in the flange area

$$c = 0.45\left(\frac{b_f}{2}\right) = \frac{0.45(9.96)}{2} = 2.24 \text{ in}$$

$\therefore$ Use $c = 2\frac{1}{4}$-in cut, (2.25 in/9.96 in) $= 0.23b_f < 0.25b_f$ . . . *o.k.*

$$R = \frac{4c^2 + b^2}{8c} = \frac{4(2.25)^2 + 22^2}{8(2.25)} = 28.0 \text{ in (radius of the flange cut)}$$

The plastic hinge may be assumed to occur at the center of the curved cut such that

$$s_h = (15.2/2) + 6.0 + (22/2) = 24.6 \text{ in}$$

and

$$L = 28.0 \text{ ft}$$

$$\therefore L' = 28 - 2(24.6/12) = 23.9 \text{ ft}$$

The length between the plastic hinges, L' (see Figure 3-8), is used to determine forces at the critical sections for connection analysis.

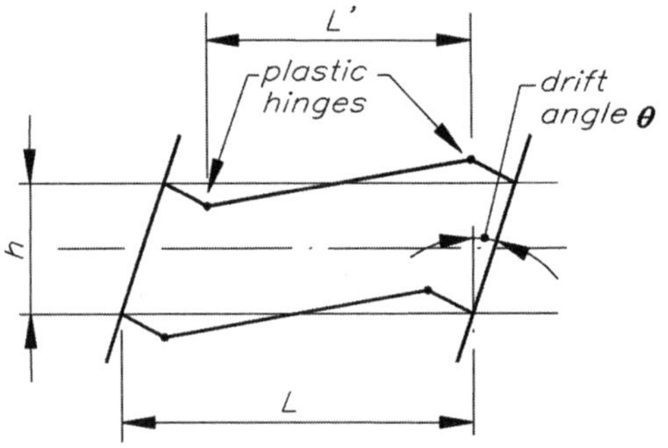

Figure 3-8. Plastic hinge locations

6b. Determine plastic section modulus at the reduced beam section — AISC 358, §5.8 Step 2

The plastic section modulus at the center of the reduced beam section is calculated as

$$Z_e = Z_x - 2ct_{bf}(d - t_{bf})$$ AISC 358, Eq 5.8-4

$$2c = 2(2.25) = 4.5 \text{ in}$$

$$Z_e = 244 - [4.5(0.64)(26.7 - 0.64)] = 168.95 \text{ in}^3$$

6c. Determine probable maximum moment at the reduced beam section — AISC 358, §5.8 Step 3

Next, the probable plastic moment at the reduced beam section, M_{pr}, is calculated as

$$M_{pr} = C_{pr}R_yF_yZ_e$$ AISC 358, Eqs 2.4.3-1, 5.8-5

The factor, C_{pr}, accounts for peak connection strength, including strain hardening, local restraint, additional reinforcement, and other connection conditions.

$$C_{pr} = \frac{F_y + F_u}{2F_y} \leq 1.2 \qquad \text{AISC 358, Eq 2.4.3-2}$$

$$C_{pr} = \frac{50 \text{ ksi} + 65 \text{ ksi}}{2(50 \text{ ksi})} = 1.15$$

$$M_{pr} = (1.15)(1.1)(50)(168.95) = 10{,}686 \text{ kip-in}$$

The value for M_{pr} must be such that the projected moment demand at the face of the column M_f, is less than the expected strength of the full beam section; this condition is verified in Part 6g below.

6d. Compute the shear force at the center of each RBS

AISC 358; §5.8, Step 4

Determine the shear force at the center of the reduced beam sections at each end of the beam. AISC 358, §5.8 requires that the shear force at the center of the reduced beam section be determined by a free body diagram of the portion of the beam between the centers of the reduced beam sections assuming the moment at the center of each RBS is M_{Pr}, and shall include gravity loads acting on the beam based on a specific load combination: $1.2D + f_1L + 0.2S$ (where $f_1 = 0.5$). See Figure 3-9.

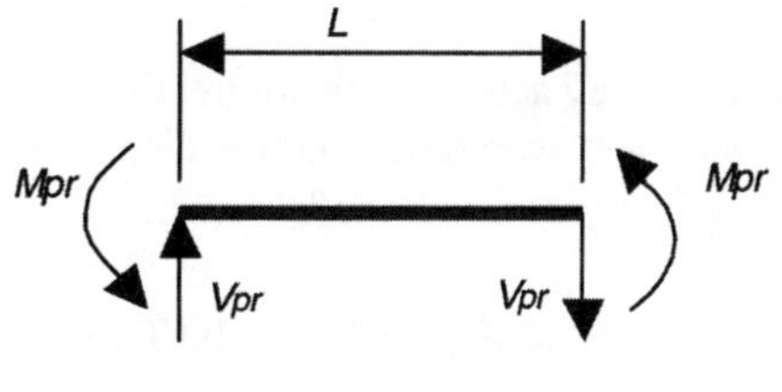

Figure 3-9. Beam equilibrium under the probable plastic moment M_{pr}

$$V_{pr} = \frac{2M_{pr}}{L'} = \frac{2(10{,}686)}{12(23.9)} = 74.5 \text{ kips}$$

The positive sense of V_{pr} results in the larger of the two values of shear force at the center of the reduced beam section at each end of the beam [defined as V_{RBS} in AISC 358 (5.8-6)].

$$V_p = 1.2(V_D) + 0.5(V_L) + (V_{pr})$$

$$\therefore V_p = 1.2(16.4) + 0.5(13.3) + (74.5) = 101 \text{ kips} = V_{RBS}$$

6e. Compute the probable maximum moment at the face of the column — AISC 358, §5.8 Step 5

$M_f = M_{pr} + V_{RBS}S_h$ — AISC 358, Eq 5.8-6

$M_f = 10{,}686 + 101(6 + 22/2) = 12{,}403$ kip-in (positive sense)

$M_f = -10{,}686 + [(-48)(6 + 22/2)] = -11{,}502$ kip-in (negative sense)

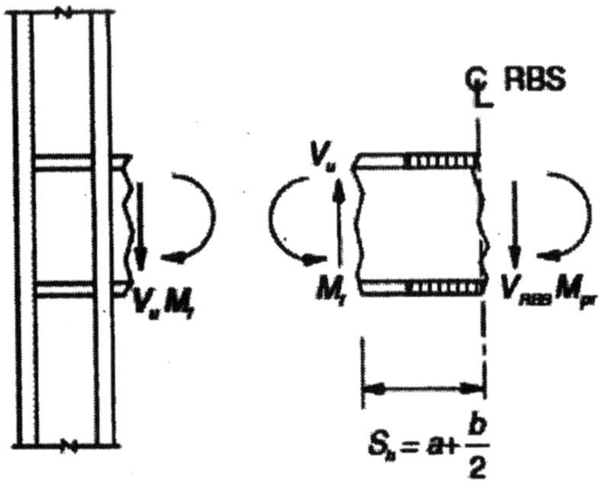

Figure 3-10. Free-body diagram between center of RBS and face of column. (AISC 358, Fig. 5.2)

In this example (as in many actual applications) the moment attributable to the gravity load applied between the plastic hinge and the face of the column flange is negligible (<0.5%), therefore it is omitted and briefly considered when comparing M_{pe} to M_f.

6f. Compute the expected plastic moment of the beam — AISC 358, §5.8, Step 6

To compute the plastic moment of the beam based on the expected yield stress of the beam material, compute M_{pe} as

$$M_{pe} = Z_b R_y F_y = (244)(1.1)(50) = 13{,}420 \text{ kip-in}$$

6g. Check that M_f does not exceed $\phi_d M_{pe}$ — AISC 358, §5.8, Step 7

Check the value of M_f against $\phi_d M_{pe}$ as follows

$$M_f \le \phi_d M_{pe} = 12{,}403 < 13{,}420 \ \ldots \ o.k.$$

If M_f exceeds $\phi_d M_{pe}$, the depth of cut at the reduced beam section (c) should be increased (not to exceed a 50-percent total reduction of the beam flange). The difference between M_{pe} and M_f is much greater than 0.5 percent, so the omission of the gravity load between the plastic hinge and the face of the column flange is acceptable.

6h. Determine the required shear strength, V_u AISC 358, §5.8, Step 8

Determine the required shear strength of the beam and beam web-to-column connection from the following equation

$$V_u = \frac{2M_{pr}}{L'} + V_{\text{gravity}} \qquad \text{AISC 358, Eq 5.8-9}$$

where

V_u = required shear strength of beam and beam web-to-column connection

L' = distance between the centers of the reduced beam sections

V_{gravity} = beam shear force resulting from $1.2D + f_1L + 0.2S$ (where $f_1 = 0.5$)

$$V_u = \frac{2M_{pr}}{L'} + (V_{RBS} - V_{pr}) = 101 \text{ kips}$$

The design shear strength of the beam is checked in accordance with AISC 360, Chapter G. This calculation was performed as part of the beam design (Part 5c). The shear strength of the W27×84 beam was determined to be 331 kips. Therefore, the beam is adequate to resist the shear demand at any location along the beam length as the calculation considers the beam web only.

6i. Design the beam web-to-column connection according to AISC 358 §5.6 and §5.8, Step 9

This check references AISC 358, §5.6. Section 5.6 indicates that the strength of the beam web-to-column connection strength must be determined in accordance with Equation 5.8-9. Furthermore, the following description identifies the only allowable detailing for the beam web-to-column connection:

Note: For SMF systems, the beam web shall be connected to the column flange using a CJP groove weld extending between weld access holes. The single plate shear connection shall be permitted to be used as backing for the CJP groove weld. The thickness of the plate shall be at least ⅜ in. (10 mm). Weld tabs are not required at the ends of the CJP groove weld at the beam web. Bolt holes in the beam web for the purpose of erection are permitted.

Because the beam web-to-column connection is made with a CJP groove weld, the shear capacity of the weld is greater than or equal to the shear capacity of the beam, so no further checks are required to verify the adequacy of this condition.

6j. Check continuity plate requirements according to AISC 358 Chapter 2 and §5.8, Step 10

This check references AISC 358, Chapter 2, §2.4.4. Section 2.4.4 identifies the following equation for use in determining the need for continuity plates and the thickness of the plates if they are required by calculation.

When the beam flange connects to the flange of a wide-flange or built-up I-shaped column having a thickness that satisfies Equations 2.4.4-1 and 2.4.4-2, continuity plates need not be provided

$$t_{cf} \geq 0.4\sqrt{1.8 b_{bf} t_{bf} \frac{F_{yb} R_{yb}}{F_{yc} R_{yc}}} = 0.4\sqrt{(1.8)(9.96)(0.64)\frac{(50)(1.1)}{(50)(1.1)}} = 1.35 \text{ in}$$

AISC 358, Eq 2.4.4-1 and Eq 2.4.4-2

$$t_{cf} \geq \frac{b_{bf}}{6} = 1.66 \text{ in}$$

where t_{cf} = minimum required thickness (inches) of column flange when no continuity plates are provided.

b_{bf} = beam flange width, in

t_{bf} = beam flange thickness, in

F_{yb} = specified minimum yield stress of the beam flange, ksi

F_{yc} = specified minimum yield stress of the column flange, ksi

R_{yb} = ratio of the expected yield stress of the beam material to the specified minimum yield stress, per AISC 341

R_{yc} = ratio of the expected yield stress of the column material to the specified minimum yield stress, per AISC 341

t_{cf} = 1.31 in < 1.35 in and 1.66 in

∴ continuity plates are required

Per AISC 358, §2.4.4a the thickness of the plates is determined as follows:

(a) For one-sided (exterior) connections, continuity plate thickness shall be at least one-half of the thickness of the beam flange.

(b) For two-sided (interior) connections, the continuity plate thickness shall be at least equal to the thicker of the two beam flanges on either side of the column.

Continuity plates shall also conform to the requirements of Section J10 of the AISC Specification.

The requirements of AISC 360, §J10 pertain to detailing/sizing of the continuity plates. The thickness requirements listed above combined with the detailing requirements of AISC 360, §J10 ensures that the continuity plate has adequate strength.

The detailing for the continuity plates dictates some of the verifying calculations associated with the continuity plate, including the welded connections between the continuity plate and the column flanges and web. AISC 358, §3.6 indicates the following detailing provisions.

Note: Along the web, the corner clip shall be detailed so that the clip extends a distance of at least 1½ inches (38 mm) beyond the published "k" detail dimension for the rolled shape. Along the flange, the plate shall be clipped to avoid interference with the radius of the rolled shape and shall be detailed so that the clip does not exceed a distance of ½ inch (12 mm) beyond the published "k_1" detail dimension. The clip shall be detailed to facilitate suitable weld terminations for both the flange weld and the web weld. When a curved clip is used, it shall have a minimum radius of ½ inch (12 mm).

Using these requirements, the projected contact area between the edge of the continuity plate and the column flange and column web are

$$A_{pb} = (W_{pb})(t_{\text{cont-}p1})$$

$$W_{pb} = b_{\text{cont-}p1} - (\text{“}k_{1\text{col}}\text{”} + 0.25 \text{ in})$$

The continuity plate width ($W_{pb\text{-flange}}$) can be determined considering the required strength at the projected bearing surface as identified in AISC 360, §J7

$$\phi R_{ct} = (0.9)1.8F_y A_{pb}$$ AISC 360, J7-1

$$\phi R_{ct} = (0.9)(1.8)F_y A_{pb} \geq \left(\frac{M_f}{d_b - t_{bf}} = \frac{12{,}403}{(26.7 - 0.64)} = 476 \text{ kips} \right)$$

$$A_{pb} = \frac{476 \text{ kips}}{(0.9)(1.8)(F_y)} = \frac{476 \text{ kips}}{(0.9)(1.8)(50)} = 5.88 \text{ in}^2$$

If the width available within the column section (between the column web and the edge of the column flange) limits the total available width for bearing, the following can be used to size the thickness of the continuity plate

$$W_{pb(\max)} = \frac{b_{cf} - t_{cw}}{2} = 7.435 \text{ in}$$

$$W_{pb} = W_{pb(\max)} - (\text{“}k_{1\text{col}}\text{”} + 0.25 \text{ in}) = 7.435 - (1.625 + 0.25) = 5.56 \text{ in}$$

$$A_{pb} = (W_{pb})(t_{\text{cont-}p1}):\ t_{\text{cont-}p1} = \frac{A_{pb}}{W_{pb}} = \frac{5.88}{5.56} = 1.06 \text{ in (say } t_{\text{cont-}p1} = 1.0 \text{ in)}$$

Therefore, use 2 pairs of 1 × 7½-in continuity plates in the column aligned with the top and bottom beam flanges.

In addition to the size of the continuity plate, the attachment/welding of the continuity plate shall meet the criteria established in AISC 358, §2.4.4b.

Note: Continuity plates, if provided, shall be welded to column flanges using CJP groove welds. Continuity plates shall be welded to column webs using groove welds or fillet welds. The required strength of the sum of the welded joints of the continuity plates to the column web shall be the smallest of the following:

(a) The sum of the design strengths in tension of the contact areas of the continuity plates to the column flanges that have attached beam flanges.

(b) The design strength in shear of the contact area of the plate with the column web.

(c) The design strength in shear of the column panel zone.

(d) The sum of the expected yield strengths of the beam flanges transmitting force to the continuity plates.

Per AISC 358, §2.4.4b, continuity plates shall be welded to column flanges with CJP groove welds, so no design calculations are required for this portion of the connection. However, the connection between the continuity plate and column web should be calculated to determine an appropriate weld for this connection.

The maximum contact area between the continuity plate and the column web is

$$A_{pw} = d_c - 2t_{cf} - 2(k + 1.5 \text{ in}) = [15.2 - 2(1.31) - 2(1.91 + 1.5)] \times 1.0 \text{ in} = 5.76 \text{ in}^2$$

The tension strength of the continuity plate is limited by the connection strength between the edge of the continuity plate and the face of the column flange. This condition is expressed in the limitations for welding required between the continuity plate and the column web. The following calculations identify the controlling case (minimum value of the four cases) for the required strength of the continuity plate-to-column web welds

$$\Sigma\phi F_y A_{pb} = 2(0.9)(50)(5.56) = 500 \text{ kips} \qquad \text{AISC 358, 2.4.4b (a)}$$

$$\phi V_{nw} = \phi(0.6)F_y A_{pw} = (1.0)(0.6)(50)(5.76) = 173 \text{ kips} \qquad \text{AISC 358, 2.4.4b (b)}$$

$$\phi R_v = 469 \text{ kips (see calculations in Part 6k)} \qquad \text{AISC 358, 2.4.4b (c)}$$

$$\Sigma\frac{\phi_d M_{pe}}{d_b - t_{bf}} = 2\left(\frac{(0.9)(13{,}420)}{26.7 - 0.64}\right) = 927 \text{ kips} \qquad \text{AISC 358, 2.4.4b (d)}$$

The smallest value is 173 kips, which will be used to design the welds between the continuity plate and the column web.

The minimum required double-sided fillet weld size to develop 173 kips follows.

$$D_{\min} = \frac{R_u}{2(1.392 \text{ k/in})[d_c - 2t_{cf} - 2(k + 1.5 \text{ in})]}$$

$$= \frac{173 \text{ k}}{2(1.392 \text{ k/in})(5.76 \text{ in})}$$

$$= 10.8$$

One could use double sided ¾-in fillet welds to connect the continuity plates to the column web, but considering costs associated with filler metal and welding, it may be more economical to use a PJP that develops the strength of the plate or a CJP groove weld between the continuity plate and the column web.

6k. Check column panel zone according to AISC 358 §5.4 and §5.8, Step 11

The panel zone strength is calculated below using the provisions of AISC 341,§9.3. The panel zone shear calculation is derived by projecting the expected moments at the beam plastic hinges to the face of the column, as follows

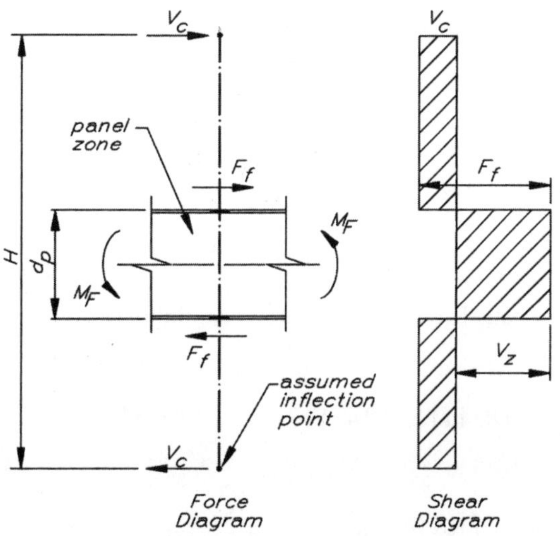

Figure 3-11. Panel zone forces

$d_p = 26.7 - 0.64 = 26.06$ in

M_f = 12,403 kip-in and 11,502 kip-in (see Part 6e)

$$R_u = \frac{\Sigma M_f}{d_p} = \frac{12{,}403 + 11{,}502}{26.06} = 917 \text{ kips}$$

The panel zone shear strength is determined from AISC 341, §9.3a, which refers to AISC 360, §J10.6

$$\phi R_v = \phi 0.6 F_y d_c t_w \left[1 + \frac{3 b_{cf} t_{cf}^2}{d_b d_c t_w}\right]$$ AISC 360, §J10-11

where:

ϕ = 1.0 per AISC 341, §9.3a

b_{cf} = width of the column flange

d_b = depth of the beam

d_c = column depth

t_{cf} = thickness of the column flange

t_q = total thickness of the panel zone, including doubler plates

For the W14×176 column, the panel zone shear strength is

$$\phi R_v = 1.0(0.6)(50)(15.2)(0.83)\left[1+\frac{3(15.7)(1.31)^2}{(26.7)(15.2)(0.83)}\right] = 469 < 917 \text{ kips} \quad \ldots \textit{ no good}$$

The minimum panel zone thickness, t_z, is also checked per AISC 341, §9.3b

$$t_z \geq (d_z + w_z)/90 \qquad \text{AISC 341, §9-2}$$

where:

d_z = panel zone depth between continuity plates

w_z = panel zone width between column flanges

t_z = 0.83 in for W14×176

$t_z = 0.83 \text{ in} \geq [(26.06) + [15.2 - 2(1.31)]]/90 = 0.43 \text{ in} \quad \ldots$ *o.k.*

Therefore, the W14×176 column panel zone strength (without doubler plates) is not adequate when matched with the W27×84 beam. When doubler plates are used in lieu of increasing the column size, then compliance with AISC 341, §9.3c is required. Considering the thickness of the column web and a reasonable plate thickness, a ½-in plate is chosen, providing a panel zone shear strength of 925 kips, which is greater than the required strength of 917 kips. Detailing associated with the doubler plates, as required by AISC 341, §9.3c is shown in Figure 3-12.

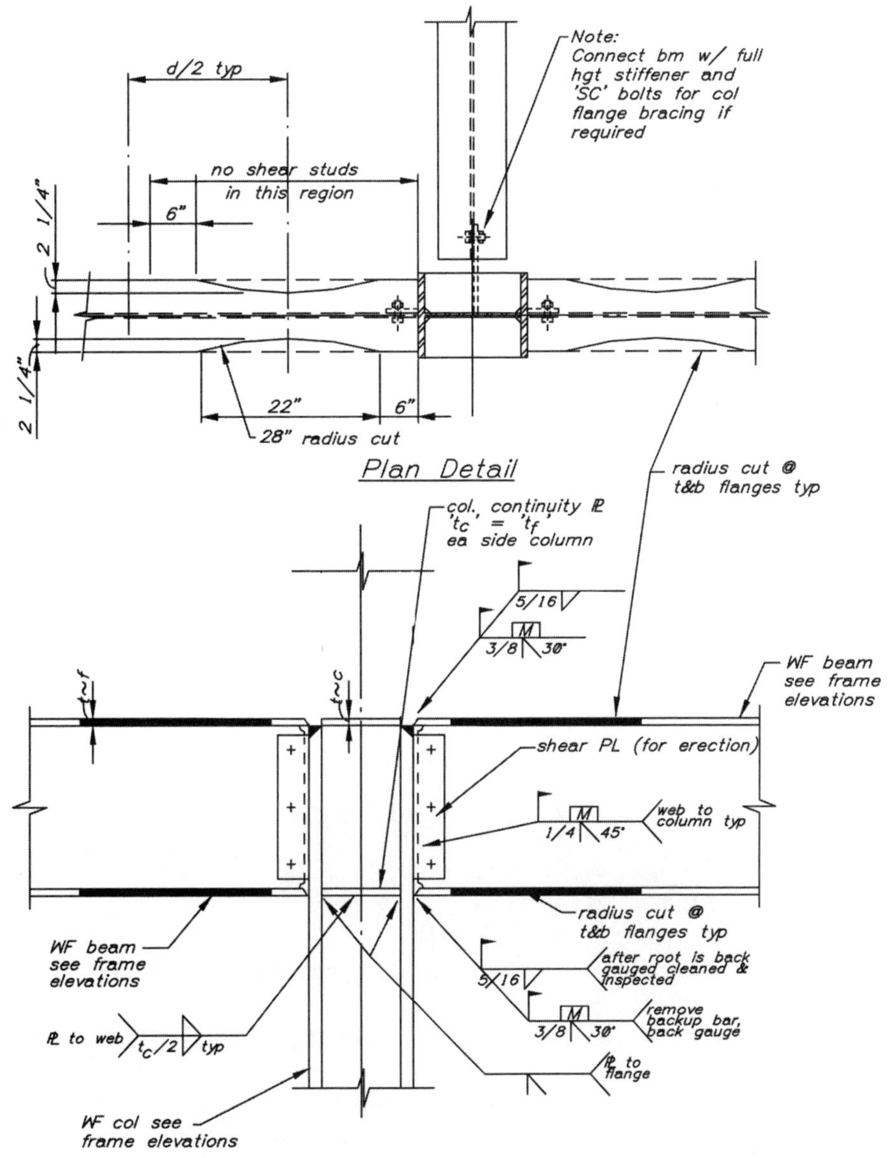

Figure 3-12. Reduced beam section connection detail

6l. Check column-beam moment ratio according to AISC 358 §5.4 and §5.8, Step 12

Per AISC 358, §5.4:

Note: For SMF systems, the column-beam moment ratio shall conform to the requirements of the AISC Seismic Provisions. The value of ΣM^*_{pb} shall be taken equal to $\Sigma (M_{pr} + M_v)$, where M_{pr} is computed according to Equation 5.8-5, and where M_v is the additional moment due to shear amplification from the center of the reduced beam section to the centerline of the column. M_v can be computed as

$V_{RBS}(a + b/2 + d_c/2)$, where V_{RBS} is the shear at the center of the reduced beam section computed per Step 4 of §5.8, a and b are the dimensions shown in Figure 5.1, and d_c is the depth of the column.

$$\frac{\Sigma M^*_{pc}}{\Sigma M^*_{pb}} \geq 1.0$$ AISC 341, Eq 9-3

Per AISC 341, §9.6, the sum of the column strengths may be taken as:

$$\Sigma M^*_{pc} = \Sigma Z_c(F_{ye} - P_u/A_g)$$

From Part 5d above, the maximum column axial demand is

$P_u/A_g = 224/51.8 = 4.32$ ksi, and for the W14×176 column, $Z_x = 320$ in^3

$$\Sigma M^*_{pc} = 2[(320)(50 - 4.32)] = 29{,}232 \text{ in-kip}$$

Per AISC 358, §5.4, the sum of the beam strengths may be taken as

$$\Sigma M^*_{pb} = \Sigma(M_{pr} + M_v)$$

$$\Sigma M_{pr} = 2(10{,}686) = 21{,}372$$

$$\Sigma M_v = \Sigma[(V_{RBS})(a + b/2 + d_c/2)]$$

$$= (101 + 48)(6 + 11 + 15.2/2) = 3665$$

$$\Sigma M^*_{pb} = 21{,}372 + 3665 = 25{,}037 \text{ in-kip}$$

$$\frac{\Sigma M^*_{pc}}{\Sigma M^*_{pb}} = \frac{29{,}232}{25{,}037} = 1.17 > 1.0$$

Therefore, the beam-column ratios (a.k.a. strong column-weak beam) are satisfied.

6m. Evaluate lateral bracing of columns — AISC 358, §5.3.2

Per AISC 358, §5.3.2, lateral bracing of columns shall conform to §9.7 or §10.7 for SMF or IMF, as applicable, in the AISC Seismic Provisions (AISC 341).

To preclude SMF column members from lateral torsional buckling under AISC 341 §9.7, §2.9.2 specifies requirements for column flange bracing. The W14×176 column has a perpendicular beam framing into it at each level providing out-of-plane joint restraint. At the beam top flange, the concrete slab effectively provides bracing for the column flange. The column flanges therefore need to be laterally braced at the beam bottom flange only if AISC 341, Equation 9-3 is not greater than or equal to 2.0.

$$\frac{\Sigma M^*_{pc}}{\Sigma M^*_{pb}} \geq 2.0$$

From Part 6l of this design example

$$\frac{\Sigma M^*_{pc}}{\Sigma M^*_{pb}} = \frac{29{,}232}{25{,}037} = 1.17 < 2.0 \text{ ... } \textit{no good}$$

The column flanges, therefore, need lateral bracing only at the moment beam bottom flange. This bracing may be provided by perpendicular beams connected to full-depth stiffeners and column continuity plates. The bracing must be designed for a required strength that is equal to 2 percent of the available beam flange strength, as appropriate.

6n. Detailing considerations

AISC 341 and AISC 358 require that the reduced beam section design entails a few unique fabrication and construction considerations (excerpts from AISC standards are italicized).

- *The reduced beam section shall be made using thermal cutting to produce a smooth curve. The maximum surface roughness of the thermally cut surface shall be 500 microinches (13 microns) in accordance with ANSI B46.1, as measured using AWS C4.1-77 Sample 4 or similar visual comparator. All transitions between the reduced beam section and the unmodified beam flange shall be rounded in the direction of the flange length to minimize notch effects due to abrupt transitions. Corners between the reduced section surface and the top and bottom of the flanges shall be ground to remove sharp edges, but a minimum chamfer or radius is not required.*

- *Thermal cutting tolerances shall be plus or minus 1/4 in. (6 mm) from the theoretical cut line. The beam effective flange width at any section shall have a tolerance of plus or minus 3/8 in. (10 mm).*

- *Gouges and notches that occur in the thermally cut RBS surface may be repaired by grinding if not more than 1/4 in. (6 mm) deep. The gouged or notched area shall be faired by grinding so that a smooth transition exists, and the total length of the area ground for the transition shall be no less than five times the depth of the removed gouge on each side of the gouge. If a sharp notch exists, the area shall be inspected by MT after grinding to ensure that the entire depth of notch has been removed. Grinding that increases the depth of the RBS cut more than 1/4 in. (6 mm) beyond the specified depth of cut is not permitted.*

- *Gouges and notches that exceed 1/4 in. (6 mm) in depth, but not to exceed 1/2 in. (12 mm) in depth, and those notches and gouges where repair by grinding would increase the effective depth of the RBS cut beyond tolerance, may be repaired by welding. The notch or gouge shall be removed and ground to provide a smooth radius of not less than 1/4 in. in preparation for welding. The repair area shall be preheated to a temperature of 150°F or the value specified in AWS D1.1 Table 3.2, whichever is greater, measured at the location of the weld repair. Notches and gouges exceeding 1/2 in. (12 mm) in depth shall be repaired only with a method approved by the engineer of record.*

- *A "protected zone" is established for the connection and shall be defined as the area from the face of the column flange to one-half of the beam depth beyond the theoretical hinge point. Bolt holes in beam webs, when detailed in accordance with the individual connection provisions of this standard, shall be permitted. With concrete fill over composite metal deck, welded shear studs should be omitted within the protected zone.*

6o. Welding parameters

Parameters for pre-qualified welded joints are presented in AISC 358, Chapter 3. The requirements identified in AISC 358 reference AISC 341 §7 and Appendix W. The requirements for welding are clearly outlined in both standards. Eventually, AISC 341, Appendix W will be superseded by requirements of AWS D1.8.

Specific protocols associated with weld backing and weld tabs are described in AISC 358, Chapter 3. All requirements are applicable to the fabrication of an RBS connection.

6p. Quality assurance and quality control

Quality assurance and quality control requirements are outlined in AISC 358, §3.7; referring to AISC 341, Appendix Q. Observation, inspection and documentation are tabulated in Appendix Q and should be accurately applied to project contract documents as applicable. Most Appendix Q requirements apply to the RBS connection. Eventually, Appendix Q will be superseded by requirements of AWS D1.8.

References

AISC, 1997, 1999, 2000. *Seismic Provisions for Structural Steel Buildings*. American Institute of Steel Construction, April 1997 with Supplement No. 1, February 1999 and Supplement No. 2, November 2000.

AISC, 2005. *Seismic Provisions for Structural Steel Buildings, Including Supplement No. 2*. American Institute of Steel Construction, Chicago, Illinois.

AISC, 2005. *Specification for Structural Steel Buildings*. American Institute of Steel Construction, Chicago, Illinois.

AISC/SSEC, 2006. *Seismic Design Manual*. American Institute of Steel Construction, Chicago, Illinois.

ASCE 2005. ASCE 7, *Minimum Design Loads for Buildings and Other Structures*, American Society of Civil Engineers, Miami, Florida.

AWS, 2005. D1.8 *Structural Welding Code—Steel*. American Welding Society, Miami, Florida.

Bozorgnia, Y., Bertero, V., 2004. *Earthquake Engineering: From Engineering Seismology to Performance-Based Engineering*. CRC Press, LLC, Danvers, Masachussetts.

Englehardt, M., 1998. Design Recommendations for Radius Cut Reduced Beam Section *Moment Connections*. University of Texas, Austin.

Englehardt, M., et al., 1996. "The Dogbone Connection, Part II," *Modern Steel Construction*. American Institute of Steel Construction, Chicago, Illinois.

FEMA-350, 2000. *Recommended Seismic Design Criteria for New Steel Moment Frame Buildings*, Federal Emergency Management Agency, Washington, D.C.

FEMA-353, 2000. *Recommended Specifications and Quality Assurance Guidelines for Steel Moment-Frame Construction for Seismic Applications*, Federal Emergency Management Agency, Washington, D.C.

FEMA-355D, 2000. *State of the Art Report on Connection Performance*, Federal Emergency Management Agency, Washington, D.C.

ICC, 2009. *International Building Code*, International Code Council, Washington, DC.

Moore, K.S., et al., 1999. "Design of Reduced Beam Section (RBS/Moment Frame Connections," *Steel Tips*. Structural Steel Educational Council, Moraga, California.

Moore, K, S., Feng, J.Y., 2007. "Design of RBS Connections for Special Moment Frames," *Steel Tips*, June. Structural Steel Educational Council, Moraga, CA.

SEAOC, 2002. *Commentary and Recommendations on FEMA-350*, Structural Engineers Association of California, Sacramento, CA.

Design Example 4
Special Plate Shear Wall

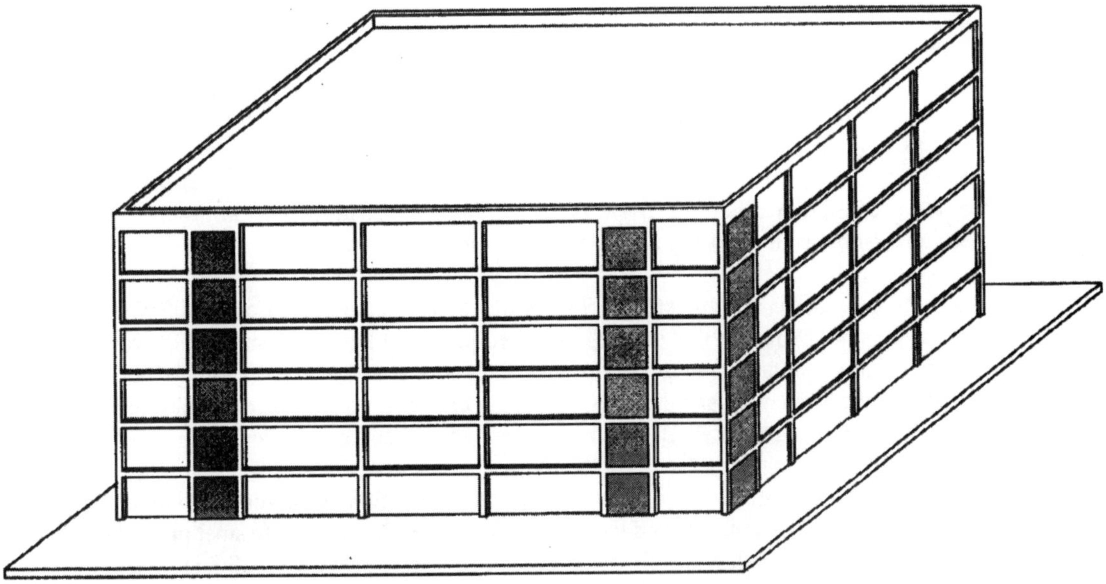

Overview

Special plate shear walls (SPSW) consist of a plate bounded at the sides by columns, also referred to as vertical boundary elements (VBE), and at the floor levels by beams, also referred to as horizontal boundary elements (HBE). The alternate nomenclature for beams and columns emphasizes the boundary element's role of resisting the tension field developed in the plate.

Steel plate shear walls have been used as early as the 1970s in the United States (Anon 1978, Anon 1977, Troy 1979). These early forms of steel plate shear walls were designed so that the web plate did not buckle under shear loading. To preclude buckling, the web plates had to either be fairly thick, or be stiffened at intermediate points across a shear panel. This type of system is commonly referred to as a stiffened steel plate shear wall. Another manifestation of the stiffened steel plate shear wall is the Composite Steel Plate Shear Wall (C-SPW) that uses concrete to brace the plate.

On the other hand, the fact that web plates have post-buckling capacity has been known and applied for decades when it comes to plate girders (Basler 1961, Porter 1975). Upon buckling, a diagonal tension field develops in the plate, which, along with the boundary members, creates a truss-like system that can resist significant shear. In some ways SPSW are similar to a plate girder turned vertical. The major difference between plate girders and SPSW is the greater strength and stiffness of the SPSW column compared to the plate girder flange (Berman 2004). Unstiffened SPSW design is centered around keeping the boundary elements elastic while the web plate yields.

Since the 1980s, tension field action has been utilized in steel plate shear walls, and is the basis of the AISC 341 provisions for SPSW. Since these systems allow thinner plates and do not require stiffeners, they are commonly cheaper, and thus have become more popular than their stiffened counterparts. Steel plate shear walls that utilize post-buckling capacity of the shear plate are commonly referred to as thin or unstiffened. U.S. code provisions apply to unstiffened steel plate shear walls that utilize the post-buckling capacity of the web plate.

The 2005 edition of AISC 341 contains the first U.S. code provisions for SPSW. Special plate shear walls represent another high-ductility lateral resisting system option. Engineers can further choose whether to design the SPSW as a dual system with special moment frames capable of taking 25 percent of the lateral forces ($R = 8.0$) or as a SPSW with ordinary moment connections ($R = 7.0$). AISC 341 requires moment connections between the HBE and VBE in both cases, because they fill out the hysterisis loops and provide resistance during the transition between tension fields in opposing directions during load reversal.

As discussed later in this design example, one of the defining characteristics of SPSW is the significant amount of strength that can be developed in a short length of wall. SPSW can be especially useful in building configurations with open floor plans and little room to hide braced frames. Of course, with more slender lateral resisting elements comes correspondingly higher overturning forces and greater likelihood that deflection will control the design.

This example is tailored to the use of unstiffened SPSW in high-seismic regions with all the ductile detailing requirements implied therein. Stiffened SPSW, composite steel plate shear walls, and low seismic applications for SPSW are not addressed. See the AISC Design Guide 20 (Sabelli and Bruneau 2006) for more information on these topics. Other references of note are the MCEER report on design of perforated steel plate shear walls (Purba and Bruneau 2007), the MCEER report by Vian and Bruneau (2007) which gives test results for panels with perforations and low yield strength steel, the paper by Berman and Bruneau (2008) which discusses capacity design of VBE's, and the paper by Qu and Bruneau (2010) for discussion of HBE capacity design.

Outline

1. **Building geometry and loads**
2. **Calculation of the design base shear and load combinations**
3. **Vertical and horizontal distribution of load**
4. **Preliminary sizing of the web plate and boundary members**
5. **Analysis and design (iterative)**
6. **Detailing and design of connections**

Given Information

- 120 × 150 ft in plan with typical floor and roof framing shown in Figure 4-1
- Six stories as shown in Figure 4-2
- Office occupancy on all floors
- Located at 1st St. and Howard St. in San Francisco, CA
- Site Class D

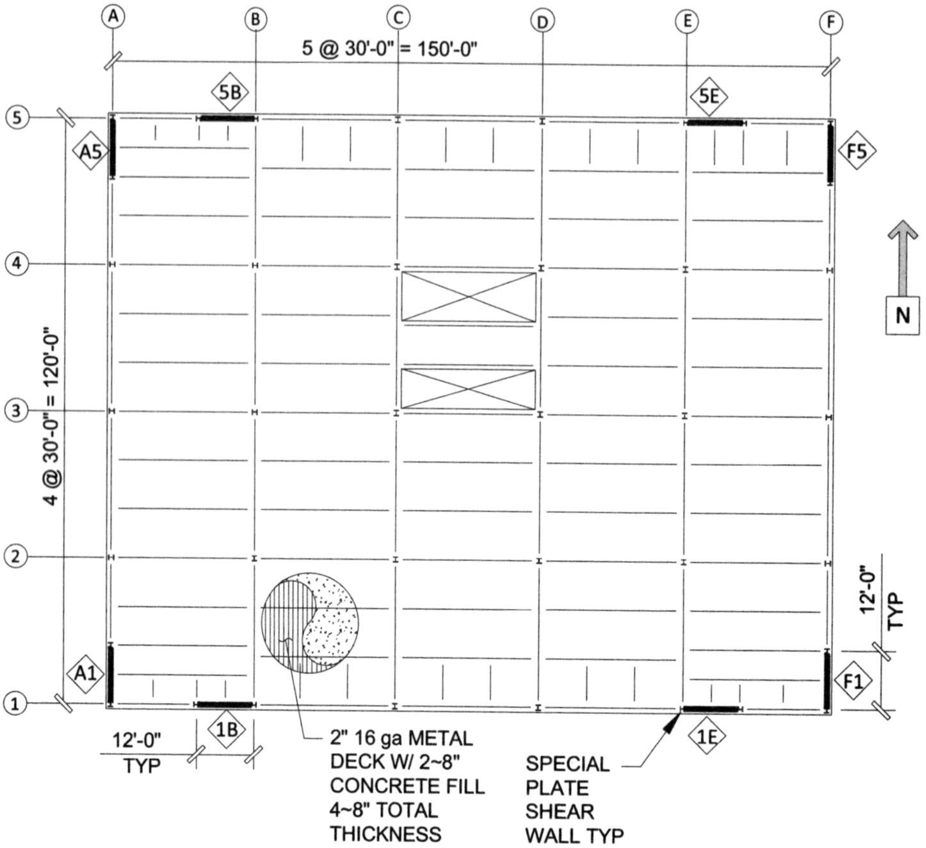

Figure 4-1. Typical floor and roof framing plan

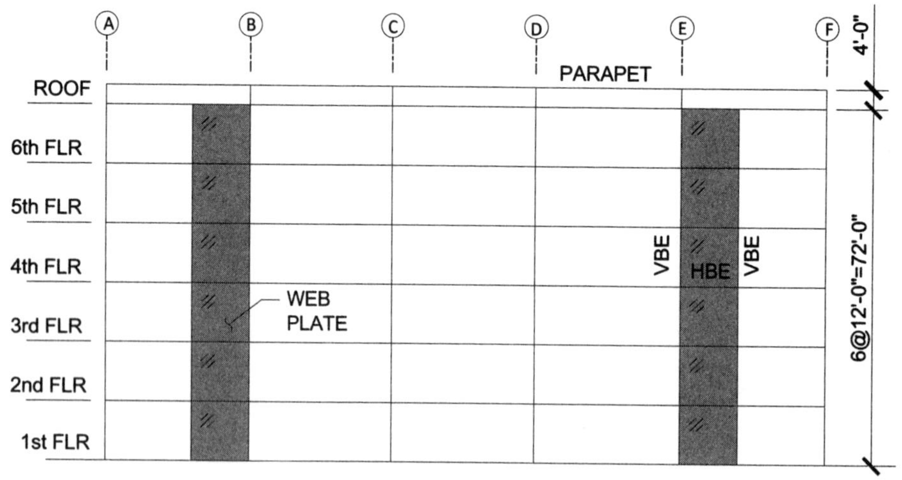

Figure 4-2. Building elevation

Calculations and Discussion | **Code Reference**

1. Building geometry and loads

1a. Assembly weights

Floor Dead Loads	Gravity Load	Effective Seismic Wt[3]
Floor Finish	5 psf	5 psf
Concrete on Metal Deck (4½″ thick)	43 psf	43 psf
Steel Framing	6 psf	6 psf
Mechanical/Plumbing/Electrical	4 psf	4 psf
Ceiling	4 psf	4 psf
Partitions	See LL below	10 psf[4]
Miscellaneous (includes SPSW plate)	3 psf	3 psf
Total Dead Load	65 psf	75 psf

Office Building Live Loads[1]	Live Load
Lobbies and Stairs = 100 psf	100 psf
Offices = 50 psf + 15 psf Partition[2]	65 psf
Corridors abv 1st Flr = 80 psf	80 psf

Roof Dead Loads	Gravity Load	Effective Seismic Wt[3]
Built-Up Roof	6 psf	6 psf
Insulation	2 psf	2 psf
Metal Roof Deck	4 psf	4 psf
Steel Framing	4 psf	4 psf
Mechanical/Plumbing/Electrical	4 psf	4 psf
Ceiling	4 psf	4 psf
Partitions	0 psf	5 psf[4]
Miscellaneous (includes SPSW plate)	1 psf	1 psf
Total Dead Load	25 psf	30 psf

Roof Live Loads[1]	Live Load
Ordinary Flat Roof	20 psf

Ext. Wall Dead Loads	Gravity Load	Effective Seismic Wt[3]
Cladding	5 psf	5 psf
½″ Plywood	2 psf	2 psf
Metal Studs	2 psf	2 psf
Insulation	2 psf	2 psf
⅝″ Gypsum Board	3 psf	3 psf
Miscellaneous	6 psf	6 psf
Total Dead Load	20 psf	20 psf

1. From ASCE/SEI 7-05 Table 4-1.
2. ASCE/SEI 7-05 §4.2.2 specifies a 15 psf **live load** where partitions will be erected or rearranged. This is a departure from previous codes that specified a 20 psf dead load for partitions.
3. ASCE/SEI 7-05 §12.7.2 describes the loads that are included in the Effective Seismic Weight.
4. Per section 12.7.2, 10 psf is included for partitions where partition load is required per ASCE/SEI 7-05 §4.2.2.

Table 4-1. Building mass calculation

Level	Assembly	Unit Wt (psf)	Area (ft^2)	Weight (kips)	Floor Wt (kips)
Second to Sixth	Floor	75	18000	1350	1480
	Ext Wall	20	6480	130	
Roof	Roof	30	18000	540	648
	Ext Wall/Parapet	20	5400	108	
Total Building Weight, W = 5 × 1480 + 648 = 8048 kips					

2. Calculation of the design base shear and load combinations

2a. Classify the structural system and determine spectral accelerations

There are two options in ASCE/SEI 7-05 Table 12.2-1 for using special plate shear walls

Table 4-2. System factors for SPSW from ASCE 7-05 Table 12.2-1

	Seismic Force-resisting System	Response Mod. Coefficient, R	System Overstrength Factor, Ω_o	Deflection Amplification Factor, C_d	Height Limits by Seismic Design Category				
					B	C	D	E	F
B27	SPSW[2]	7.0	2.0	6.0	NL	NL	160	160	100
D13	SPSW— SMF Dual System[1,2]	8.0	2.5	6.5	NL	NL	NL	NL	NL

1. Dual system with special moment frames capable of resisting at least 25% of prescribed seismic forces.
2. 2009 IBC §2205.2 references AISC 341 for determining requirements. In seismic design categories A, B, and C, it is not required to follow the provisions of AISC 341 if using $R = 3$, ($\Omega_o = 3$, and $C_d = 3$ from part H of Table 12.2-1.

For the purposes of this example, a special plate shear wall without the dual system will be used

$R = 7.0$	$\Omega_o = 2.0$	$C_d = 6.0$

2b. Find design spectral accelerations

The building is located at 1st and Howard St in San Francisco, CA. The maximum considered earthquake (MCE) spectral accelerations for 0.2 sec, S_s, and 1.0 sec, S_1, can be found using ASCE 7-05 or IBC §1613 Figures 22-1 to 22-14. Figures 22-3 and 22-4 show an enlarged map of California with S_s and S_1 values respectively. Values obtained from these figures are

$S_s = 150\%g$ F 22-3

$S_1 \approx 68\%g$ (interpolated between 60% and 75%) F 22-4

Alternatively, the internet can be used to determine the spectral accelerations at this location. Websites such as http://www.worldatlas.com/aatlas/latitude_and_longitude_finder.htm and http://itouchmap.com/latlong.html can be used to find the longitude and latitude of the site:

Latitude 37°47′, Longitude −122°23′, which could also be written as: 37.788°, −122.394°

The longitude and latitude are then entered into the USGS website at https://geohazards.usgs.gov/secure/designmaps/us/application.php to get the S_s and S_1 values interpolated for this site. The output values are

$S_s = 1.50g$	$S_1 = 0.607g$

Note that in near fault areas, the gradations on Figures 22-3 and 22-4 are so close together it is difficult to interpolate accurately. S_1 as determined from Figure 22-4 was 8 percent higher than the value obtained from the USGS website. For this example, the more accurate interpolation provided by the USGS website will be used. The site class is D, so the factors to modify the MCE spectral accelerations are

$F_a = 1.0$ T 11.4-1

$F_v = 1.5$ T 11.4-2

These values are used to modify the spectral accelerations

$S_{MS} = F_a S_s = 1.0 \times 1.50 = 1.50g$ Eq 11.4-1

$S_{M1} = F_v S_1 = 1.5 \times 0.067 = 0.911g$ Eq 11.4-2

The spectral accelerations to be used in design are

$$S_{DS} = \frac{2}{3} S_{MS} = \frac{2}{3} \times 1.50 = 1.00g$$ Eq 11.4-3

$$S_{D1} = \frac{2}{3} S_{M1} = \frac{2}{3} \times 0.911 = 0.607g$$ Eq 11.4-4

$S_{DS} = 1.00g$	$S_{D1} = 0.607g$

2c. Construct the design response spectrum

Determine the approximate fundamental building period using §12.8.2.1

$C_t = 0.02$ and $x = 0.75$ T 12.8-2

$T_a = C_t h_n^x = 0.02 \times 72^{0.75} = 0.49$ sec (See discussion below) Eq 12.8-7

$T = 0.49$ sec

$$T_o = 0.2\frac{S_{D1}}{S_{DS}} = 0.2\frac{0.607}{1.00} = 0.121 \text{ sec}$$

$$S_a = S_{DS}\left(0.4 + 0.6\frac{T}{T_o}\right) = 0.4 + 4.96T \text{ for } T < T_o$$ Eq 11.4-5

$$T_s = \frac{S_{D1}}{S_{DS}} = \frac{0.607}{1.00} = 0.607 \text{ sec}$$ §11.4.5

$$S_a = \frac{S_{D1}}{T} = \frac{0.607}{T} \text{ for } T > T_s$$ Eq 11.4-6

The long period equation for S_a does not apply here because the long period transition occurs at 12 sec (from Figure 22-16).

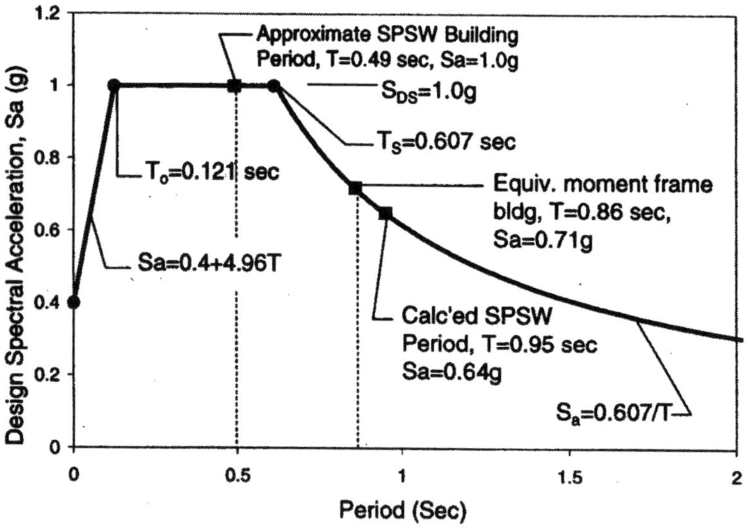

Figure 4-3. Design Response Spectrum for the example building

As shown in Figure 4-3, the design spectral acceleration is between T_o and T_s, so the design spectral acceleration S_a is 1.0g. It is not required to construct the design response spectrum when using the equivalent lateral force procedure, because the response spectrum is implicit in the calculation of C_s in §12.8.1.1.

The response spectrum demonstrates the effect of the assumptions used in the calculation of building period. Values of $C_t = 0.02$ and $x = 0.75$ were selected as specified for shear wall buildings, which result in an approximate period, $T_a = 0.49$ sec. As shown later in this example, drifts are close to, if not at, the story drift limits of §12.12.

The period of the structure can be established through structural analysis of the SPSW. Section 12.8.2, however, limits the period that can be used to calculate spectral acceleration to a value of $T_{max} = C_u \times T_a$, where C_u is a factor found in Table 12.8-1. In this case $T_{max} = 1.4 \times 0.49 = 0.69$ sec. It is recommended that this period, T_{max}, be used to design the SPSW and that the period of the building be verified later in the design process. The first mode period of the final design shown in Figure 4-16 was calculated to be 0.95 sec. Therefore, the initial assumption of $T = 0.69$ sec is shown to not only be valid, but still quite conservative.

2d. Determine seismic design category

Occupancy Category = II — IBC §1604.5

Importance Factor = 1.0 — T 11.5-1

Short Period Seismic Design Category = D (for $S_{DS} = 1.0g$) — T 11.6-1

1-sec Seismic Design Category = D (for $S_{D1} = 0.607g$) — T 11.6-2

According to §11.6, the seismic design category is the more severe of the results of Tables 11.6-1 and 11.6-2. Both resulted in Seismic Design Category D, so

Seismic Design Category = D — §11.6

2e. Examine the building for Horizontal Irregularities — T 12.3-1

1a. Torsional Irregularity—A torsional irregularity exists when the maximum story drift computed, including accidental torsion, is more than 1.2 times the average story drift. From the rigid diaphragm analysis later in this example:

$$\delta_{max} = 0.27F_i/K$$

$$\delta_{avg} = 0.25F_i/K$$

$$\frac{\delta_{max}}{\delta_{avg}} = \frac{0.27F_i/K}{0.25F_i/K} = 1.08 < 1.2 \rightarrow \text{NO TORSIONAL IRREGULARITY}$$

1b. to 5 By inspection, the building does not qualify for any of these horizontal structural irregularities.

NO HORIZONTAL STRUCTURAL IRREGULARITIES

2f. Examine the building for Vertical Irregularities — T 12.3-2

1a. to 5b. By inspection, the building does not qualify for any of the vertical structural irregularities.

NO VERTICAL STRUCTURAL IRREGULARITIES

2g. Select lateral force procedure — T 12.6-1

1. Simplified Alternative Structural Design Criteria According to §12.14.1.1 this analysis procedure can be used for SPSW, but not for buildings over 3 stories—NOT PERMITTED

2. Equivalent Lateral Force Analysis—According to Table 12.6-1, since $T < 3.5T_s$ (0.69 sec < 2.12 sec), and the building is regular and Occupancy Category II—PERMITTED

3. Modal Response Spectrum Analysis—PERMITTED

4. Seismic Response History Procedures—PERMITTED

USE EQUIVALENT LATERAL FORCE ANALYSIS

2h. Calculate the seismic response coefficient, C_s, and total base shear

$$C_s = \frac{S_{DS}}{\left(\frac{R}{I}\right)} = \frac{1.00}{\left(\frac{7.0}{1.0}\right)} = 0.14 \le \left[\frac{S_{D1}}{T\left(\frac{R}{I}\right)} = \frac{0.607}{0.69\left(\frac{7.0}{1.0}\right)} = 0.126\right]$$ Eqs 12.8-2 and 12.8-3

Also,

$$C_s \ge 0.01, \text{and } C_s \ge \frac{0.5S_1}{\left(\frac{R}{I}\right)} = \frac{0.5 \times 0.607}{\left(\frac{7.0}{1.0}\right)} = 0.043$$ Eq 12.8-6

$$C_s > 0.044S_{DS}I = 0.044(1.0)g(1.0) = 0.044g$$ Eq 12.8-5 Supplement 2

$$\boxed{C_s = 0.126}$$

$$V = C_s W = 0.126 \times 8048 = 1014 \text{ kips}$$ Eq 12.8-1

$$\boxed{V = 1014 \text{ kips}}$$

2i. Check wind

This is a simple check to verify that wind doesn't control over seismic loads. If the seismic base shear (multiplied by 0.7 as done in ASD load combinations) is divided by the area of the larger face of the building, an equivalent wind pressure can be calculated

$$P_{\text{equiv}} = \frac{0.7V}{\text{width} \times \text{height}} = \frac{710}{150 \times 72} = 66 \text{ psf}$$

It is safe to assume that the average wind pressure will be well below 66 psf. As expected, wind does not control the lateral loads.

2j. Calculate the redundancy factor

According to §12.3.4 the redundancy factor should be calculated for each principal axis. The redundancy factor is 1.3 unless either §12.3.4.2(a) or §12.3.4.2(b) is shown to be true in which case the Redundancy Factor can be taken as 1.0. Section 12.3.4.2(b) requires that there be two bays of SPSW on each perimeter line. The Redundancy Factor is, therefore, 1.0 for both the East-West and North-South directions.

$\rho = 1.0$ for East-West Direction

$\rho = 1.0$ for North-South Direction

2k. Load combinations

$E = E_h \pm E_v$ — Eq 12.4-1 and Eq 12.4-2

$E_h = \rho Q_E = 1.0Q_E$ — Eq 12.4-3

$E_v = 0.2S_{DS}D = 0.2 \times 1.0D = 0.2D$, therefore: — Eq 12.4-4

$E = 1.0Q_E \pm 0.2\text{D}$

Load combinations of consequence are — §2.3.2

$1.4D$ — Load Comb. 1

$1.2D + 1.6L + 0.5L_r$ — Load Comb. 2

$1.2D + 1.0E + 0.5L = 1.4D + 0.5L + 1.0Q_E$ — Load Comb. 5 (modified)

$0.9D + 1.0E = 0.7D + 1.0Q_E$ — Load Comb. 7 (modified)

3. Vertical and horizontal distribution of load §12.8.3

3a. Vertical distribution of shear

Table 4-3. Vertical Distribution of Shear

Level	w_i kips	h_i (ft)	$w_i h_i^k$	C_{vx}	F_x (kips)
Roof	648	72	70,041	0.16	158
6th	1480	60	131,020	0.29	296
5th	1480	48	102,617	0.23	232
4th	1480	36	74,888	0.17	169
3rd	1480	24	48,039	0.11	108
2nd	1480	12	22,489	0.05	51
Total	9576		449,094		1014

The terms used in this table are defined in Section 12.8.3. Since the period = 0.69 > 0.5 sec, the value for k is interpolated between a value of 1.0 for $T = 0.5$ sec and 2.0 for $T = 2.5$ sec. In this example, $k = 1.095$. The distribution of story shear is carried out using

$$F_x = C_{vx}V, \text{ where, } C_{vx} = \frac{w_x h_x^k}{\sum_{i=1}^{n} w_i h_i^k}$$ Eq 12.8-11 and Eq 12.8-12

3b. Horizontal distribution of story shear

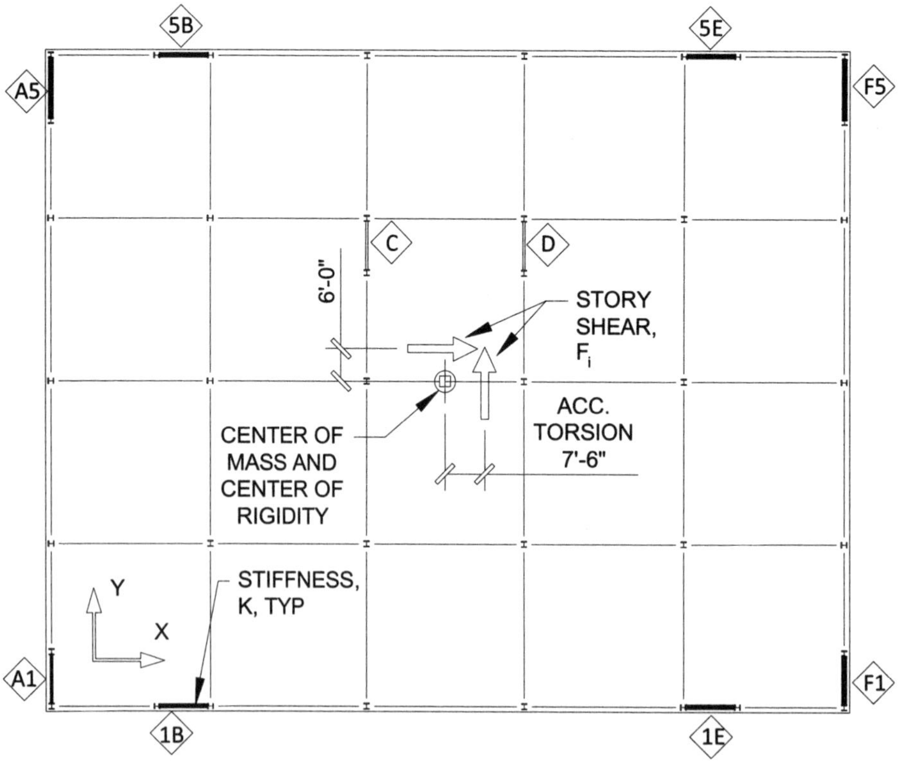

Figure 4-4. Rigid diaphragm analysis to distribute story shear to SPSW

As shown in Figure 4-4, the center of mass and center of rigidity coincide at the middle of the building. To distribute the load to each SPSW, the story shear, F_i, is applied in the X and Y directions. Per §12.8.4.2, the point of application of the story shear is offset 5 percent to account for accidental eccentricity. Since all of the SPSW are the same, a generic stiffness, K, is used for all walls.

Table 4-4. Rigid diaphragm analysis to distribute shear to SPSW

Wall	x	y	$R_y x$	$R_x y$	Rd^2	X-Direction			Y-Direction		
						V_{direct}	V_{tors}	V_{total}	V_{direct}	V_{tors}	V_{total}
A1	–75		–75 K		5625 K	0.25 F_i	–0.012 F_i	0.24 F_i	0	–0.015 F_i	–0.015 F_i
A5	–75		–75 K		5625 K	0.25 F_i	–0.012 F_i	0.24 F_i	0	–0.015 F_i	–0.015 F_i
F1	75		75 K		5625 K	0.25 F_i	0.012 F_i	0.26 F_i	0	0.015 F_i	0.015 F_i
F5	75		75 K		5625 K	0.25 F_i	0.012 F_i	0.26 F_i	0	0.015 F_i	0.015 F_i
1B		–60		–60 K	3600 K	0	–0.01 F_i	–0.01 F_i	0.25 F_i	–0.012 F_i	0.24 F_i
1E		–60		–60 K	3600 K	0	–0.01 F_i	–0.01 F_i	0.25 F_i	–0.012 F_i	0.24 F_i
5B		60		60 K	3600 K	0	0.01 F_i	0.01 F_i	0.25 F_i	0.012 F_i	0.26 F_i
5E		60		60 K	3600 K	0	0.01 F_i	0.01 F_i	0.25 F_i	0.012 F_i	0.26 F_i

$\Sigma Rd^2 = 36{,}900\ K$

The following calculations find the force in SPSW 5E for shear in the Y-Direction.

$$V_{direct} = F_i \frac{R_y}{\sum R_y} = 0.25F_i$$

$$M \quad = F_i \times (0.05 \times 150) = 7.5F_i$$

$$V_{tors} \quad = \frac{M \times R_x y}{\sum Rd^2} = \frac{7.5F_i \times 60K}{36{,}900K} = 0.012F_i$$

$$V_{total} \quad = V_{direct} + V_{tors} = 0.25F_i + 0.012F_i = 0.26F_i$$

From the above table it is shown that the maximum design shear for any wall is 0.26 times the story shear. The rest of this example focuses on the design of SPSW 5E.

Table 4-5. Story forces applied to SPSW 5E

Level	Incremental Shear	Cumulative Shear
Roof	41 kips	41 kips
6th	77	118
5th	60	178
4th	44	222
3rd	28	250
2nd	13	263

4. Preliminary sizing of the web plate and boundary members

4a. SPSW layout

One of the defining characteristics of steel plate shear walls is the significant amount of strength that can be developed in relatively thin plate in a short length of wall. The magnitude of overturning forces or deflection will often control the SPSW design. To demonstrate this fact, consider using only two 12 ft, 0-in-long special plate shear walls in each direction. Using the method described later, the plate sizes given in Table 4-6 can be shown to satisfy strength requirements, even considering the redundancy factor, $\rho = 1.3$.

Table 4-6. Example of web plate thicknesses for only two 12-ft shear walls in each direction

Level	Plate Material	Plate Thickness (inches)	Shear Capacity, ϕV_n
6th	A1011	⅛	187
5th	A36	¼	407
4th	A36	5⁄16	509
3rd	A36	7⁄16	713
2nd	A36	½	815
1st	A36	½	815 kips

Although structural engineers may be more used to specifying plate thicknesses like these, rather than those given in Table 4-7, the resulting overturning and deflection become unmanageable. Even the largest W14 section is not sufficient to resist the VBE overturning and tension field forces. The uplift forces are likewise large, and even though deflection was not checked for this scenario, it would likely be a problem. For these reasons, and because only two lateral resisting elements is not desirable from a redundancy standpoint, four walls in each direction were used.

Even with four SPSW in each direction, note how little wall length is required: four 12-ft wall segments in each direction = 48 ft versus a comparable braced frame design which might use four 30-ft bays in each direction = 120 ft. Also, it has been shown that SPSW require less steel than moment frames (Eatherton, 2006). SPSW, therefore, are especially well suited for open floor plans with maximum programming flexibility and few permanent solid walls to hide lateral resisting elements. In these cases, the SPSW system is a cheaper alternative to moment frames.

4b. SPSW configuration

AISC 341 limits the aspect ratio of SPSW panels to be representative of panels that have been tested

$$0.8 < \frac{L}{h} \le 2.5, \text{ for this example, } \frac{L}{h} = \frac{12}{12} = 1.0 \rightarrow \text{OK}$$ AISC 341 §17.2b

L = distance between VBE centerlines

h = distance between HBE centerlines

4c. Preliminary web plate design

Shear capacity of a SPSW panel is given by

$$V_n = 0.42 F_y t_w L_{cf} \sin(2\alpha) \text{ with } \phi = 0.9 \qquad \text{AISC Eq 17-1}$$

where:

F_y and t_w are the yield strength and thickness of the plate, respectively

L_{cf} = clear distance between VBE = 12 ft × 12 in/ft −17 in = 127 in (assuming d_c = 17 in)

α = angle of inclination of tension field relative to vertical given by the equation

$$\tan^4\alpha = \frac{1 + \dfrac{t_w L}{2A_c}}{1 + t_w h\left(\dfrac{1}{A_b} + \dfrac{h^3}{360 I_c L}\right)} \qquad \text{AISC Eq 17-2}$$

Definitions for these terms are given in AISC §17.2a. The equation for α is a function of the vertical and horizontal boundary member sizes, but since the VBEs and HBEs depend on the plate thickness, an initial guess for α is necessary. α usually varies between 35° and 45°, so 40° is an adequate preliminary approximation. Although α will be verified later, the web plate design is based on $\sin(2\alpha)$, and thus is not so sensitive to α that iterations are required. The components of the tension field force and the corresponding boundary member designs and connections are more sensitive since they are based on factors of $\cos^2\alpha$ and $\sin^2\alpha$. Iterations in plate thickness may, however, be required to limit story drifts.

Size the web plate at the 1st floor. Try ³⁄₁₆-in-thick A36 plate.

$V_u = 263$ kips — From T 4-5 in part 3b

$\phi V_n = 0.9 \times 0.42 \times 36 \times 3/16 \times 127 \sin(2 \times 40) = 319 \text{ kips} \rightarrow \text{OK}$

$\phi V_n = 319$ kips

The rest of the web plate designs are tabulated in Table 4-7.

Table 4-7. Preliminary web plate design at each level

Level	Plate Mat'l	Yield Stress, F_y (ksi)	Plate Thick.	Design Thick., t_w (inches)	R_y[2, 3]	Clear Dist., L_{cf} (inches)	Shear, V_u (kips)	Shear Capacity, ϕV_n (kips)	Demand to Capacity Ratio
6th	A1011[1]	33	14 ga	0.075	1.1	127	41	117	0.35
5th	A1011[1]	33	12 ga	0.105	1.1	127	118	164	0.72
4th	A1011[1]	33	10 ga	0.134	1.1	127	178	209	0.85
3rd	A1011[1]	33	7 ga	0.179	1.1	127	222	279	0.80
2nd	A36	36	3⁄16	0.1875	1.3	127	250	319	0.78
1st	A36	36	3⁄16	0.1875	1.3	127	264	319	0.82

1. For sheet steel, AISC 341 section 6.1 allows A36 or A1011 HSLAS Gr. 55. However, lower yield strength sheet steel is specified here to keep the plate from getting too thin. See the discussion about plate thickness and material for more information.
2. Used later in the design example.
3. R_y is the factor applied to the design yield stress to estimate actual material yield stress. AISC 341 Table I-6-1 gives values for R_y for various materials, including A1011 HSLAS grade 55. A1011 grade 33 is not included in Table I-6-1, but the actual yield stress can be verified. See the section on sheet steel availability for recommendations.

4d. Note about plate thickness and material

At the upper floors, even 3⁄16-in A36 plate provides more capacity than is needed. AISC 341 does not limit plate slenderness or plate thickness. Instead, the commentary §C17.2b states that story drift limitations will indirectly limit the plate slenderness. This will be exhibited to some degree later in this example, but the question remains as to how thin a plate is appropriate. 12-gage and 14-gage plate have been used successfully in SPSW projects (Eatherton, 2006). In research, SPSW utilizing 20-gage (Berman 2005), and even 22-gage sheet steel (Elgaaly 1997) have been successfully tested and have demonstrated favorable response. For the purposes of this example, 14-gage plate was used as a minimum to avoid difficulty with welding very thin web plates.

Another consideration when choosing plate thickness is the material specification. Section 6.1 of the AISC 341 limits the materials that can be used in the seismic load resisting system. Listed plate material includes A36 and A1011 Grade 55 (in either SS or HSLAS). A1011 is hot rolled sheet steel, and as such should have more pronounced yielding and more ductility than cold rolled sheet metal (A1008). It would be preferable, however, to use A1011 Grade 30 or 33 sheet steel as it has a lower yield and more elongation. The AISC 341 commentary §C17.2 suggests that special highly ductile low-yield steel having specified minimum yield in the range of 12 to 33 ksi can be used. A1011 Grade 33 was selected for the sheet steel web plates in this design example.

4e. Note about sheet steel availability

Manufacturers such as U.S. Steel and Nucor are capable of producing a wide range of sheet thicknesses in many different grades of A1011. Typically, however, material ordered direct from the steel mills has to be in quantities on the order of 50 metric tons. Steel suppliers buy these large quantities of sheet steel and sell it in smaller amounts to the fabricators. Availability will vary, but a survey of three suppliers in northern California found that 7-gage to 16-gage steel sheet was stocked in coils up to 72 inches wide. This sheet steel is available in A1011 (or called by the old designation of A569), but is almost exclusively commercial grade (A1011 CS). Most clients don't require structural sheet steel (SS, or HSLAS), so the suppliers did not pay the cost premium that comes with having the material further graded. Because the mill certification comes with chemical composition, though, a supplier can select commercial grade A1011 sheet steel that is likely to have a certain minimum yield stress.

When specifying A1011 sheet steel, there are notes that could be added to the construction documents to prevent the web plate from becoming an expensive special order. Regardless of which grade is specified (between 33 ksi and 55 ksi may be the most realistic range), if mill certification demonstrating the specified grade is not available, as would be the case for A1011 CS, then relatively inexpensive coupon tests can be required to verify the yield stress and elongation. The results of the coupon tests could be used as acceptability criteria for the web plate, and may save money by allowing the use of more readily available material. For this example, A1011 Grade 33 is specified. If mill certifications demonstrating Grade 33 are not available, several coupon tests per coil will be required to verify that the yield stress is between 33 ksi and 37 ksi (R_yF_y = 37 ksi), and the elongation is at least 20 percent.

4f. Discussion of SPSW shear capacity equation

The configuration shown in Figure 4-5 represents an intermediate floor of a multistory SPSW frame. Thorburn and Kulak simplified the tension field into a series of inclined tension strips (Thorburn et al 1983). Using the tension strip model, the shear, V_y, that causes the entire plate to yield along its base can be derived $V_y = \frac{1}{2}F_y t_w L_{cf} \sin(2\alpha)$. The derivation of this equation is based on the shear stress, F_{12}, on a horizontal section as illustrated in Figure 4-7. AISC Equation 17-1 for design shear strength of a SPSW panel is identical, with a factor of 1.2 applied to represent the ratio of full yield to first yield.

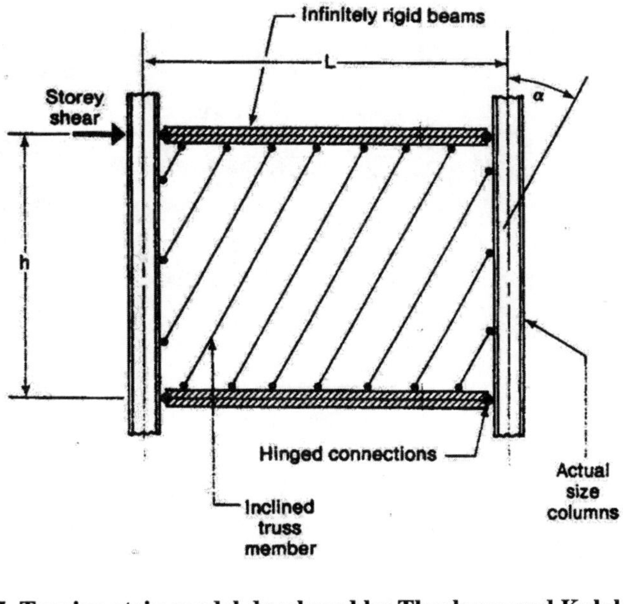

Figure 4-5. Tension strip model developed by Thorburn and Kulak (1983).

As shown in Figure 4-5, the beam-to-column connections are considered as pinned in the derivation of shear strength. For the SPSW system not using a dual system, the moment connections are added to achieve higher ductility and energy absorption, rather than as a means to increase strength.

4g. HBE design

For boundary-element design, AISC 341 §17.4a simply states that boundary members are to be designed for the forces corresponding to the expected yield strength, in tension, of the web calculated at an angle α. The definition of the expected yield strength consists of multiplying F_y by R_y to get the expected yield stress. The values of R_y for different materials are given in Table I-6-1 and are tabulated for this design example in Table 4-7. Note that R_y for A36 has been increased in recent editions of AISC 341 to 1.3 (for A36 plate), whereas A1011 Grade 55 is assumed to have $R_y = 1.1$. Boundary elements have to be designed, therefore, for approximately 18 percent more load when using A36 web plate over a comparable A1011 plate.

Figure 4-6 shows the application of the expected yield strength of the plate at angle α

$$F = F_y R_y t_w \rightarrow \text{Expected yield strength at angle } \alpha$$

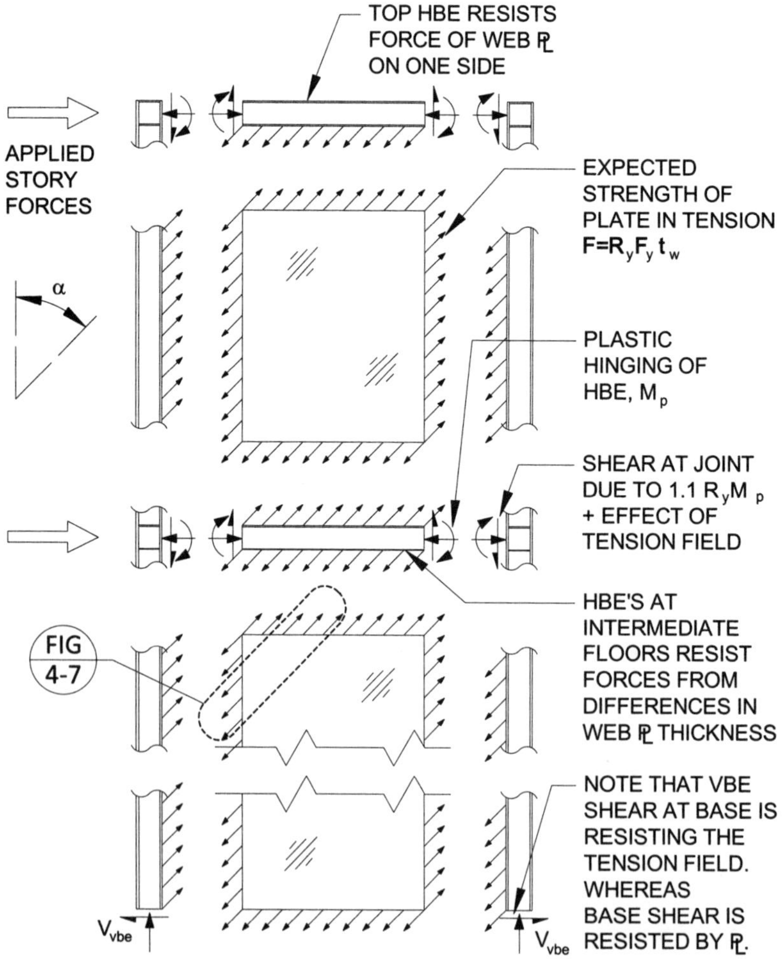

Figure 4-6. SPSW divided into parts to show forces

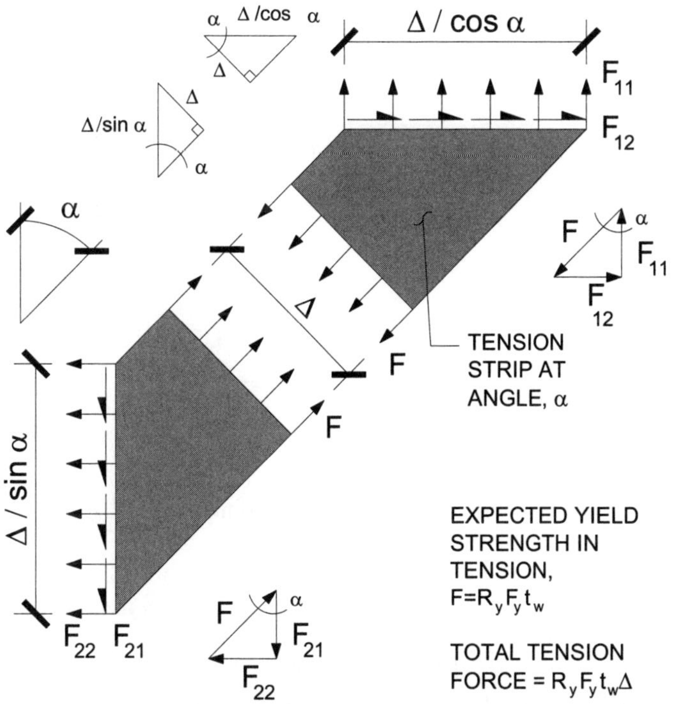

Figure 4-7. Components of expected web strength applied to HBE and VBE (based on Sabelli 2006)

For the design of the HBE, however, it is necessary to separate the force F into components. Figure 4-7 shows a representative strip of web plate acting at an angle α. Sabelli and Bruneau (2006) present a derivation of the resulting horizontal and vertical forces acting on the HBE and VBE

$$F_{11} = \frac{\text{Force}}{\text{Length}} = \frac{R_y F_y t_w \Delta \cos\alpha}{\Delta / \cos\alpha} = R_y F_y t_w \cos^2\alpha$$

$$F_{12} = \frac{R_y F_y t_w \Delta \sin\alpha}{\Delta / \cos\alpha} = R_y F_y t_w \sin\alpha\cos\alpha = R_y F_y t_w \left[\frac{1}{2}\sin(2\alpha)\right]$$

$$F_{21} = \frac{R_y F_y t_w \Delta \cos\alpha}{\Delta / \sin\alpha} = R_y F_y t_w \sin\alpha\cos\alpha = R_y F_y t_w \left[\frac{1}{2}\sin(2\alpha)\right]$$

$$F_{22} = \frac{R_y F_y t_w \Delta \sin\alpha}{\Delta / \sin\alpha} = R_y F_y t_w \sin^2\alpha$$

For the 6th floor, the forces F_{11} and F_{12} are calculated below. All tension field components are given in Table 4-8.

$$F_{11\text{-}6} = F_{y6} R_{y6} t_{w6} \cos^2\alpha_6 = 33 \times 1.1 \times 0.075 \cos^2(40) = 1.60 \text{ kip/in}$$

$$F_{12\text{-}6} = F_{y6} R_{y6} t_{w6} \left[\frac{1}{2}\sin(2\alpha_6)\right] = 33 \times 1.1 \times 0.075[0.5\sin(2 \times 40)] = 1.34 \text{ kip/in}$$

Table 4-8 Components of the expected strength of the tension field

Level	F_{11} (k/in)	F_{12} (k/in)	F_{22} (k/in)	F_{21} (k/in)
6th	1.60	1.34	1.12	1.34
5th	2.24	1.88	1.57	1.88
4th	2.85	2.40	2.01	2.40
3rd	3.81	3.20	2.68	3.20
2nd	5.15	4.32	3.63	4.32
1st	5.15	4.32	3.63	4.32

The HBE is designed for the vertical force, F_{11}, and the shear force, F_{12}. Figure 4-8 shows the forces acting on the roof HBE and the 6th floor HBE. Note the following:

- The roof HBE resists the force of the tension field on one side only. In some cases this will result in a heavier top HBE than in floors below.
- At intermediate floors, if the plate is the same thickness above and below, the HBE does not resist vertical forces other than gravity loads. When the plates are of different thickness above and below, the HBE is designed for the forces associated with the difference.
- The HBE is not required to be designed for end moments, unless using a dual system where the moment frame is to be designed to resist 25 percent of the load.

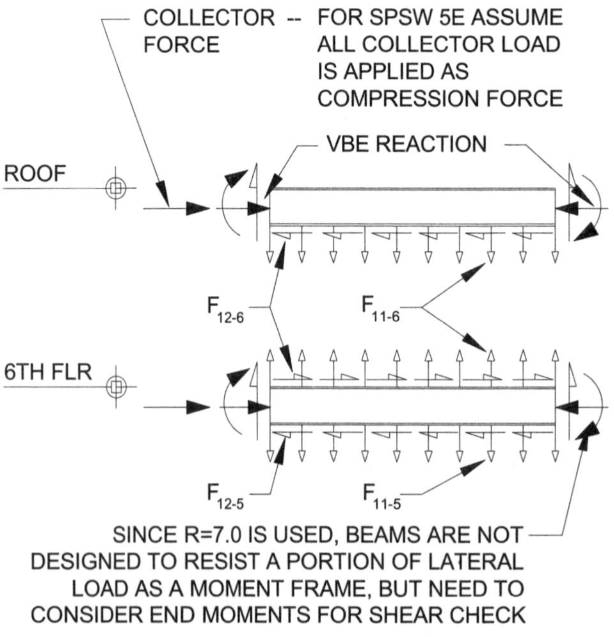

Figure 4-8. HBE design forces

4h. Find the roof HBE moment

Assume that the HBE will be braced at mid-span with a beam. Find the moment on the roof HBE for SPSW 5E

$$\begin{Bmatrix} w_{D1} \\ w_L \end{Bmatrix} = 6\text{ ft}\begin{Bmatrix} 25\text{ psf } D \\ 20\text{ psf } L_r \end{Bmatrix} = \begin{matrix} 0.15\text{ k/ft } D \\ 0.12\text{ k/ft } L_r \end{matrix}$$ Line Loads

$w_{D2} = (6\text{-ft wall} + 4\text{-ft parapet}) \times 20 = 0.20\text{ kip/ft } D$ Uniform load due to exterior wall

$F_{11\text{-}6} = 1.60\text{ k/in} \times 12\text{ in/ft} = 19.2\text{ kip/ft } Q_E$

Using load combination number 5:

$$M_u = 1.4\left(\frac{(w_{D1} + w_{D1})L_{cf}^2}{4}\right) + 1.0\left(\frac{F_{11\text{-}6}L_{cf}^2}{4}\right)$$ Load Comb. 5 (modified)

This equation is based on work by Bruneau (2011) that shows that HBEs subjected to end moments. Uniform load should be designed for a maximum moment of $w_L{}^2/4$ to prevent in-span plastic hinging.

$$M_u = 1.4\left(\frac{(0.15 + 0.20)(127/12)^2}{4}\right) + 1.0\left(\frac{19.2 \times (127/12)^2}{4}\right)$$

$$M_u = 13.7 + 537.6 = 551\text{ kip-ft}$$

$$\boxed{M_u = 551\text{ kip-ft}}$$

4i. Find the roof HBE compression force

Next, find the compression force in the roof HBE for SPSW 5E. The roof HBE resists two components of compression:

1. Compression from the story shear collected to the wall, P_{12} (for SPSW 5E, the collector forces are delivered mainly from one side)
2. Horizontal reaction from the VBE that resists the inward pull of the tension field, P_{22}

The compressive load from collecting load to the SPSW is limited by the shear that can be developed at the top of the panel:

$$P_{12} = F_{12\text{-}6}L_{cf} = 1.34 \times 127 = 170\text{ kips}$$

Compare this load to the collector force §12.10.1.1

$$F_{px\text{-roof}} = \frac{\sum_{i=x}^{n} F_i}{\sum_{i=x}^{n} w_i} w_{px} = 158 \text{ kips}$$

$$0.2S_{DS}Iw_{px} \le F_{px\text{-roof}} \le 0.4S_{DS}Iw_{px}$$

$$130 \text{ kips} \le F_{px\text{-roof}} \le 260 \text{ kips}$$

$$F_{px\text{-roof}} = 158 \text{ kips}$$

$$P_{\text{collector}} = \Omega_o \frac{F_{px\text{-root}}}{4 \text{ frames}} = \frac{2.0 \times 158}{4} = 79 \text{ kips}$$

The design collector load, $P_{\text{collector}}$, is less than the force required to fully yield the web plate (if the moment connections are ignored), P_{12}. The intent of the AISC 341 provisions is clear, in that the boundary elements are to be able to resist the effects of the expected yield strength of the web plate. The HBE should, therefore, be designed for the larger of $P_{\text{collector}}$ or P_{12} plus the horizontal reaction from the VBE calculated below.

The horizontal reaction from the VBE is the load required to resist the component, F_{22}, for the height of the column tributary to the top beam. This force accounts for the tendency of the columns to bow inward from tension field action. The HBEs brace the column against this inward pull.

$F_{22\text{-}6} = 1.12$ kip/in from Table 4-8

$$P_{22} = F_{22}\frac{h - d_b}{2} = 1.12 \times \frac{12 \times 12 - 18}{2} = 71 \text{ kips (assuming a W18 beam)}$$

$$P_u = P_{12} + P_{22} = 241 \text{ kips}$$

$$\boxed{P_u = 241 \text{ kips}}$$

4j. Choose a section and check width-thickness ratio for the roof HBE

W18 sections are used for the preliminary HBE sections in this example. Choose a W18×86. For this example it is desired to only have one point of lateral bracing between the columns. Check the maximum spacing for lateral bracing this section.

$$L_{b\text{-max}} = 0.086 r_y \frac{E}{F_y}$$

$$L_{b\text{-max}} = 0.086 \times 2.63 \frac{29{,}000}{50} = 131 \text{ in} = 10 \text{ ft-11 in}$$ §17.4d

The HBE is laterally braced at 6 ft, 0 in → OK

$$\boxed{\text{HBE lateral bracing is OK}}$$

Because it is necessary for the HBEs to develop plastic hinges for the full mechanism to occur, AISC 341 §17.4c requires that the HBEs meet the width-thickness ratios in Table I-8-1

$\lambda_f \le (0.3\sqrt{E/F_y} = 7.22)$ $\quad \lambda_f = 7.2 \rightarrow \text{OK}$ T I-8-1

$C_a = P_u/\phi P_n = 241/1078 = 0.22$ T I-8-1

$\lambda_w \le 1.12\sqrt{E/F_y}(2.33 - C_a) = 56.9$ $\quad \lambda_w = 33.4 \rightarrow \text{OK}$ T I-8-1

Beam width-thickness ratios are OK

4k. Check combined flexure and axial for the roof HBE

For a W18×86 roof HBE

$$\frac{KL}{r} = \frac{1.0 \times (6\text{ ft} \times 12\text{ in/ft})}{2.63} = 27 \text{ for weak axis buckling}$$

$$\frac{KL}{r_x} = \frac{1.0 \times (12\text{ ft} \times 12\text{ in/ft})}{7.77} = 19 \text{ for strong axis buckling}$$

$$F_e = \frac{\pi^2 E}{(KL/r)^2} = 393 \text{ ksi}$$ Eq E3-4

Limit on inelastic buckling, $4.71\sqrt{\dfrac{E}{F_y}} = 113$

Because $\dfrac{KL}{r} < 4.71\sqrt{\dfrac{E}{F_y}}$, $\quad F_{cr} = \left[0.658^{F_y/F_e}\right]F_y = 47.4 \text{ ksi}$ Eq E3-2

$\phi P_n = 0.90 \times F_{cr} A_g = 0.90 \times 47.4 \times 25.3 = 1079 \text{ kips}$ Eq E3-1

$\phi P_n = 1079 \text{ kips}$

$L_p = 9.3 \text{ ft}$ $\quad L_r = 28.6 \text{ ft}$ Eq F2-5 and Eq F2-6
(can also be found in AISC 13th Table 3-2)

$L_b = 6$ ft-0 in, so L_b is between L_p and L_r

$M_n = M_p = F_y Z_x = 50 \text{ ksi } (186 \text{ in}^3)/12 \text{ in/ft} = 775 \text{ kip-ft}$

$\phi M_n = 0.9 \times 775 \text{ kip-ft} = 698 \text{ kip-ft}$

$\phi M_n = 698 \text{ kip-ft}$

Since $P_r/P_c > 0.2$

$$\frac{P_r}{P_c} + \frac{8}{9}\left(\frac{M_{rx}}{M_{cx}} + \frac{M_{ry}}{M_{cy}}\right) \leq 1.0 \qquad \text{Eq H1-1a}$$

$$\frac{P_u}{\phi P_n} + \frac{8}{9}\left(\frac{M_u}{\phi M_n}\right) = \frac{241}{1079} + \frac{8(551)}{(9 \times 698)} = 0.93 \rightarrow \text{OK}$$

Combined axial and flexure = 0.93

4l. Check shear for the 6th floor HBE

The shear at the face of the column is the sum of the shear associated with plastic hinging in the beam and the effect of the expected yield strength of the tension field. The following calculations show that the factored shear in the 6th floor beam is less than the factored nominal shear capacity. The beam is a W18×65.

$$V_{u-6} = \frac{2 \times 1.1 R_y M_p}{L_{cf}} + (F_{11-5} - F_{11-6})\frac{L_{cf}}{2} = \frac{2 \times 1.1 \times 1.1 \times 554}{\frac{127}{12}} + (2.24 - 1.60)\frac{127}{2}$$

$$V_{u-6} = 127 + 41 = 168 \text{ kips}$$

For the W18×65:

$$\frac{h}{t_w} = 35.7 < 2.24\sqrt{\frac{E}{F_y}} = 53.9 \text{ so:} \qquad \text{Section G2.1}$$

$$\phi V_n = \phi_v 0.6 F_y A_w C_v = 1.0 \times 0.6 \times 50 \times 18.35 \times 0.45 \times 1.0 = 248 \text{ kips} \rightarrow \text{OK} \qquad \text{Eq G2-1}$$

$V_{u-6} = 168$ kips $< \phi V_n = 248$ kips

4m. Find the 6th-floor HBE moment

The design of the other HBEs is similar to the design of the roof HBE. The differences for HBEs at intermediate floor levels are discussed in this section and then a table giving design information for the other HBEs follows.

There are vertical forces associated with tension field acting above and below the 6th floor HBE, as shown in Figure 4-8. The net vertical force due to the tension field is, therefore, the difference in the effects of tension field above and below.

$F_{11\text{-net}} = F_{11-5} - F_{11-6} = 2.24 - 1.60 = 0.64$ kip/in $= 7.68$ kip/ft (See Table 4-8 for F_{11})

$$\begin{Bmatrix} w_{D1} \\ w_L \end{Bmatrix} = 6\text{ ft}\begin{Bmatrix} 65\text{ psf } D \\ 65\text{ psf } L \end{Bmatrix} = \begin{matrix} 0.39\text{ k/ft } D \\ 0.39\text{ k/ft } L \end{matrix} \qquad \text{Line Loads}$$

$w_{D2} = 12\text{ ft} \times 20 = 0.24$ kip/ft D Uniform load due to exterior wall

Using load combination number 5:

$$M_u = 1.4\left(\frac{(w_{D1}+w_{D2})L_{cf}^2}{4}\right)+0.5\left(\frac{w_L L_{cf}^2}{4}\right)+1.0\left(\frac{F_{11\text{-net}}L_{cf}^2}{4}\right) \qquad \text{Load Comb. 5 (modified)}$$

$$M_u = 1.4\left(\frac{(0.39+0.24)\times(127/12)^2}{4}\right)+0.5\left(\frac{0.39\times(127/12)^2}{4}\right)+1.0\left(\frac{7.68\times(127/12)^2}{4}\right)$$

$$M_u = 24.7 + 5.5 + 215 = 245 \text{ kip-ft}$$

$$\boxed{M_u = 245 \text{ kip-ft}}$$

4n. Find the 6th-floor HBE compression force

The compression force in the 6th-floor HBE consists of two components: 1) The portion of the 6th-floor story shear collected to the wall and 2) The horizontal reaction due to bracing the VBE against the inward pull of the tension field. For the purposes of calculating the 6th-floor HBE compression, the first component is taken as the larger of the diaphragm collector force, $P_{collector}$, or the horizontal force developed due to the difference in the 5th- and 6th-floor web plate thicknesses, P_{12}, assuming both the 5th- and 6th-floor web plates are fully yielded.

$$F_{12\text{-net}} = F_{12\text{-}5} - F_{12\text{-}6} = 1.88 - 1.34 = 0.54 \text{ kip/in} \quad (\text{See T 4-8 for } F_{12})$$

$$P_{12} = F_{12\text{-net}}L_{cf} = 0.54 \times 127 = 69 \text{ kips}$$

$$F_{px\text{-}6} = \frac{\sum_{i=x}^{n} F_i}{\sum_{i=x}^{n} w_i} w_{px} = 316 \text{ kips} \qquad \text{Eq 12.10-1}$$

$$F_{px\text{-}6} \geq 0.2S_{DS}Iw_{px} = 0.2 \times 1.0 \times 1.0 \times 1480 = 296 \text{ kips}$$

$$F_{px\text{-}6} = 316 \text{ kips}$$

$$P_{collector} = \Omega_o \frac{F_{px\text{-roof}}}{4 \text{ frames}} = \frac{2.0\times 316}{4} = 158 \text{ kips}$$

$$F_{22\text{-}6} = 1.12 \text{ kip/in} \quad (\text{See T 4-8 for } F_{22})$$

$$F_{22\text{-}5} = 1.57 \text{ kip/in}$$

$$P_{22} = F_{22\text{-}5}\frac{h-d_b}{2} + F_{22\text{-}6}\frac{h-d_b}{2} = 169 \text{ kips}$$

$$P_u = \max(P_{12}, P_{collector}) + P_{22} = 158 + 170 = 327 \text{ kips}$$

$$\boxed{P_u = 327 \text{ kip-ft}}$$

4o. Size the remaining HBEs

A W18×65 is chosen for the 6th-floor HBE, and can be shown to satisfy width-thickness limitations and combined axial/flexure interaction requirements. Table 4-9 gives the preliminary sizes for the HBEs with key design information.

Table 4-9. Summary of preliminary HBE designs

Level	VBE Reaction (kips)	Collector Force (kips)	P_u (kips)	M_u (k-ft)	Size	Lateral Bracing, L_{bmax} (ft)	ϕP_n (kips)	ϕM_n (k-ft)	Combined Axial and Flexure[1]
Roof	71	170	241	551	W18×86	10.9	1078	698	0.93
6th	170	158	328	245	W18×65	7.0	753	499	0.87
5th	226	148	373	238	W18×65	7.0	753	499	0.92
4th	295	148	443	352	W18×86	10.9	1078	698	0.86
3rd	396	148	544	479	W18×97	11.0	1215	791	0.99
2nd	456	148	604	30	W18×60	7.0	694	461	0.93

Note: 1. Using Eq H1-1a of AISC 360.

4p. VBE design

Section 17 of AISC 341 has several requirements for the VBE. Similar to HBEs, the VBE should be designed to resist the forces associated with the expected yield strength of the plate, $F_yR_yt_w$, acting at an angle α (still assumed to be 40°). The difference for the VBE is that it is subjected to six floors of web plates developing their expected yield strength. This is not dissimilar to the eccentrically braced frame or buckling restrained braced frames that require the columns to be designed for the loads associated with each floor's energy dissipating element at its capacity. Unlike those systems, however, columns in SPSW are required to resist considerable moment due to the tension field pulling inward, in addition to the overturning forces. In this sense, the columns are doing more work in SPSW than in other systems and large structural sections are often required.

For VBE design, the moment and compression forces are separated and examined individually and then interaction is considered in accordance with Chapter H of AISC 360. The design checks for the VBE include the following.

1. Combined axial and flexure considering
   - Moment due to expected yield strength of the web plate acting at the angle α in addition to plastic hinging of the HBE
   - Compression / tension due to overturning loads
2. Width-thickness ratio limits
3. Minimum moment of inertia requirement
4. Strong-column / weak-beam check

4q. Find the moment in the VBE

Figure 4-9 shows the forces that are transmitted to the VBE by the web plate and HBEs. Moment in the VBE results from the horizontal component of the web plate expected strength, F_{22}, and plastic hinging of the HBE, M_p. Figure 4-9 shows these forces applied to the VBE. The HBE and base connection are modeled as pinned supports.

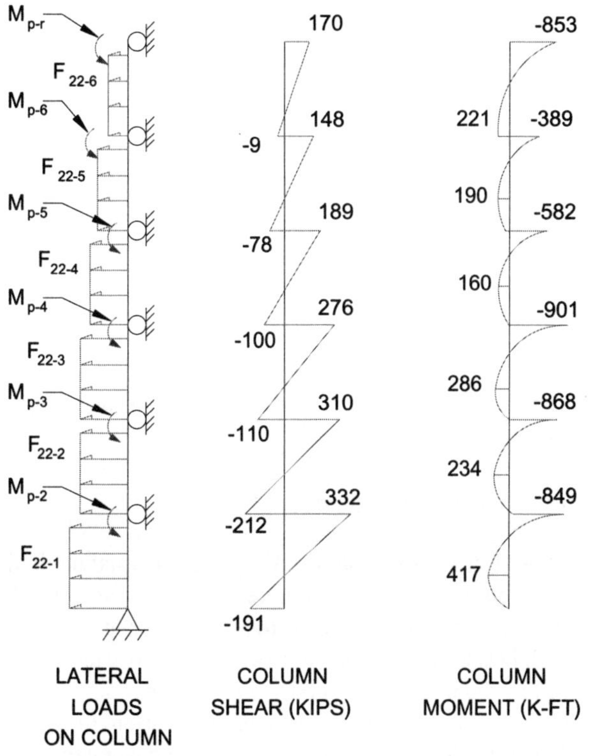

Figure 4-9. Moment on the VBE (method from Sabelli and Bruneau 2006)

The moment distribution can be found using a simple 2d model of a continuous beam over multiple supports. The forces applied at the 6th floor and roof, are given by:

$$F_{22\text{–}6} = 1.12 \text{ kip/in} \quad \text{(See Table 4-8 for } F_{22}\text{)}$$

$$M_{P\text{–roof}} = R_y F_y Z_{x\text{-roof}} = 1.1 \times 50 \times 186/12 \text{ in/ft} = 853 \text{ kip-ft}$$

Note that the beam plastic hinge moments can be reduced if HBE axial forces are considered. See the AISC Design Guide 20 for more information (Sabelli and Bruneau 2006). The rest of the loads applied to the 2d model of the column are given in Table 4-10, and the resulting moments are given graphically in Figure 4-9.

Table 4-10. Loads on VBE to be used in 2d model to determine moment

Level	Tension Field, F_{22} (k/in)	HBE Plastic Hinge, M_p (k-ft)
Roof	—	853
6th	1.12	610
5th	1.57	610
4th	2.01	853
3rd	2.68	967
2nd	3.63	564
1st	3.63	—

4r. Determine column compression

The columns should be designed for compression loads associated with the web plate yielding. The commentary to AISC 341 suggests three methods for determining the overturning forces:

1. Nonlinear-pushover analysis
2. Combined Linear Elastic Computer Programs and Capacity Design Concept (LECP&CD concept)
3. Indirect Capacity Design Approach

The method used in this example is a simplified rational method to determine the upper bound compression from a nonlinear-pushover analysis. In essence, forces are calculated based on a complete mechanism with all web plates yielding and plastic hinges at each beam-column connection. This is consistent with the design of other components and the intent of the code. If a nonlinear pushover analysis was stopped at a certain level of story drift or plastic hinge rotation, the compression forces considered here would not be realized, hence the designation as an upper bound. This method yields the maximum compressive forces that can be transferred to the column by the SPSW system.

The methods for calculating column compression based on the LECP&CD concept and the Indirect Capacity Design Approach are not discussed in detail here, but are fairly easy to compute and are shown in Figure 4-10 alongside the upper bound on the nonlinear pushover analysis. The reader is directed to Berman and Bruneau (2008) for more information. Both methods begin with a 2d model of the frame with web elements to determine compression due to code level forces (also shown in Figure 4-10). The LECP&CD applies Ω_o to the compression forces associated with code level story forces above the floor under consideration. The effect of the web plate yielding at the floor in question is then added. This method is consistent with the definition of Ω_o. and for this example produced VBE compression just less than that determined by the upper bound on the nonlinear pushover analysis. See the section on strong column-weak beam check for additional discussion when using the LECP&CD concept.

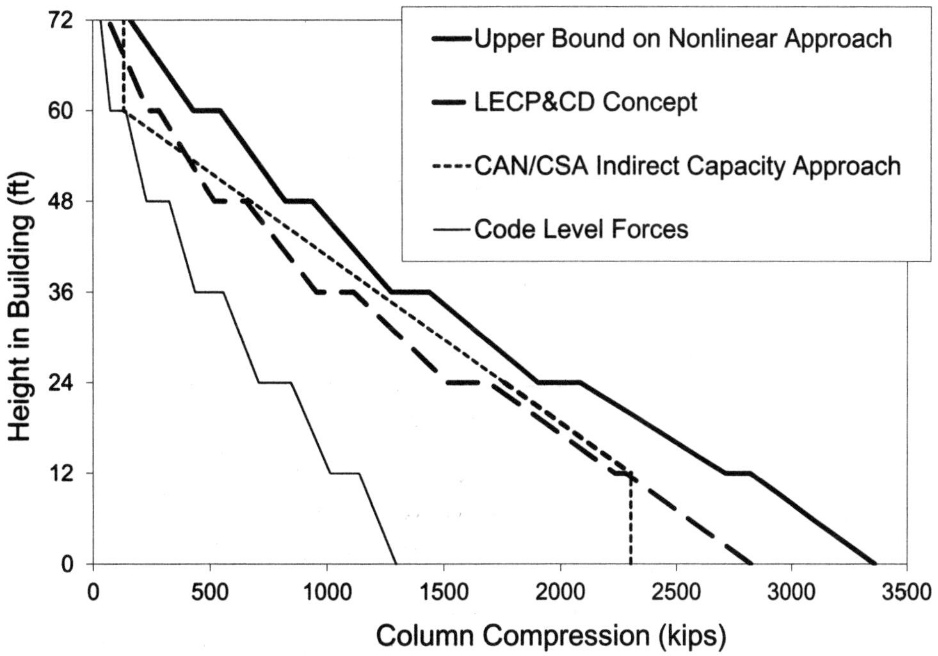

Figure 4-10. Column compression due to each of the three methods

The Indirect Capacity Design Approach applies a factor B to the code level overturning at the base which then reduces linearly to B times the overturning force at the 6th floor. The factor B is defined as the shear that causes full yielding in the 1st floor web plate (discounting the moment connections) over the code level shear, V_u. Because the beams are moment connected to the columns, however, it would be possible to develop more compression in the columns than this method predicts, especially at the base of the building. Because it is unlikely that every plate will yield and every beam develop plastic hinges, however, it is possible that this level of compression is appropriate. More research is required to determine which method is most appropriate. For the purposes of this example, the more conservative upper bound on the nonlinear pushover analysis was used.

The column compression calculated by the upper bound nonlinear approach consists of: downward components of the tension field, beam shear due to tension field, and beam shear due to plastic hinging (shown in Figure 4-11). The forces applied at the roof and 6th floor are given by

$F_{11\text{-}6} = 1.60$ kip/in (See Table 4-8 for F_{11})

$V_{11\text{-}r} = F_{11\text{-}6} \times L_{cf}/2 = 1.60 \times 127/2 = 102$ kips

$$V_{p\text{-}r} = \frac{2M_{p\text{-}r}}{L_{cf}} = \frac{2 \times 853 \times 12 \text{ in/ft}}{127} = 161 \text{ kips}$$

$F_{21\text{-}6} = 1.34$ kip/in (See Table 4-8 for F21)

The rest of the applied compression loads are given in Table 4-11, and the resulting compression forces are given graphically in Figure 4-11.

Table 4-11. Loads on VBE contributing to compression

Level	Tension Field, F_{21} (k/in)	Beam Shear due to Plate, V_{11} (kips)	Beam Shear due to Hinge, V_p (kips)
Roof	—	102	161
6th	1.34	41	115
5th	1.88	39	115
4th	2.40	61	161
3rd	3.20	85	183
2nd	4.32	0	107
1st	4.32	—	—

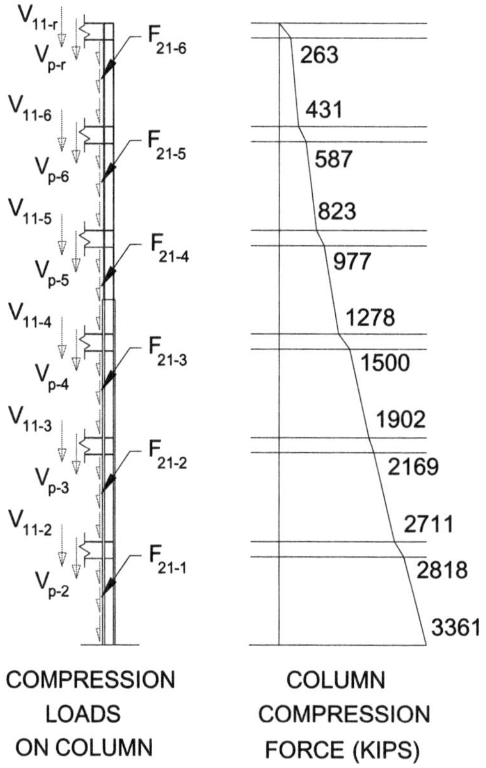

Figure 4-11. VBE compression forces for the Upper Bound on Nonlinear Pushover Analysis method

4s. Check width-thickness and moment of inertia for VBE

With the moment distribution given in Figure 4-9 and the compression forces given in Figure 4-11, a wide flange section is chosen that satisfies the compactness criteria, that satisfies the minimum moment of inertia, and that can resist the combined forces. Note that the column is to be spliced above the 4th floor. Choose a W14×342 for the 1st-floor column section. Check width-thickness limitations

$\lambda_f \leq (0.38\sqrt{E/F_y} = 9.15)$ $\quad \lambda_f = 3.31 \rightarrow \text{OK}$ — AISC 341 T I-8-1

$C_a = P_u/\phi P_n = 3361/4177 = 0.80$ — AISC 341 T I-8-1

$\lambda_w \leq 1.12\sqrt{E/F_y}(2.33 - C_a) = 41.3$ $\quad \lambda_w = 7.4 \rightarrow \text{OK}$ — AISC 341 T I-8-1

VBE width-thickness ratios are OK

Section 17.4g places a minimum on the moment of inertia of the VBE. The limit is intended to require a stiff enough VBE such that the tension field can develop uniformly across the panel.

$$I_{min} = \frac{0.00307 t_w h^4}{L} = 1719 \text{ in}^4 \quad < I_c = 4900 \text{ in}^4 \rightarrow \text{OK}$$ §17.4g

VBE moment of inertia is OK

4t. Check combined flexure and axial forces on VBE

$$\frac{KL}{r_y} = \frac{1.0 \times (12 \text{ ft} \times 12 \text{ in/ft})}{4.24} = 34.0 \text{ for minor axis}$$

$$\frac{KL}{r_x} = \frac{1.0 \times (12 \text{ ft} \times 12 \text{ in/ft})}{6.98} = 20.6 \text{ for major axis}$$

$$F_e = \frac{\pi^2 E}{(KL/r)} = 248 \text{ ksi}$$ Eq E3-4

Limit on inelastic buckling, $4.71\sqrt{\frac{E}{F_y}} = 113$

Because $\frac{KL}{r} < 4.71\sqrt{\frac{E}{F_y}}$, $F_{cr} = \left[0.658^{F_y/F_e}\right]F_y = 46.0 \text{ ksi}$ — Eq E3-2

$\phi P_n = 0.90 \times F_{cr} A_g = 0.90 \times 46 \times 101 = 4177 \text{ kips}$ — Eq E3-1

$\phi P_n = 4177$ kips

$L_p = 1.76 r_y \sqrt{\frac{E}{F_y}} = 15.0 \text{ ft}$ — Eq F2-5

$L_b = 12$ ft, 0 in, so L_b is less than L_p

$\therefore M_n = 0.9 F_y Z_x = 0.9 \times 50 \times 672 = 2520 \text{ kip-ft}$ — Eq F2-1

$\phi M_n = 2520$ kip-ft

The worst-case combined axial and flexure might occur at one of three locations: the base of the column, the maximum moment in the 1st story, or at the 2nd-floor level since the moment is larger than below. All locations were checked, as given in Table 4-12, but the worst case was found to be at the 2nd-floor level

Since $P_r/P_c > 0.2$

$$\frac{P_r}{P_c}+\frac{8}{9}\left(\frac{M_{rx}}{M_{cx}}+\frac{M_{ry}}{M_{cy}}\right)\leq 1.0 \qquad \text{Eq H1-1a}$$

$$\frac{P_u}{\phi P_n}+\frac{8}{9}\left(\frac{M_u}{\phi M_n}\right)=\frac{2818}{4177}+\frac{8}{9}\frac{849}{2520}=0.97 \rightarrow \text{OK}$$

Combined axial and flexure = 0.97

4u. Summary of VBE design

The design of the upper column is carried out in a similar manner. The resulting design and key design parameters are given in Table 4-12.

Table 4-12. Summary of the VBE design

Location	Parameter	1st Floor Column	4th Floor Column
1. At bottom of column	Compression (kips)	3361	1278
	Corresponding Moment (k-ft)	0	48
	Combined Loading	0.80	0.71
2. Max moment between HBEs	Compression (kips)	3144	11.58
	Corresponding Moment (k-ft)	417	160
	Combined Loading	0.90	0.74
3. At next higher HBE	Compression (kips)	2818	977
	Corresponding Moment (k-ft)	849	582
	Combined Loading	0.97	0.99
	Section	W14×342	W14×145
	Compression Capacity (kips)	4177	1746
	Moment Capacity (k-ft)	2520	975
	Worst Case Combined Flexure and Axial	0.97	0.99

4u. Strong column — weak beam check

AISC 341 §17.4a requires that the beam-column moment ratio provisions in §9.6 be met for all HBE to VBE intersections without the consideration of web plates. The intent of the strong column-weak beam check of §9.6 is to keep the column from hinging before the beam, and in this way preclude the development of a story mechanism. Carrying out this design check in the absence of the web plate tension will greatly underestimate the amount of moment applied to the column, thereby undermining the intent of this provision.

With the upper bound on a nonlinear pushover analysis method of column design, however, the strong-column weak beam moment ratio is implicitly satisfied, considering the effect of the web plate yielding. The column is designed to resist the full compression and moment due to the web plate yielding concurrent with plastic hinging of the beams. If the upper bound on nonlinear pushover analysis method is used to design the columns, the strong column/weak-beam check is unnecessary.

If the LECP&CD concept is used to determine column compression, the strong-column/weak-beam check should be carried out. Note, however, that this check will not explicitly preclude a story mechanism because the web plate is not considered.

5. Analysis and design (iterative)

5a. Check angle of inclination

Now that the web plates, HBEs, and VBEs have been sized, the accuracy of the initial estimate for α, can be assessed. The equation for the angle of inclination is

$$\tan^4 \alpha = \frac{1 + \frac{t_w L}{2A_c}}{1 + t_w h\left(\frac{1}{A_b} + \frac{h^3}{360 I_c L}\right)} \qquad \text{Eq 17-2}$$

t_w = web plate thickness

L = distance between VBE centerlines

h = distance between HBE centerlines

A_c = cross-sectional area of VBE

A_b = cross-sectional area of HBE

I_c = moment of inertia of VBE perpendicular to plate

For the 6th-floor level this works out to

$$\tan^4 \alpha = \frac{1 + \frac{0.075 \times 144}{2 \times 46.7}}{1 + 0.075 \times 144\left(\frac{1}{25.3} + \frac{144^3}{360 \times 1900 \times 144}\right)} = 0.64; \alpha = \tan^{-1}(0.64)^{0.25} = 41.8°$$

$$\boxed{\alpha_6 = 41.8°}$$

The angle α, was calculated for all floors and the results are given in Table 4-13.

Table 4-13. Calculated angle of inclination

Level	Angle of Inclination of the Tension Field, α
6th	41.8°
5th	40.3°
4th	39.5°
3rd	39.9°
2nd	40.1°
1st	38.5°

The estimate of α = 40° proved to be fairly accurate. For the final design, the calculated angle of inclination, α, will be used, but first, deflections are checked.

5b. Check deflection

Calculating deflection requires the use of a 2d computer model. There are multiple ways to model the web plate, of which two of the most useful are shown in Figure 4-12. The Canadian steel code (CAN/CSA S16-01) recommends the tension strip model. In this model, the web plate is divided into at least ten tension strips per panel, oriented at the angle α. The tension strip section is merely the tributary width of plate times the thickness of the plate, and is pinned at each end. CAN/CSA S16-01 uses this model to determine the design capacity of the wall as the shear that causes a strip to reach its factored tension capacity. The elastic 2d model is not appropriate, however, for boundary element design. In order to use the 2d model to design the boundary elements, it would be necessary to account for nonlinearities such as the plate yielding and plastic hinging of the beams.

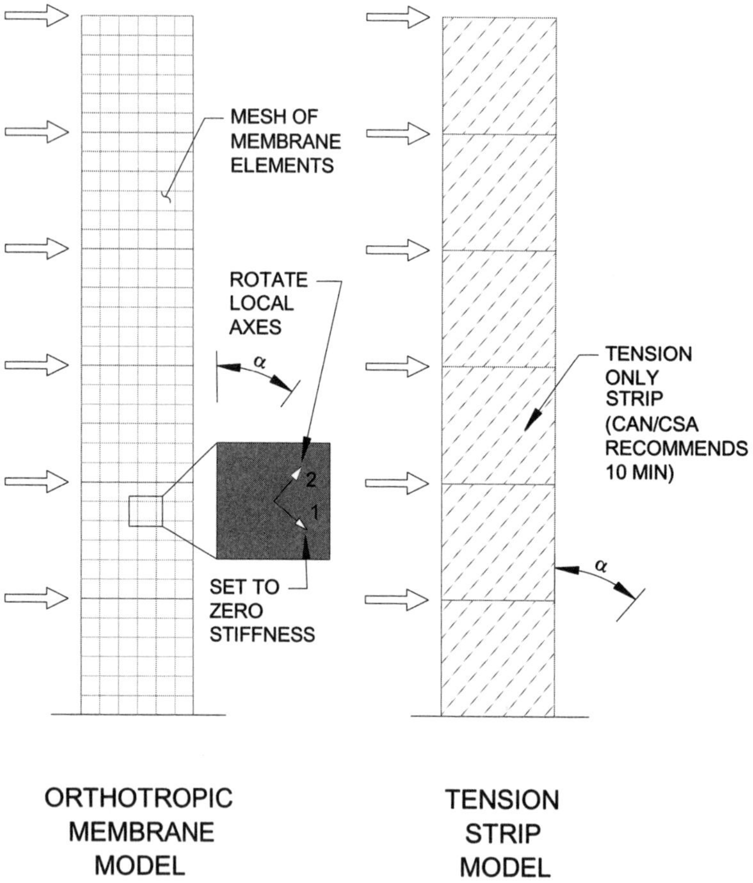

Figure 4-12. Modeling options to check deflection

Another modeling option is the orthotropic membrane model. The web plate is meshed into a number of membrane elements. The local axes of these elements are rotated to the angle of inclination, a. The modulus of elasticity along local axis 1 and the shear modulus are set to zero (or as near zero as the modeling software allows). The modulus of elasticity along local axis 2 is set to 29,000 ksi for steel. The membrane thickness is then set to the design web plate thickness. Advantages to this type of modeling include

1. The angle of inclination, α, can be adjusted without reconfiguring the model geometry.

2. The membrane can be meshed automatically, whereas, the tension strips have to be added individually.

For these reasons, the orthotropic membrane model was used for this design example. As shown in Figure 4-12, each panel was divided into 36 elements, 2 ft, 0 in on a side. The resulting elastic deflections are given in Table 4-14. The elastic story drifts are the difference in the elastic deflections at the floor above and below. The predicted inelastic story drift is given by the following for the 5th floor

$$\delta_x = \frac{C_d \delta_{xe}}{I} = \frac{6.0 \times 0.47}{1.0} = 2.82 \text{ in} \qquad \text{Eq12.8-15}$$

The limit on story drift is given in Table 12.12-1 based on building system and the occupancy category

$$\Delta_a = 0.020 h_{sx} = 0.020 \times 144 = 2.88 \text{ in} \qquad \text{T 12.12-1}$$

$$\boxed{\delta_x \le \Delta_a \rightarrow \text{drift is OK}}$$

Table 4-14. Calculated deflection for preliminary design

Level	Elastic Deflection (in)	Elastic Story Drift, δ_{xe} (in)	Predicted Inelastic Story Drift, δ_x (in)	Story Drift Limitation, Δ_a (in)
Roof	2.34	0.37	2.22	2.88
6th	1.97	0.44	2.64	2.88
5th	1.54	0.47	2.82	2.88
4th	1.07	0.39	2.34	2.88
3rd	0.68	0.34	2.04	2.88
2nd	0.34	0.34	2.04	2.88

5c. Iterate to satisfy deflection

The drifts were found to satisfy the story drift limits. However, if the inelastic story drift had exceeded the story drift limit it would have been necessary to either increase the size of HBEs and VBEs to increase the stiffness of the moment frame or increase the web plate thickness. If the capacity design approach is used, the column moments are the only parameters not easily calculated by closed form equations (although they can be estimated). It is possible, therefore, to create a spreadsheet or Mathcad sheet to carry out most, if not all, of the required calculations. Using one of these computer tools can significantly simplify the iteration process.

5d. Final plate thicknesses and boundary members

Since all design checks were satisfied, the final design for the web plate, HBEs and VBEs are shown in Figure 4-13

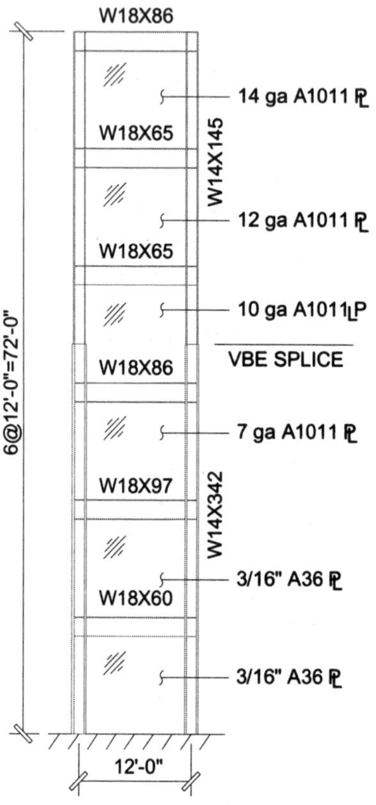

Figure 4-13. Final plate and boundary member design

6. Detailing and design of connections

6a. Panel zone

AISC 341 §17.4f references Section 9.3 for a check of the panel zone shear for the panel zones next to the top and base HBE of the SPSW. The roof HBE and an HBE at the foundation are expected to experience larger shear forces because there is web plate on only one side of the HBE. For this design example, check the column panel zone at the VBE (W14×145) roof connection with the HBE (W18×86).

$$V_u = F_f - V_c = 581 - 142 = 439 \text{ kips}$$

where,

$$F_f = \frac{M_{pr}}{h_o} = \frac{R_y Z_x F_y}{d_b - t_{fb}} = \frac{10{,}230}{17.6} = 581 \text{ kips}$$

$$V_c = \frac{2M_{pr}}{h} = \frac{2 \times 10{,}230}{(12' \times 12)} = 142 \text{ kips}$$

Since $\frac{P_u}{\phi P_n} = \frac{263}{1910} = 0.14 < 0.75$, then: AISC 360 §J10.6

$$\phi R_n = 1.0 \times 0.60 F_y d_c t_w [1 + 3b_{cf} t_{cf}^2/(d_b d_c t_w)] = 443 \text{ kips}$$ Eq J10-10

and the minimum panel zone thickness is determined from Section 9.3b as:

$$t_{\min} \geq \frac{(d_z + w_z)}{90} = \frac{(18.4 - 2 \times 0.77) + (15.0 - 2 \times 1.19)}{90} = 0.327''$$ AISC 341 Eq 9-2

$V_u = 439$ kips	$\phi R_n = 443$ kips
$t_{\min} = 0.328''$	$t_{wc} = 0.745''$

6b. HBE-to-VBE connection

AISC 341 §17.4b requires that the HBE to VBE connections satisfy the requirements of §11.2, beam-to-column connections for ordinary moment frames. Criteria for removal of backing bars, backgouging, and reinforcement with fillet welds are given. These criteria should be called out on the construction documents. Additionally, where weld access holes are provided they need to match the figure given in Section 11.2. For this example, the HBE web and flanges will get complete joint penetration welds to the VBE flange. Since the HBE web will get a complete joint penetration weld to the VBE flange, a check of the HBE shear will suffice. The shear in the HBE is checked in a previous section of this design example.

6c. Plate to boundary elements

In general, tolerances make it impractical to weld the plate directly to HBEs and VBEs. Either a plate or angle is required to provide a flat surface, against which the plate can be placed and welded. For this example a plate will be used. The connecting plate is sometimes referred to as a "fish plate." Section 17.3 specifies that the connection of the plate to the HBE and VBE shall resist the forces associated with the expected yield strength of the plate in tension, oriented at the angle, α. In this case it is most convenient to express this force in terms of F_1 and F_2 — the effect of the tension field acting on the connection at the angle, α. Figure 4-14 and the following equations demonstrate the decomposition of the tension field force into these components.

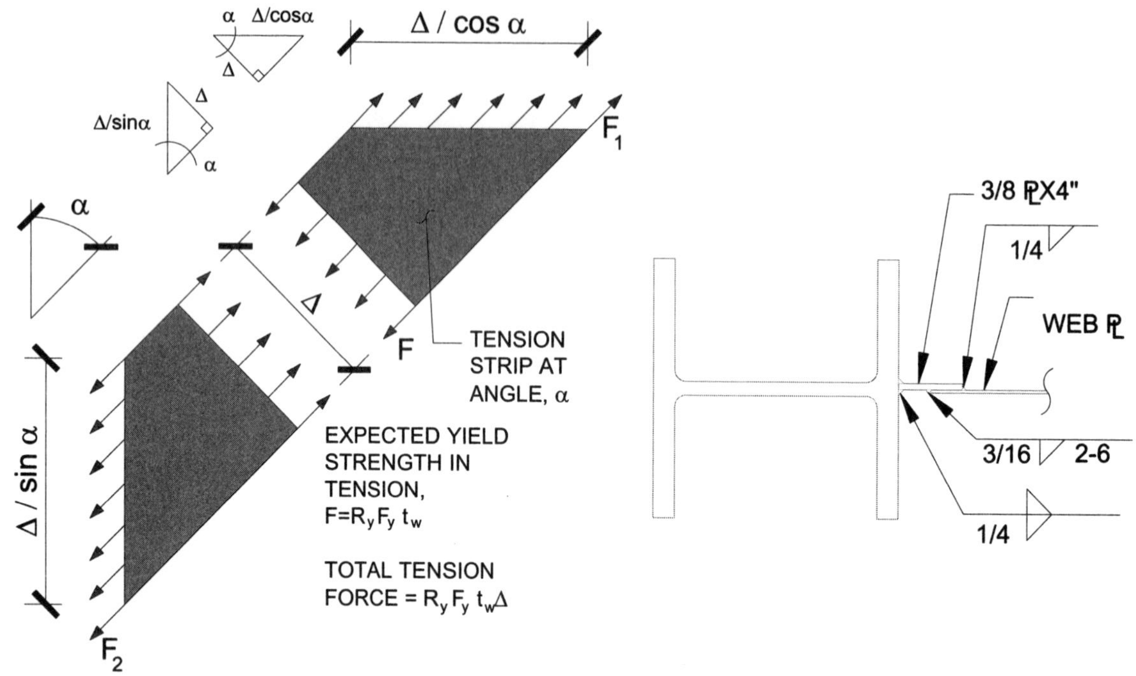

Figure 4-14. Web plate to boundary element connection

$$F_1 \frac{\text{force}}{\text{length}} = \frac{R_y F_y t_w \Delta}{\Delta / \cos\alpha} = R_y F_y t_w \cos\alpha$$

$$F_2 \frac{\text{force}}{\text{length}} = \frac{R_y F_y t_w \Delta}{\Delta / \sin\alpha} = R_y F_y t_w \sin\alpha$$

For the 1st floor plate, this becomes

$$F_1 = 1.3 \times 36 \times 0.1875 \times \cos(38.5) = 6.87 \text{ kip/in}$$

$$F_2 = 1.3 \times 36 \times 0.1875 \times \sin(38.5) = 5.46 \text{ kip/in}$$

Design the weld of the plate for the worst of the two, 6.86 kip/in, and use it for both conditions. Try ¼-inch continuous weld on one side and 3⁄16-inch weld intermittent at 2 inches every 6 inches on the other side. The equation for weld capacity comes from §J2 of the LRFD specification and includes a factor for the angle of loading relative to the longitudinal axis of the weld, θ, which effectively increases the weld capacity.

$$\phi R_n = 0.75 \times 0.6 F_{EXX}(1.0 + 0.50 \sin^{1.5}\theta)\, A_{eff} \qquad \text{Eqs J2-4 and J2-5}$$

$$\phi R_n = 0.75 \times 0.6 \times 70 \times \frac{1}{\sqrt{2}}\left(\frac{1}{4} \times \frac{6}{6} + \frac{3}{16} \times \frac{2}{6}\right)(1 + 0.5(\sin 38.5)^{1.5}) = 8.67 \text{ kip/in} \rightarrow o.k.$$

$V_u = 6.87$ kips/in	$\phi R_n = 8.67$ kips/in

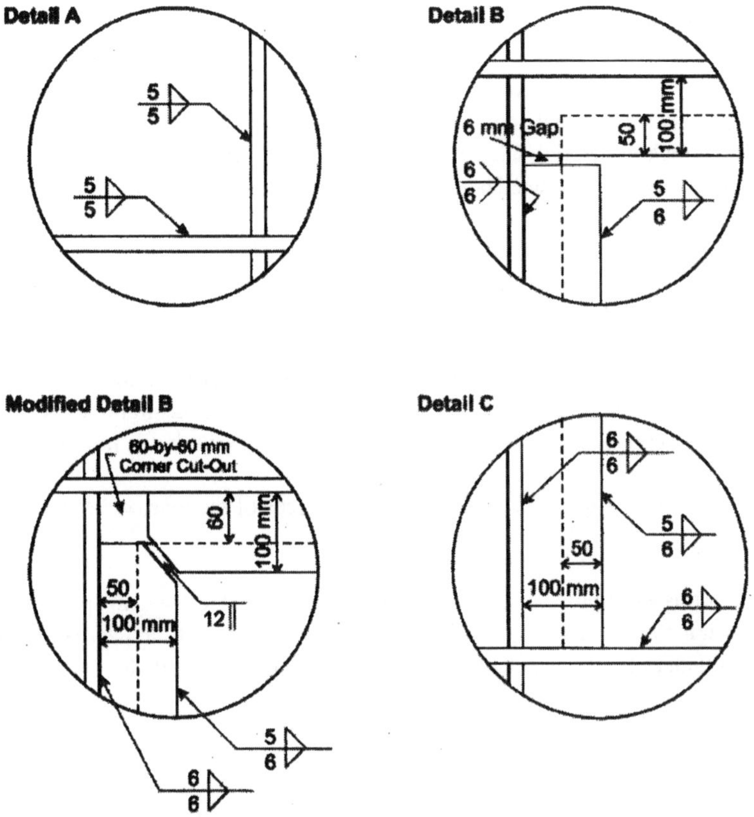

Figure 4-15. Corner details tested by Schumacher — all were found to be acceptable (from Schumacher 1999)

At the corners of the panels, it must be decided how to treat the intersection of the two fish plates. Schumacher, Grondin, and Kulak (Schumacher et al, 1999), studied the effect of different ways of joining the fish plates. Figure 4-15 shows the four types of corner connections tested. All four corner details exhibited similar response and were found to be acceptable for use in the SPSW system.

6d. Column splice

AISC 341 §17.4e references §8.4 for splices in the VBE. As stated previously, columns in the SPSW system do more work than other lateral resisting systems. As shown in Figure 4-15 there is considerable moment that is developed in the VBE. Figure 4-15 also shows that the maximum moment occurs at the floor levels, and although it is far from zero at a typical splice height, it is not going to be subjected to inelastic demands. For this example the column splice is accomplished with complete joint penetration welds at the flanges and web, so the connection will have at least the strength of the smaller section.

6e. Plate splice

A1011 sheet steel is manufactured in long lengths and then wrapped onto a coil for easy storage and shipping. Coil widths range from 60 to 72 inches for 16-gage and thicker material. Clearly, with a maximum width of 72 inches, a splice will be necessary in the plate. There are two ways to treat the splice.

1. Add an intermediate HBE at mid-floor height that not only acts as a mechanism to splice the plate, but also serves as a strut to resist the inward pull of the columns. This requires a more substantial element, but can lead to a more economical VBE design. It is recommended that if this approach is used, the connection of the intermediate HBE to the VBE transfer the axial force without developing fixity. It is not desirable to develop moment connections with inelastic demands at a point that is not laterally braced.

2. Use a splice plate to transfer the forces of the tension field that does not restrain the columns.

For this example, the second option is chosen, and a ⅜ × 4-in splice plate is used to transfer the tension field stresses. The procedure outlined above in §6.2, Plate to Boundary Elements, can be used to design the welds from the web plate to the splice plate.

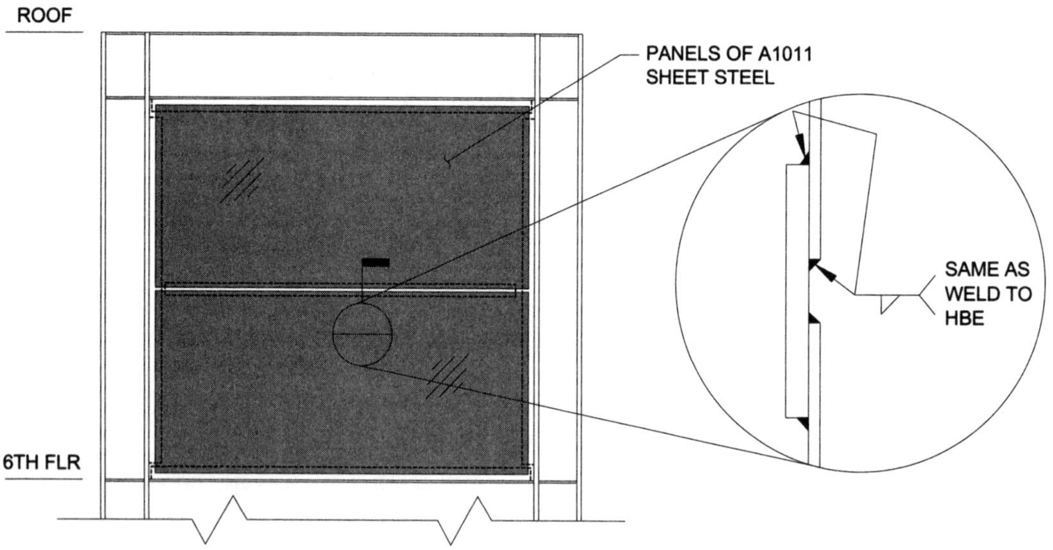

Figure 4-16. Splice in light gage plate

6f. Base of wall to foundation

There are different approaches that can be used to anchor the 1st floor web plate to the foundation. It is sometimes preferable to use a wide flange beam. See the AISC Design Guide 20 for an example (Sabelli 2006). For this example, a WT is used to transfer the significant uplift and shear to the foundation. The anchor bolts will be designed to resist the combined shear and tension. The AISC Design Guide 1 (DeWolf and Ricker 1990) suggests that if anchor bolts are used to resist shear, they either be designed for friction or include a weld washer. The weld washer allows for the use of oversized holes in the WT and tolerances in anchor bolt placement, while still encouraging uniform bearing on each bolt.

For an anchor bolt spacing of 12 inches determine the loads to each bolt

$$F_{11} = 1.3 \times 36 \times 0.1875 \cos^2(38.5) = 5.37 \text{ kips/in}$$

$$F_{12} = 1.3 \times 36 \times 0.1875[0.5 \sin(2 \times 38.5)] = 4.28 \text{ kips/in}$$

Bolt tension without prying action, $R_{ut} = \dfrac{F_{11} \times 12\text{-in spacing}}{2 \text{ bolts}} = 32.2$ kips

Bolt shear, $V_u = \dfrac{F_{12} \times 12\text{-in spacing}}{2 \text{ bolts}} = 25.7$ kips

Prying action will amplify the bolt tension. Chapter 9 of the AISC Manual 13th Edition gives the method to determine the added bolt tension due to prying action

Bolt gage, $g = 6$ in

WT flange width, $b_f = 10.0$ in, so a = 2.0 in, and b = 2.8 in   AISC Manual Ch 9

Using 1⅛-in bolts, a′ = 1.5 in and b′ = 2.2 in

The additional tension force can be calculated to be

$$q = 15.3 \text{ kips}$$

The resulting factored tension is

$$T_u = r_{ut} + q = 32.2 + 15.3 = 47.5 \text{ kips}$$

Try two 1⅛-inch A1554 Grade 105 bolts at 12-in spacing. $A_b = 0.99$ in^2, $F_u = 125$ ksi

$$\phi T_n = 0.75 \times 0.75 F_u A_b = 0.75 \times 0.75 \times 125 \times 0.99 = 69.6 \text{ kips} \quad \text{T J3.2}$$

$$\phi V_n = 0.75 \times 0.40 F_u A_b = 0.75 \times 0.40 \times 125 \times 0.99 = 37.1 \text{ kips} \quad \text{T J3.2}$$

$$\left(\frac{T_u}{\phi T_n}\right)^2 + \left(\frac{V_u}{\phi V_n}\right)^2 \le 1.0 \quad \text{Eq C-J3-5a}$$

$$\left(\frac{47.5}{69.6}\right)^2 + \left(\frac{25.7}{37.1}\right)^2 = 0.94$$

Combined tension and shear = 0.94

Figure 4-17 shows the detail at the base of the wall.

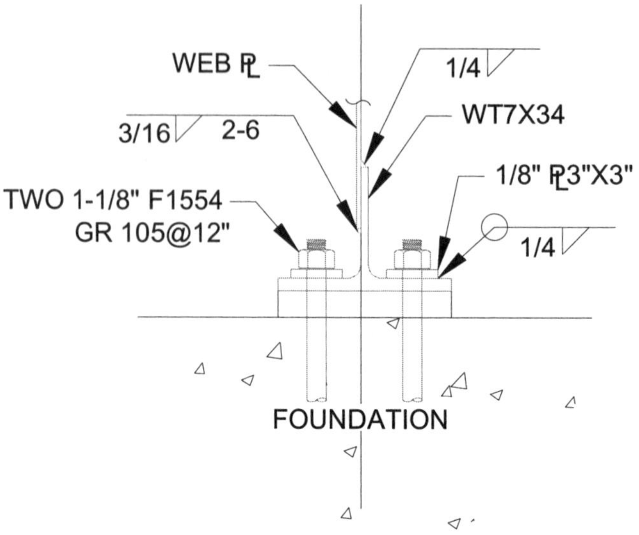

Figure 4-17. Web plate connection to foundation

Additional design checks for the WT that should be examined include ensuring the stem is stronger than the SPSW web plate, bolt bearing, bending of the WT between bolts, and local bending of the flange.

6g. Lateral bracing

Because the HBE is expected to undergo inelastic demands, it is necessary to adequately brace the HBE against lateral movement. §17.4d gives the requirements for laterally bracing the HBE. The maximum spacing for lateral bracing has already been checked in conjunction with the HBE design.

Lateral bracing is required at the HBE to VBE and at the midpoint of the HBE. The connection at the 2nd floor (W18×60) mid-span lateral bracing should be designed to resist the following load at the bottom flange:

$$P_{br} = 0.02F_y b_f t_f = 0.02 \times 50 \times 7.56 \times 0.695 = 5.2 \text{ kips} \qquad \text{§17.4d}$$

The required lateral bracing stiffness for the second floor HBE (W18×60) is given by

$$\beta_{br} = \frac{1}{\phi}\left(\frac{10M_u C_d}{L_b h_o}\right) = 7.6 \text{ kip/in} \qquad \text{LRFD App. 6, Eq A-6-8}$$

where

ϕ = 0.75

M_u = 30 kip-ft = 360 kip-in (calculated using the same procedure as in part 4m)

C_d = 2.0 for double curvature

L_b = 72 in

h_o = distance between flange centroids = 17.5 in

Floor beams are shown framing into the SPSW panels at the ends and at the mid-span of the HBE in Figure 4-1. The mid-span floor beam will attach to a full-depth stiffener in the HBE web. The lateral bracing detail is not pursued further here, but the mid-span floor beam and its attachment could be designed to satisfy the minimum strength and stiffness requirements.

6h. Column continuity plates

Section 17 of AISC 341 does not specifically require column continuity plates. Requirements for ordinary moment frame beam-to-column connections are used for the HBE to VBE connection per §17.4b, but panel zones are designed to special moment frame connection standards. The choice of which type of provisions to follow for SPSW continuity plates is made simple, because §9.5 specifies that continuity plates for special moment frame connections be consistent with the prequalified connection being used. Since the ordinary moment connection is not prequalified, provisions relating to ordinary moment frame provisions should be followed. §11.5 should be checked to determine if continuity plates are required.

6i. Column base connection

The VBE, in addition to resisting considerable compression, will experience large uplift forces associated with seismic forces. Uplift forces can be computed by the capacity design approach similar to that employed to determine compression. It is shown in the AISC Design Guide 20 that tension forces are best calculated using this type of capacity design approach because it can yield less uplift than the other three methods (Sabelli and Bruneau 2006). Vertical forces on the tension column include: the beam shear from plastic hinging, V_p; beam shear from web plate yielding, V_{11}; and the vertical component of the expected web plate strength acting on the column, V_{21}. The capacity design approach would produce the uplifts given in Table 4-15.

Table 4-15. Column Uplift Computed by the Capacity Design Approach

Level	Plate Uplift on Column, V_{21} (kips)	Beam Shear due to Plate, V_{11} (kips)	Beam Shear due to Hinge, V_p (kips)	Incremental Uplift (kips)	Cumulative Uplift, T (kips)
Roof	0	101	−161	60	−60
6th	−169	41	−115	−244	−303
5th	−236	39	−115	−312	−616
4th	−302	61	−161	−402	−1018
3rd	−403	85	−183	−501	−1519
2nd	−544	0	−107	−651	−2170
1st	−544	0	0	−544	−2714

The uplift at the base of the column is found to be 2714 kips. Compare this to the design uplift computed by multiplying the uplift due to code level forces by Ω_o:

Uplift due to code level forces found using 2d model, $T_{\text{code}} = 992$ kips.

$T_u = \Omega_o T_{\text{code}} = 2.0 \times 992 = 1984$ kips

The forces computed this way are approximately 27% less than the capacity design approach. For consistency, this example will use the maximum uplift forces that can be transferred to the base connection, $T_u = 2714$ kips. A similar method can be used to estimate the maximum shear at the base of the column as:

$$V_u = F_{22-1}\frac{(h-d_b)}{2} + \frac{M_{p-2}}{h} = 3.63\frac{(144-18.24)}{2} + \frac{5.64}{12} = 275 \text{ kips}$$

Try fourteen, 2-inch ϕ F1554 Gr. 105 bolts. The nominal tensile and shear stress are obtained from AISC 360 Table J3.2 for threaded parts with threads not excluded from the shear planes.

$\phi T_n = 0.75 \times 0.75 F_u A_b N_b = 0.75 \times 0.75 \times 125 \times 3.14 \times 14 = 3091$ kips Table J3.2

$\phi V_n = 0.75 \times 0.40 F_u A_b N_b = 0.75 \times 0.40 \times 125 \times 3.14 \times 14 = 1649$ kips Table J3.2

$$\left(\frac{T_u}{\phi T_n}\right)^2 + \left(\frac{V_u}{\phi V_n}\right)^2 \leq 1.0 \qquad \text{Eq C-J3-5a}$$

$$\left(\frac{2714}{3091}\right)^2 + \left(\frac{275}{1649}\right)^2 = 0.75$$

Combined tension and shear = 0.75

The column base plate is designed not to resist moment as reflected by the zero moment at the base of the column in Figure 4-9. Design checks for the base plate are required, but not shown here.

References

Anonymous, 1977. "Shear walls and slipforming speed Dallas' Reunion project" Engineering News Record, pp 20-21, July 28.

Anonymous, 1978. "Patent problems, challenge spawn steel seismic walls" Engineering News Record, pp 17, January 26.

Anonymous, 1978. "Quake-proof hospital has battleship-like walls" Engineering News Record, pp 62-63, Sept. 21.

Anonymous, 1978. "California hospital is quake-proof, high-rise" Engineering News Record, pp 20-21, August 10.

Basler, Konrad 1961. "Strength of Plate Girders in Shear" Journal of the Structural Division, ASCE, Vol. 87, No. ST7 October.

Berman, Jeffrey W., Celik, Oguz C., and Bruneau, Michel 2005. "Comparing Hysteretic Behavior of Light-Gauge Steel Plate Shear Walls and Braced Frames" Engineering Structures 27 pp 475-485.

Berman, Jeffrey W. and Bruneau, Michel 2004. "Steel Plate Shear Walls are Not Plate Girders" AISC Engineering Journal, Third Quarter.

Berman, Jeffrey W. and Bruneau, Michel 2008. "Capacity Design of Vertical Boundary Elements in Steel Plate Shear Walls" AISC Engineering Journal, First Quarter.

Bruneau, M., Uang, C.M., and Sabelli, R. 2011. *Ductile Design of Steel Structures*. New York: McGraw-Hill.

CAN/CSA S16-012001. "Limit States Design of Steel Structures" published by Canadian Standards Association. With CAN/CSA S16S1-05 "Supplement No. 1 to CAN/ CSA-S16-01."

DeWolf, John T., and Ricker, David T. 1990. "AISC Design Guide 1— Column Base Plates," Published by the American Institute of Steel Construction, AISC.

Eatherton, Matthew 2006. "Design and Construction of Steel Plate Shear Walls" Proceedings of the Eighth U.S. National Conference on Earthquake Engineering 8NCEE San Francisco, CA April 18-22.

Elgaaly, Mohamed and Liu, Yinbo 1997. "Analysis of Thin-Steel-Plate Shear Walls" Journal of Structural Engineering, ASCE, Vol. 123, No. 11, November.

Porter, D.M., Rockey, K.C. and Evans, H.R. 1975. "The collapse behavior of plate girders loaded in shear," The Structural Engineer, London England, Vol. 53, No. 8., Aug.

Purba, Ronny, and Bruneau, Michel 2007. "Design Recommendations for Perforated Steel Plate Shear Walls" Technical Report MCEER-07-0011

Qu, Bing, and Bruneau, Michel 2010. “Capacity Design of Intermediate Horizontal Boundary Elements of Steel Plate Shear Walls” Journal of Structural Engineering, Vol. 136, No. 6.

Sabelli, Rafael and Bruneau, Michel 2007. “AISC Design Guide 20 — Steel Plate Shear Walls,” Published by the American Institute of Steel Construction, AISC

Schumacher, A., Grondin, G.Y. and Kulak, G.L. 1999. “Connection of Infill Panels in Steel Plate Shear Walls” Canadian Journal of Civil Engineering, Vol. 26.

Thorburn, L. Jane, Kulak, G.L.and Montgomery, C.J. 1983. “Analysis of Steel Plate Shear Walls” Structural Engineering Report No. 107, Department of Civil Engineering — University of Alberta at Edmonton, May.

Vian, Darren, and Bruneau, Michel 2005. “Steel Plate Shear Walls for Seismic Design and Retrofit of Building Structures” Technical Report MCEER 05-0010.

Design Example 5
Reinforced Concrete Wall

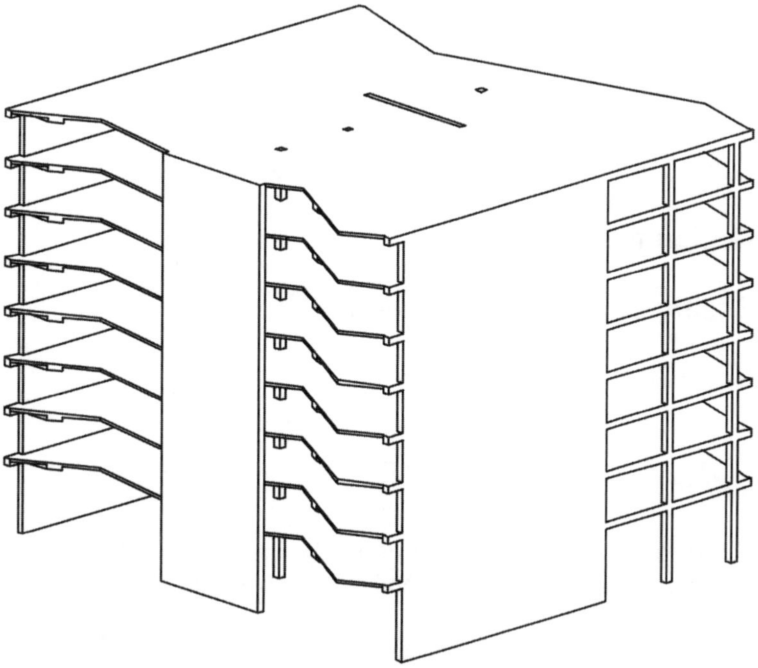

Figure 5-1. Eight-story reinforced concrete parking garage

Overview

The structure in this design example is an eight-story parking garage with load-bearing reinforced concrete walls as its seismic-force-resisting system, as shown in Figure 5-1. This design example focuses on the design and detailing of one of the 30-foot 6-inch-long walls running in the transverse building direction.

The purpose of this design example is twofold:

1. To demonstrate the design of a solid reinforced concrete wall for flexure and shear, including bar cut-offs and lap splices.

2. To demonstrate the design and detailing of wall boundary zones.

The design example assumes that design lateral forces have already been determined for the structure, and that the forces have been distributed to the walls of the structure by a hand or computer analysis. The SEAOC Blue Book (2009, Article 09.05.010) discusses design considerations for concrete parking structures, such as the participation of ramps, which are assumed to be included where appropriate in the building analysis. This analysis has provided the lateral displacements corresponding to the design lateral forces.

Outline

This design example illustrates the following parts of the design process

1. **Load combinations for design**
2. **Preliminary sizing of wall**
3. **Flexural strength of wall**
4. **Lap splices and curtailment of vertical bars**
5. **Design for shear**
6. **Sliding shear (shear friction)**
7. **Boundary zone detailing**

Given Information

Mapped spectral response acceleration values may be obtained from IBC §1613.5.1 maps, ASCE/SEI 7-05 maps (Figures 22-1 through 22-14), or from the USGS website earthquake.usgs.gov/hazards:

$S_1 = 0.65g$

$S_s = 1.60g$

Site class D

Occupancy Category I

Seismic Design Category D

Reliability/redundancy factor, $\rho = 1.0$

Seismic Importance factor, $I = 1.0$

Concrete strength, $f'_c = 5000$ psi

Steel yield strength, $f_y = 60$ ksi

Figure 5-2 shows the typical floor plan of the structure. Figure 5-3 shows the wall elevation and shear and moment diagrams. The wall carries axial forces, P_D (resulting from dead load including self-weight of the wall) and P_L (resulting from live load) as shown in Table 5-1. Live loads have already been reduced according to IBC §1607. The shear, V_E, and moment, M_E, resulting from the design lateral earthquake forces are also shown in Table 5-1. The forces are from a linear static analysis.

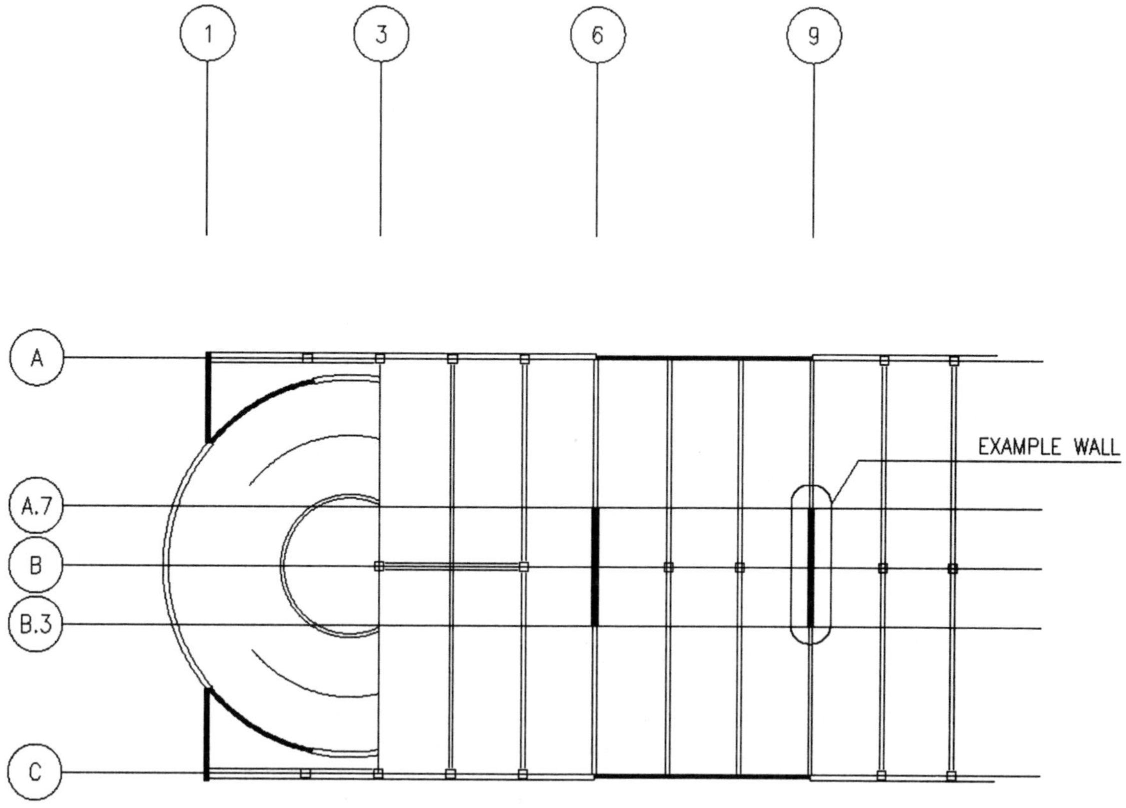

Figure 5-2. Floor plan

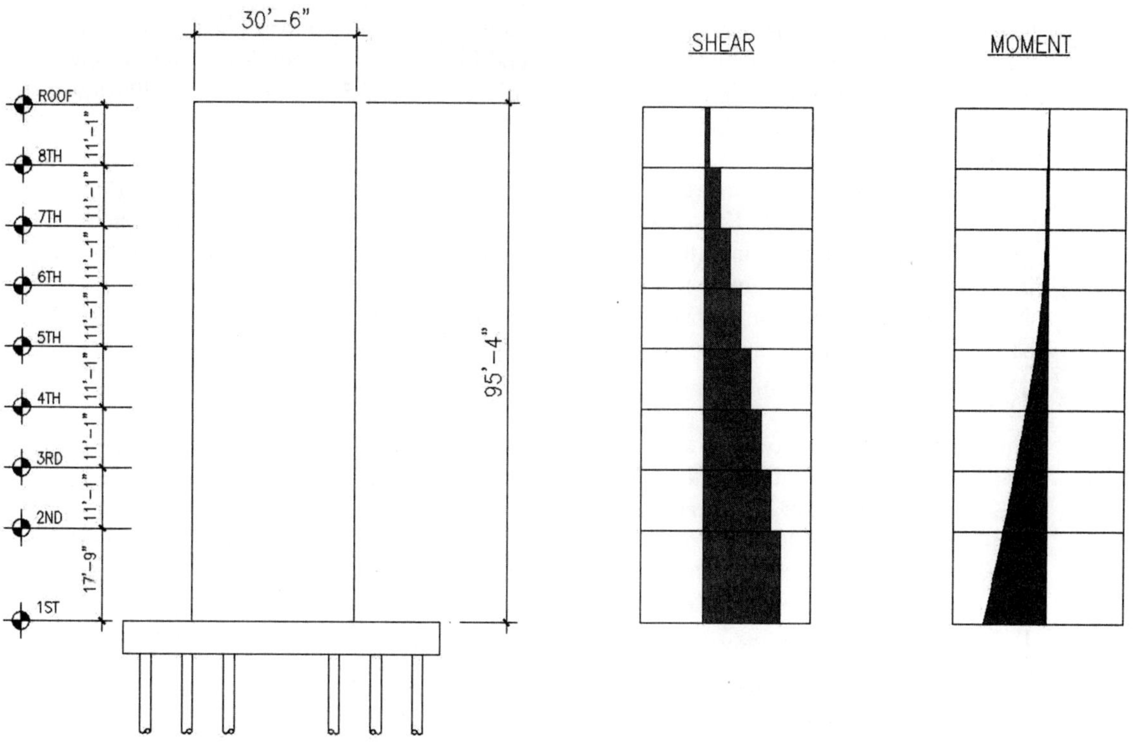

Figure 5-3. Wall elevation, shear, and moment diagram

Table 5-1. Design loads and lateral forces

Level	P_D (kips)	P_L (kips)	V_E (kips)	M_E (kip-ft)
Roof	193	37	93	0
8	388	72	253	928
7	573	108	423	3640
6	758	144	604	8250
5	945	181	794	14,900
4	1130	217	996	23,600
3	1310	253	1230	34,400
2	1540	290	1420	48,000
1				73,000

For this design example, it is assumed that the foundation system is rigid, and thus the wall is considered to have a fixed base. The fixed-base assumption is made here primarily to simplify the example. In an actual structure, the effect of foundation flexibility and its consequences on structural deformations should be considered.

The analysis uses effective section properties for the stiffness of concrete elements. Please refer to Example 6 for a discussion of effective section properties for use in analysis.

Using the fixed-base assumption and effective section properties, the horizontal displacement at the top of the wall corresponding to the design lateral forces is 1.55 inches. This displacement is needed for the detailing of boundary zones according to ACI 318 §21.9.6, which is illustrated in Part 7 of this design example.

The design and analysis of the structure is based on a response modification coefficient, R, of 5 (ASCE/SEI 7-05 Table 12.2-1) for a bearing wall system with special reinforced concrete shear walls. The deflection amplification factor, C_d, is 5. The SEAOC Blue Book (2009, Article 09.01.010) expresses the opinion that the R value for concrete bearing wall systems ($R = 5$) and that for walls in building frame systems ($R = 6$) should be the same, which may be justified based on detailing provisions. To be consistent with the current code requirements, though, this design example uses $R = 5$.

This design example follows the general building code requirements of the 2009 IBC and ASCE/SEI 7-05. Unless indicated otherwise, section numbers used in this example refer to ASCE requirements. For structural concrete design, the 2009 IBC references the American Concrete Institute Building Code (ACI 318), as indicated in §1901.2. This example follows the requirements of ACI 318-08. Discussions of the SEAOC Blue Book recommendations refer to the document *Recommended Lateral Force Recommendations and Commentary* (SEAOC, 1999) as well as the Blue Book online articles on specific topics (SEAOC, 2009) as applicable.

Calculations and Discussion **Code Reference**

1. Load combinations for design

Load combinations for the design of concrete are given in IBC §1605.2. Equations 5 and 7 of §1605.2 are the basic design load combinations to be used for concrete. For seismic design, these equations are modified by ASCE 7 §12.4.2.3 referenced in IBC §1605.1.

$$1.2D + 1.0E + L + 0.2S$$

$$0.9D + 1.0E + 1.6H$$

Load combinations for *nonseismic* loads for reinforced concrete are given in §2.3.2, Equations 1 through 4 and 6.

Horizontal and vertical components of earthquake force E

The term E in the load combinations includes horizontal and vertical components according to Equations 12.4-1 and 12.4-2 of §12.4.2.

$$E = E_h + E_v$$ Eq 12.4-1

$$E = E_h - E_v$$ Eq 12.4-2

Where E_h and E_v are defined according to Equations 12.4-3 and 12.4-4 of §12.4.2.1 as follows:

$$E_h = \rho Q_E$$ Eq 12.4-3

$$E_v = 0.2S_{DS}D$$ Eq 12.4-4

Substituting this into the basic load combinations results in seismic design load combinations:

$$(1.2 + 0.2S_{DS})D + \rho Q_E + L + 0.2S$$

$$(0.9 - 0.2S_{DS})D + \rho Q_E + 1.6H$$

Since there is no lateral earth pressure, $H = 0$, and since there is no snow load, $S = 0$, Section 12.4.2.3 permits the load factor on L in the above combination to be reduced to 0.5 where L_o in Table 1607.1 is less than or equal to 100 psf, with the exception of garages or areas occupied as places of public assembly. Because this example building is a parking garage, the reduced factor on L is not applicable.

Given site class D and $S_S = 1.60$, from IBC Table 1613.5.3(1) or ASCE 7 Table 11.4-1, $F_a = 1.0$

$$S_{MS} = F_a S_S = 1.60$$ Eq 11.4-1

$$S_{DS} = \frac{2}{3} S_{MS} = 1.07$$ Eq 11.4-3

With $S_{DS} = 1.07$ and $\rho = 1.0$, the governing load combinations for this design example become

$$1.41D + Q_E + L$$

$$0.686D + Q_E$$

Actions at base of wall

For the example wall, the dead and live loads cause axial forces only, and the earthquake forces produce shear and moment only. The second of the above combinations gives the lower bound axial force. Because the axial force is less than that which would cause the balanced strain condition, the second of the above load combinations will be the more critical for the flexural strength of the wall.

The governing axial force at the base of the wall is thus

$$P_u = 0.686P_D = 0.686(1540 \text{ kips}) = 1060 \text{ kips}$$

The governing moment and shear at the base of the wall is

$$M_u = M_E = 73{,}000 \text{ kip-ft}$$

$$V_u = V_E = 1420 \text{ kips}$$

2. Preliminary sizing of wall

2a. Shear stress and reinforcement ratio rules of thumb

The dimensions and required number of walls in a building can be selected by limiting the average shear stress in the walls, corresponding to factored lateral forces, to between $3\sqrt{f_c'}$ and $5\sqrt{f_c'}$. Walls with higher levels of shear stress are permitted by ACI 318, but shear stress within the range suggested leads to more easily constructible detailing for shear strength, sliding shear, and boundary confinement. For taller buildings with significant influence of higher modes, PEER *Guidelines for Performance-Based Seismic Design of Tall Buildings* (PEER, 2010) recommends limiting shear stresses to the range of $2\sqrt{f_c'}$ and $3\sqrt{f_c'}$ for preliminary design.

For the example wall, the maximum factored shear force equals 1420 kips. Using a $3\sqrt{f_c'}$ criterion, for a wall length of 30 feet, 6 inches, the wall thickness equals

$$\frac{1{,}420{,}000 \text{ lb}}{366 \text{ in}(3\sqrt{5000}) \text{ psi}} = 18.3 \text{ in}$$

Say $b = 20$ inches

Detailed shear design for calculated demands is discussed in Section 5.

2b. Minimum wall thickness to prevent wall buckling

For structures with tall story heights, the designer should check that the wall thickness exceeds $l_u/16$ where l_u is the unsupported height or length of the wall, whichever is shorter. The SEAOC Blue Book Commentary (C407.5.6, page 178) recommends that "the wall boundary thickness limit of $l_u/16$ be applied at all potential plastic hinge locations, regardless of whether boundary zone confinement is required."

For the example wall, the clear height at the first story is 17 feet.

Recommended thickness $= l_u/16 = 17(12)/16 = 12.8 \text{ in} < 20 \text{ in}$ . . . *o.k.*

2c. Layout of vertical reinforcement

Based on brief calculations and the preliminary sizing considerations discussed here, the wall section and reinforcement layout shown in Figure 5-4 is proposed for the base of the wall.

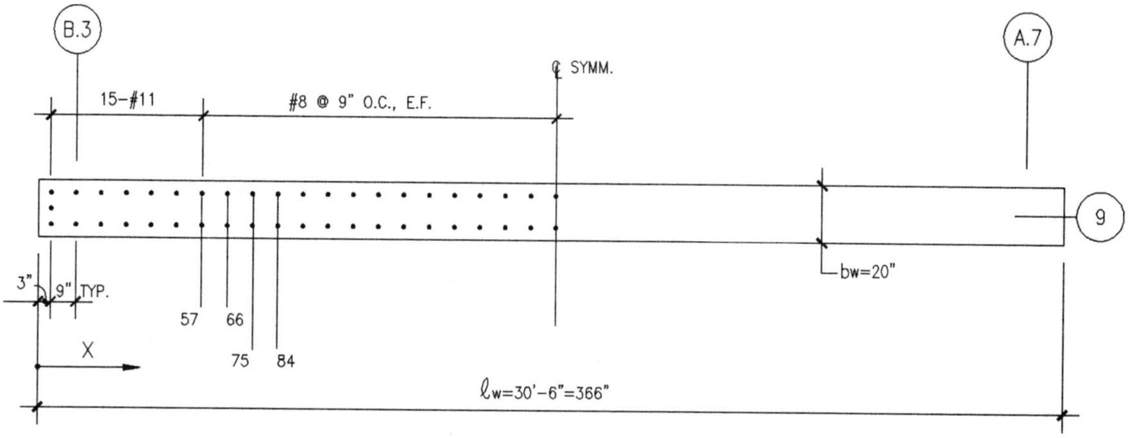

Figure 5-4. Layout of vertical reinforcement at wall base

The reinforcement layout addresses the following issues:

- Vertical bars are spaced longitudinally at 9 inches on center for ease of construction.
- The maximum permissible center-to-center spacing of vertical bars is 14 inches in boundary regions of walls where confinement is needed, according to ACI 318 §21.6.4.2, which is referenced by ACI 318 §21.9.6.4(c). This means that at the ends of the 20-inch-thick wall, three bars are used as shown in Figure 5-4.

ACI 318 §21.9.2.1 specifies a minimum reinforcement ratio of 0.0025 for both vertical and horizontal reinforcement of structural walls. For the proposed layout, at the center portion of the wall's length

$$\rho_v = A_s/bs = 1.58 \text{ in}^2/(9 \text{ in} \times 20 \text{ in}) = 0.0088 > 0.0025 \ \ldots \ o.k.$$

3. Flexural strength of wall

As required by ACI 318 §21.9.5.1, all "developed longitudinal reinforcement within effective flange widths, boundary elements, and the wall web shall be considered effective." Thus, the vertical reinforcement in the web of the wall and axial force contributions to the flexural strength of wall sections may not be neglected.

The 1995 and earlier editions of ACI 318 and the 1991 and earlier editions of the UBC required wall boundaries to carry all moment and gravity forces. This practice results in higher flexural strengths in walls, which can lead to poor earthquake performance because it makes shear failure more likely to occur. By ACI 318 §21.9.5.1, this design practice is no longer permitted.

Wall flexural strength can be computed by hand calculations, spreadsheet calculations, or a computer program such as spColumn. All three calculation approaches are demonstrated below. All of the calculation methods are based on an assumed strain distribution and an iterative calculation procedure.

3a. Assumed reinforcement strain

For cyclic loading all vertical reinforcement along the wall can be assumed to yield in either tension or compression. This assumption simplifies the hand calculation of moment capacity and is used in the hand calculations shown below.

Alternatively the reinforcement strain can be assumed to be directly proportional to distance from the neutral axis, as discussed in ACI 318 §10.2.2. This assumption is used in the spreadsheet calculations demonstrated here and is also used by the spColumn computer program.

The assumption of all reinforcement yielding results in a slightly greater flexural strength compared to the strain assumption of §10.2.2, but the difference is not significant. The two possible assumed strain distributions are illustrated in Figure 5-5.

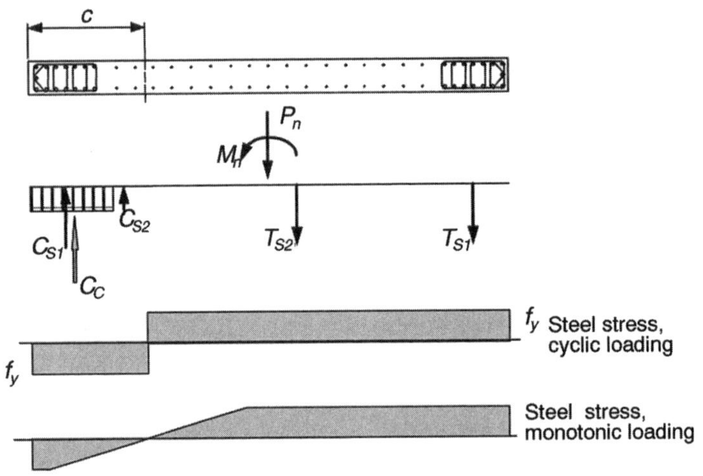

Figure 5-5. Steel stress and neutral axis depth

In calculating flexural strength, it is necessary to determine the neutral axis depth, c, as shown in Figure 5-5. A typical calculation of flexural strength is based on the following steps:

1. An initial estimate of c
2. Calculation of the steel reinforcement tensile and compressive forces
3. Balancing the forces to calculate the concrete compressive force, $C_c = (P_n + \Sigma T_s - \Sigma C_s)$
4. Calculation of the stress block length, a, which corresponds to C_c
5. Calculation of c equal to a/β_1, and a reiteration of Steps 1 through 4, if necessary

3b. Strength reduction factor, ϕ

ACI 318-08 §9.3.2 gives the requirements for determining the strength reduction factor ϕ. For reinforced concrete sections with flexure and axial force, the value of ϕ depends on whether the section is tension-controlled or compression-controlled. This procedure for determining ϕ was introduced in the 2002 edition of ACI 318 (ACI 2002).

ACI 318 §10.3.3 defines the conditions for compression-controlled sections, and §10.3.4 defines the conditions for tension-controlled sections. The definitions depend on the strain values at the cross sections at nominal strength. For Grade 60 reinforcement, sections are compression controlled if the net tensile strain in the extreme tension steel, ε_t, is less than or equal to 0.002 when the concrete in compression reaches its assumed strain limit of 0.003. Sections are tension controlled if ε_t is greater than or equal to 0.005 when the concrete in compression reaches its assumed strain limit of 0.003.

For tension-controlled sections ϕ is 0.9. For compression-controlled sections (without spiral reinforcement) ϕ is 0.65. When ε_t is between 0.002 and 0.005, the value of ϕ is linearly interpolated as indicated in §9.3.2.

3c. Hand calculation

The calculation of moment strength is based on the free-body diagram shown in Figure 5-6.

The strength reduction factor, ϕ, is determined according to ACI 318 §9.3.2. Considering the moderate amount of reinforcement in the wall section and relatively small (less than $0.10f'_cA_g$) axial force, it is reasonable to assume that the net tensile strain in the extreme tension steel, ε_t, is greater than 0.005 when the concrete in compression reaches its assumed strain limit of 0.003. Thus the section is tension controlled as defined in §10.3.4. With this assumption, ϕ is equal to 0.9 per §9.3.2.1. (We will verify this assumption and modify the value of ϕ if required in the subsequent design steps.)

$$P_n = P_u/\phi = 1060 \text{ kips}/0.9 = 1180 \text{ kips}$$

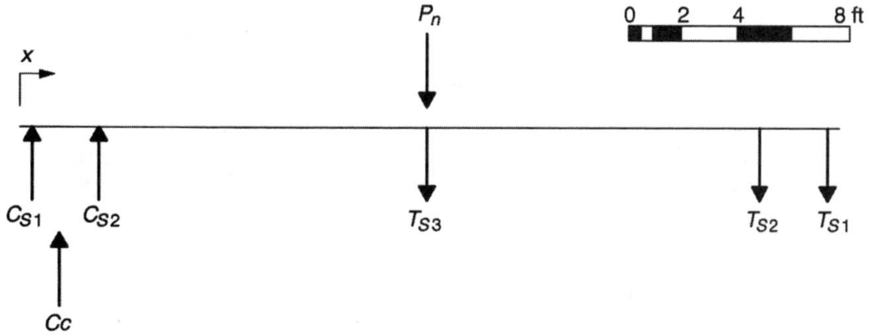

Figure 5-6. Free body diagram for flexural strength

The iterative calculation of neutral axis depth and flexural strength is shown in Tables 5-2 and 5-3.

Table 5-2. First iteration for c and M_n

Force	Reinforcement Bars	A_s (in^2)	$A_s f_y$ kips	x (in)	$A_s f_y * x$ kip-in
C_{S1}	3 #11	4.68	−281	3	−842
C_{S2}	12 #11	18.7	−1120	34.5	−38,700
T_{S3}	54 #8	42.7	2560	183	469,000
T_{S2}	12 #11	18.7	1120	332	373,000
T_{S1}	3 #11	4.68	281	363	102,000
P_n			1180	183	216,000
C_c			−3740	22.0	−82,300
			0		1,040,000 = M_n (kip-in) 86,500 = M_n (kip-ft)

Table 5-3. Second iteration for c and M_n

Force	Reinforcement Bars	A_s (in^2)	$A_s f_y$ kips	x (in)	$A_s f_y * x$ kip-in
C_{S1}	3 #11	4.68	−281	3	−842
C_{S2}	10 #11	15.6	−936	30	−28,100
T_{S3}	54 #8	42.7	2560	187	479,000
T_{S2}	12 #11	18.7	1120	332	373,000
T_{S1}	3 #11	4.68	281	363	102,000
P_n			1180	183	216,000
C_c			−3920	23.0	−90,200
			0		1,050,000 = M_n (kip-in) 87,600 = M_n (kip-ft)

First iteration, assume c = 60 in

Therefore, 15 #11 bars yield in compression, 54 #8 bars (all web vertical bars) plus 15 #11 bars yield in tension. (Assume all reinforcement yields in either tension or compression.)

Solve for C_c to balance forces, $C_c = 3740$ kips

Calculate a corresponding to C_c,
$a = C_c/(0.85 f'_c b) = 3740$ kips/(0.85(5 ksi)20 in $\quad a = 44.0$ in

Calculate $c = a/\beta_1 = 44.0/0.80 \quad c = 55.0$ in

Second iteration, assume *c* = 55 in

Therefore, 13 #11 bars yield in compression, 54 #8 bars plus 15 #11 bars yield in tension. Neglect force in 2 #11 located at x = 58.5 inches. Therefore, centroid of 10 #11 bars is at $x = 30$ inches. Assume all other reinforcement yields.

Solve for C_c to balance forces, $C_c = 3920$ kips

Calculate a corresponding to
C_c, a = $C_c/(0.85 f'_c b)$ = 3920 kips/(0.85(ksi)20 in) $a = 46.1$ in

Calculate $c = a/\beta_1$ $c = 57.6$ in   solution converged

This results in $M_n = 87{,}600$ kip-ft

Now we check the assumption that the section is tension controlled per §10.3.4. Using the strain compatibility for the section of the wall the tensile strain in the extreme tension steel can be calculated as follows:

$$\varepsilon_t = \frac{0.003}{c}(l_w - 3\text{ in} - c) = \frac{0.003}{57.6\text{ in}}(366\text{ in} - 3\text{ in} - 57.6\text{ in}) = 0.0159 > 0.005$$

Therefore $\phi = 0.9$.

$$\phi M_n = 0.9(87{,}600\text{ kip-ft}) = 78{,}800\text{ kip-ft} > M_u\ 73{,}000\text{ kip-ft}\ \ldots\ o.k.$$

3d. Calculation using a general spreadsheet

The approach used above to calculate flexural strength can be done on a spreadsheet or by hand. A more generally applicable spreadsheet to calculate wall flexural strength can also be created. Such a spreadsheet is shown in Figure 5-7.

This spreadsheet is set up so that each individual layer of reinforcement is represented by a spreadsheet row. The input variables are at the top of the spreadsheet. The user adjusts the input value of the neutral axis depth, c, on the spreadsheet until the tension and compression forces on the section are balanced, as indicated by the added notes on the section.

The spreadsheet gives a design moment capacity, ϕM_n, of the selected section equal to 78,100 kip-ft, approximately the same as that calculated by hand in the previous section.

3e. Calculation by software

Computer software tools are also available to design wall sections for combined flexural and axial forces. The example wall section is analyzed using the program spColumn, and the flexural strength obtained is approximately the same as that calculated by the hand and spreadsheet methods. The printed screen output from spColumn is shown in Figure 5-8.

Strain Compatibility for Shear Walls Job: Wall Example 5 Date: Aug-11
Limiting Concrete Strain Job No.: 8.23 By: JRM
Rectangular Stress Block (ultimate) Wall: 30.5' long rectangular wall

f'c = 5 ksi
fy = 60 ksi
l = 366 inches length Pu =0.686DL +E
b = 20 inches thickness protection password: shwall
Ag = 7320 inches ρtotal= 0.0122
x = 60 inches Compression Zone Boundary Element Length
y = 20 inches Compression Zone Boundary Element Width
cg = 183.0 inches from extreme compression fiber
Pu = 1060.0 kips
strain = 0.003 Ultimate concrete strain
c = 61.0 inches
n = 7.2
beta1 = 0.80

Layer	y (in)	strain (- is compr)	fc (ksi) or fs (ksi)	As (in^2)	Ts (kips)	Cc (kips) Cs (kips)	Moment Calculation	
compression zone	61	-0.00300	5.0			4148.0		657873
As1	363	0.01485	60.0	4.68	280.8	0.0	50544	0
As2	354	0.01441	60.0	3.12	187.2	0.0	32011	0
As3	345	0.01397	60.0	3.12	187.2	0.0	30326	0
As4	336	0.01352	60.0	3.12	187.2	0.0	28642	0
As5	327	0.01308	60.0	3.12	187.2	0.0	26957	0
As6	318	0.01264	60.0	3.12	187.2	0.0	25272	0
As7	309	0.01220	60.0	3.12	187.2	0.0	23587	0
As8	300	0.01175	60.0	1.58	94.8	0.0	11092	0
As9	291	0.01131	60.0	1.58	94.8	0.0	10238	0
As10	282	0.01087	60.0	1.58	94.8	0.0	9385	0
As11	273	0.01043	60.0	1.58	94.8	0.0	8532	0
As12	264	0.00998	60.0	1.58	94.8	0.0	7679	0
As13	255	0.00954	60.0	1.58	94.8	0.0	6826	0
As14	246	0.00910	60.0	1.58	94.8	0.0	5972	0
As15	237	0.00866	60.0	1.58	94.8	0.0	5119	0
As16	228	0.00821	60.0	1.58	94.8	0.0	4266	0
As17	219	0.00777	60.0	1.58	94.8	0.0	3413	0
As18	210	0.00733	60.0	1.58	94.8	0.0	2560	0
As19	201	0.00689	60.0	1.58	94.8	0.0	1706	0
As20	192	0.00644	60.0	1.58	94.8	0.0	853	0
As21	183	0.00600	60.0	1.58	94.8	0.0	0	0
As22	174	0.00556	60.0	1.58	94.8	0.0	-853	0
As23	165	0.00511	60.0	1.58	94.8	0.0	-1706	0
As24	156	0.00467	60.0	1.58	94.8	0.0	-2560	0
As25	147	0.00423	60.0	1.58	94.8	0.0	-3413	0
As26	138	0.00379	60.0	1.58	94.8	0.0	-4266	0
As27	129	0.00334	60.0	1.58	94.8	0.0	-5119	0
As28	120	0.00290	60.0	1.58	94.8	0.0	-5972	0
As29	111	0.00246	60.0	1.58	94.8	0.0	-6826	0
As30	102	0.00202	58.5	1.58	92.4	0.0	-7484	0
As31	93	0.00157	45.6	1.58	72.1	0.0	-6490	0
As32	84	0.00113	32.8	1.58	51.8	0.0	-5131	0
As33	75	0.00069	20.0	1.58	31.5	0.0	-3407	0
As34	66	0.00025	7.1	1.58	11.3	0.0	-1318	0
As35	57	-0.00020	5.7	3.12	0.0	17.8	0	2243
As36	48	-0.00064	18.5	3.12	0.0	44.6	0	6019
As37	39	-0.00108	31.4	3.12	0.0	84.6	0	12188
As38	30	-0.00152	44.2	3.12	0.0	124.7	0	19077
As39	21	-0.00197	57.0	3.12	0.0	164.7	0	26687
As40	12	-0.00241	60.0	3.12	0.0	173.9	0	29744
As41	3	-0.00285	60.0	4.68	0.0	260.9	0	46964
				89.5	3749	5019	240435	800794
				Pn =	1178			
					4927	5019		

Ag= 7320 inches^2
phi= 0.900
Mn = 1041229 in*kips **86769** ft*kips
phi * Mn 937106 in*kips 78092 ft*kips
phi * Pn 1060 kips

Figure 5-7. General spreadsheet to calculate flexural strength

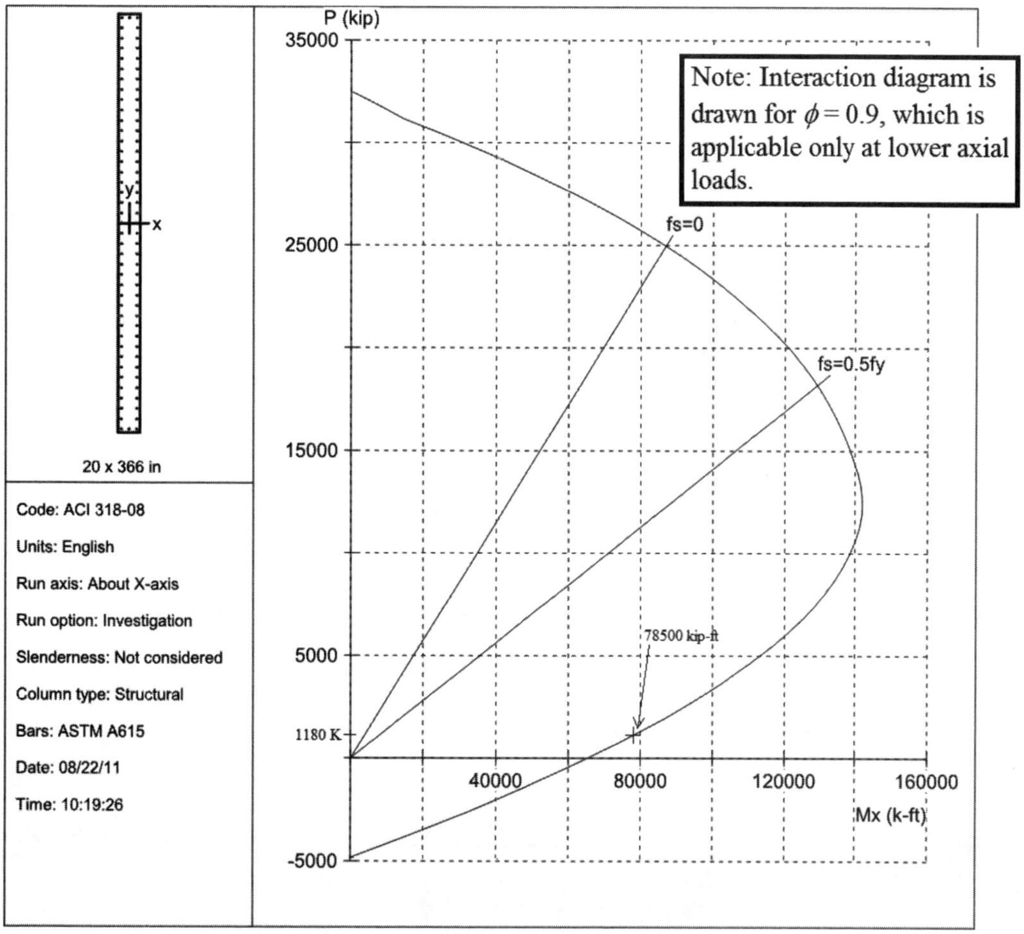

Figure 5-8. Analysis of Wall Section by spColumn

4. Lap splices and curtailment of vertical bars

4a. Bar cut-offs

ACI 318 §12.10.3 addresses the development of flexural reinforcement and states "Reinforcement shall extend beyond the point at which it is no longer required to resist flexure for a distance equal to the effective depth of the member or $12d_b$, whichever is greater." ACI 318 §21.9.2.3(a) indicates that this requirement is applicable to walls, and that the effective depth may be considered equal to $0.8l_w$. Section 402.7 of the SEAOC Blue Book also discusses this requirement.

Applying the bar cut-off requirement to the example wall, the flexural strength is reduced in two steps over the height of the wall: above Level 5 and above Level 7. The dimensions of the wall section and the number of vertical bars are unchanged at these transitions—only the bar size is reduced. The selection of vertical reinforcement sizes and cut-offs is shown in the wall elevation of Figure 5-9. A summary of flexural reinforcement and flexural strength over the wall height is given in Table 5-4.

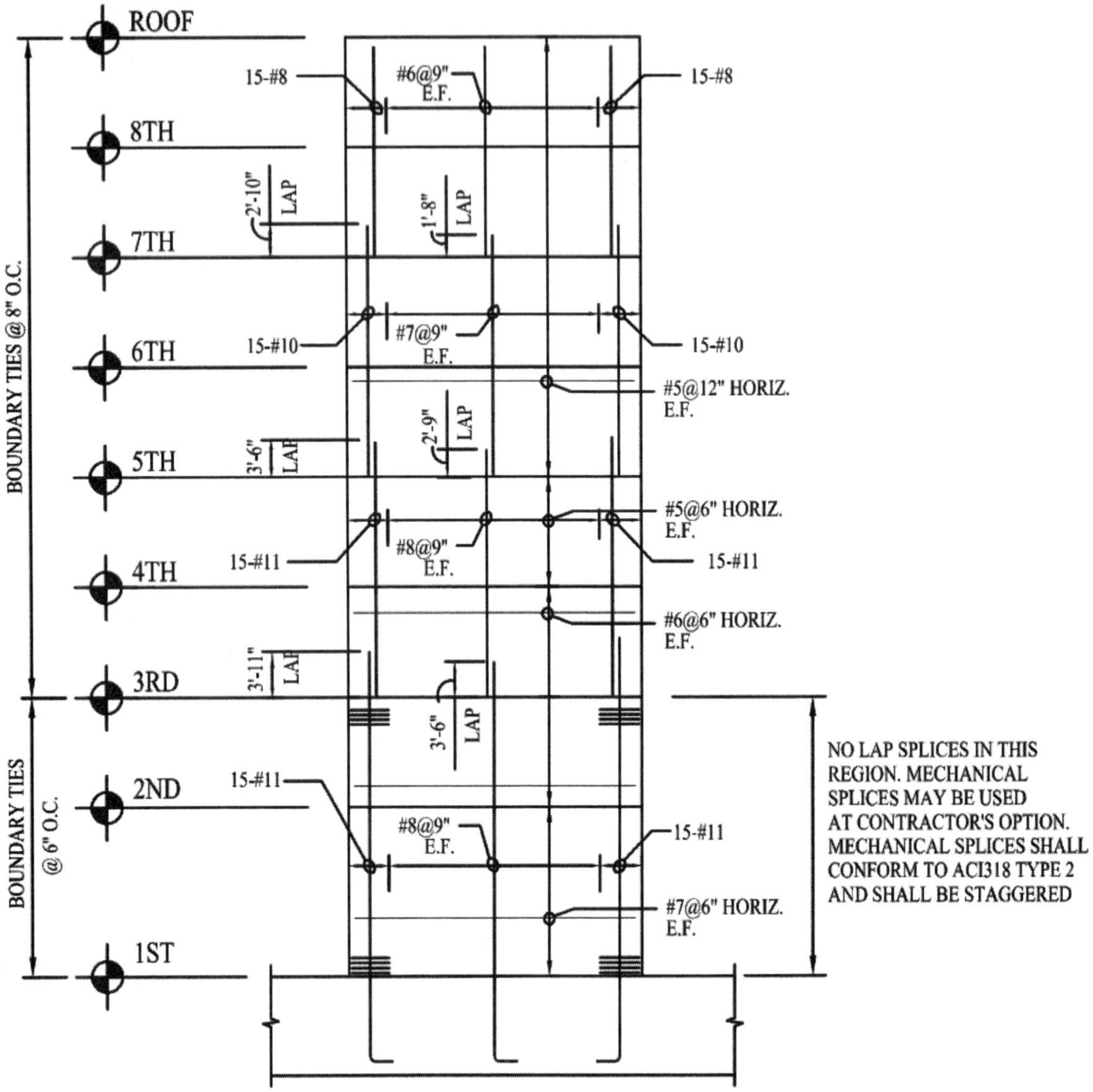

Figure 5-9. Wall elevation

Table 5-4. Boundary and vertical web reinforcement

Location	Vertical Bars Each Boundary	Web Vertical Bars	Axial Force $P_u = (0.9 - 0.2S_{DS})P_D$	Design Flexural Strength, ϕM_n
Level 1– Level 5	15 #11	54 #8	1060 kip	78,100 kip-ft
Level 5 – Level 7	15 #10	54 #7	520 kip	59,000 kip-ft
Level 7 – Level 9	15 #8	54 #6	266 kip	39,900 kip-ft

The flexural strengths for each reinforcement arrangement were calculated using the spreadsheet procedure described in Part 3d, above.

The moment demand above Level 5 is checked by the calculation below. For simplicity, the moment diagram is assumed to be linear over the building height. This also addresses higher mode dynamic response effects according to the recommendations of Paulay and Priestley (1992, Figures 5-28 and 5-29).

Height of reinforcement cut-off above base = 51 ft + 2-ft, 9-in lap splice = 53.75 ft

Height after subtracting $0.8l_w$ bar extension = 53.75 ft – 0.8(30.5 ft) = 29.4 ft

Moment demand M_u at the base of the wall = 73,000 kip-ft

Overall wall height, h, = 95.3 ft

Moment demand at h = 29.4 based on linear moment diagram = (73,000)(95.3 – 29.4)/95.3 = 50,500 kip-ft < 59,000 . . . *o.k.*

Similarly, the moment demand above Level 7 is checked by the following calculation.

Height of reinforcement cut-off above base = 73-ft, 2-in + 1-ft, 8-in lap splice = 74.8 ft

Height after subtracting $0.8l_w$ bar extension = 74.8 – 0.8(30.5) = 50.4 ft

Moment demand at h = 50.4 based on linear moment diagram = (73,000)(95.3 – 50.4)/95.3 = 34,400 kip-ft < 39,900 . . . *o.k.*

The calculations for bar cut-off locations are illustrated in Figure 5-10.

The wall reinforcement ratio, ρ_v, shall not be less than 0.0025 per ACI §21.9.2.1. For the top levels of the building, with #6 @ 9-inch web vertical bars, ρ_v is calculated as

$$\rho_v = \frac{2(0.44 \text{ in}^2)}{9 \text{ in}(20 \text{ in})} = 0.00489 \geq 0.0025 \ \ldots \ o.k.$$

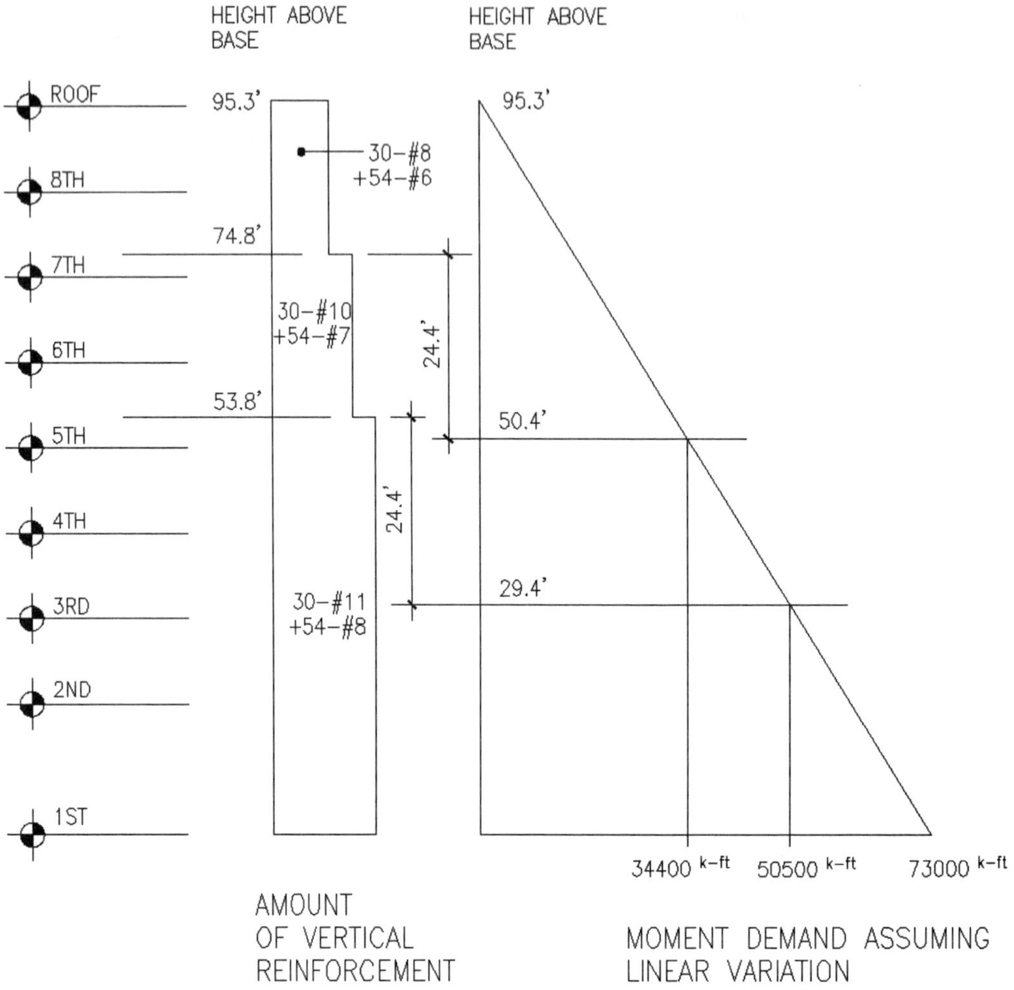

Figure 5-10. Calculation of required flexural strength at bar cut-off locations

4b. Splices of reinforcement

The lap splices of the vertical reinforcement are shown in the wall elevation of Figure 5-9.

Lap splice length

Lap splice lengths are calculated according to ACI 318 §12.15. This section indicates that Class B tension lap splices are to be 1.3 times the tension development length, l_d, from ACI 318 §12.2.

For cases with ample cover and spacing between bars or where reinforcement exists to provide $A_{tr} > 0$ per ACI 318 Chapter 12, the more explicit calculation per ACI 318 §12.2.3 will result in smaller l_d than calculations per ACI 318 §12.2.2. Given below are sample calculations for the two lap splice conditions that occur for the vertical bars just above the 5th floor.

Lap splices of the #7@9-inch vertical bars of the wall web

The calculation of l_d is according to ACI 318 Equation (12-1):

$$l_d = \left(\frac{3}{40} \frac{f_y}{\lambda \sqrt{f'_c}} \frac{\psi_t \psi_e \psi_s}{\left(\dfrac{c_b + K_{tr}}{d_b} \right)} \right) d_b$$

Because the vertical bars in the wall web are in the outside layer, there is no transverse reinforcement that "crosses the potential plane of splitting," so $A_{tr} = 0$. Thus, $K_{tr} = 0$.

Clear cover for these bars is specified on the drawings as 1½ inches. The distance c_b is defined in ACI 318 as the smaller of (a) the distance from center of a bar or wire to nearest concrete surface, and (b) one half the center-to-center spacing of bars or wires being developed. In this case, c_b, equals the cover measured from the center of the bar: 1.5 in + 0.875 in/2 = 1.94 inches.

The term $(c_b + K_{tr})/d_b$ equals 2.22, which is less than the upper limit of 2.5 from ACI 318 §12.2.3.

Other input values to equation (12-1) are

$\psi_t = 1.0$ (vertical bars)

$\psi_e = 1.0$ (uncoated reinforcement)

$\psi_s = 1.0$ (#7 bars or larger)

$\lambda = 1.0$ (normal weight concrete)

$d_b = 0.875$ in

$$l_d = \left(\frac{3}{40} \frac{60{,}000 \text{ psi}}{(1.0)\sqrt{5000 \text{ psi}}} \frac{(1.0)(1.0)(1.0)}{(2.22)} \right) d_b = 28.7 d_b$$

The required length for a Class B lap splice equals

$1.3(28.7d_b) = 37.3(0.875 \text{ in}) = 2$ ft, 9 in

Lap splices of the #10 vertical bars within boundary ties

The calculation of lap splice length for vertical boundary bars is similar, except that the ties around the bars cross the potential plane of splitting and provide a transverse reinforcement area, A_{tr}, that can reduce the required lap splice length.

ACI 318 defines A_{tr} as the "total cross-sectional area of all transverse reinforcement within spacing s that crosses the potential plane of splitting through the reinforcement being developed." The quantity n is the number of bars or wires being spliced or developed along the plane of splitting. The interpretation of how to determine A_{tr} and n is not entirely clear from ACI 318, but Figure 5-11, adapted from Jirsa et al [1979] and the commentary to the New Zealand concrete code [Standards New Zealand 1995], depicts potential splitting planes and corresponding values for A_{tr}/n.

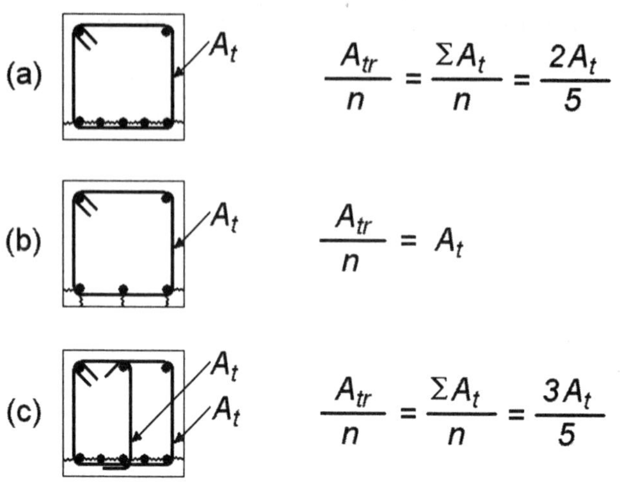

Figure 5-11. Determination of A_{tr}/n for calculation of lap splice length
(adapted from Figure 4 of Jirsa et al [1979] and Figure C7.5 of Standards New Zealand [1995])

As can be deduced from Jirsa et al, case (b) governs when spacing between bars is relatively large, as is the case for the boundaries of the example wall. Therefore,

$$A_{tr}/n = A_t = 0.20 \text{ in}^2$$

Ties around the boundary bars are spaced at 8 inches on center, as shown in Figure 5-9, so $s = 8$ in

$$K_{tr} = \frac{40A_{tr}}{sn} = \frac{(40)0.20 \text{ in}^2}{(8 \text{ in})(1)} = 1.00 \text{ in}$$

Clear cover is specified as 1½ inches to the tie, so

$$c_b = 1.5 \text{ in} + 0.5 \text{ in} + 1.27 \text{ in}/2 = 2.64 \text{ in}$$

The term $(c_b + K_{tr})/d_b$ equals 2.87, which is greater than the upper limit of 2.5 from ACI 318 §12.2.3, so the value input into ACI 318 Equation (12-1) is 2.5

$$d_b = 1.27 \text{ in}$$

$$l_d = \left(\frac{3}{40} \frac{60{,}000 \text{ psi}}{(1.0)\sqrt{5000 \text{ psi}}} \frac{(1.0)(1.0)(1.0)}{(2.5)} \right) d_b = 25.5 d_b$$

The required length for a Class B lap splice equals

$$1.3(25.5d_b) = 33.1(1.27 \text{ in}) = 3\text{ft, } 6\text{in}$$

Splice considerations in plastic hinge regions

At locations where yielding of longitudinal reinforcement is likely to occur ACI 318 §21.9.2.3 requires the development lengths of longitudinal reinforcement to be 1.25 times the values calculated for f_y in tension.

For the example wall, yielding of longitudinal reinforcement is expected (and designed to occur) at the base of the wall where the bending moment is the highest. (See Figure 5-3.) This is the intended plastic hinge region of the wall.

Where possible, lap splices of flexural reinforcement should be avoided in plastic hinge regions of walls. As indicated in 1999 Blue Book Sections C402.7 and C404.3 (and in the commentary ACI 318 §21.5.2.3, applicable to flexural members of frames), lap splices in plastic hinge regions are likely to slip unless they are surrounded by confining ties. Even well-confined lap splices that do not slip are undesirable in plastic-hinge regions because they concentrate yielding over a shorter length of the flexural reinforcement than occurs if lap splices are not present.

Paulay and Priestley (1992) note that splices in plastic hinge zones tend to progressively unzip, and that attempting to mitigate the problem by making lap splices longer than required is unlikely to ensure satisfactory performance.

Mechanical splices

Properly designed mechanical splices are preferable to lap splices in plastic hinge regions. Ideally, the mechanical splices should be able to develop the breaking strength of the bar. As a minimum, mechanical splices must be Type 2 splices according to ACI 318 §21.9.2.3 and §21.1.6.1 if they are to be used in plastic hinge regions. SEAOC recommends that mechanical splices be staggered so that no more than one half of the reinforcement is spliced at one section, and the stagger is not less than 2 feet. Staggering of the splices is not required by ACI 318.

Plastic hinge length and zone in which to exclude lap splices

The equivalent plastic hinge length, l_p, of a wall section can be taken as $0.2l_w + 0.07\ M/V$ where M/V is the moment to shear ratio at the plastic hinge location (ATC, 1999).

For the example wall, l_p is calculated as shown below:

$$l_p = 0.21l_w + 0.07M/V = 0.2\ (30.5\text{ ft}) + 0.07(73{,}000\text{ kip-ft}/1420\text{ kips})$$

$$= 6.1\text{ ft} + 3.6\text{ ft}$$

$$= 9.7\text{ ft}$$

Equivalent plastic hinge lengths, as calculated above, can be used to relate plastic curvatures to plastic rotations and displacements. The actual zone of yielding and nonlinear behavior typically extends beyond the *equivalent* plastic hinge length.

For this design example wall, the expected zone of yielding is taken as equal to approximately $2l_p$ (19.4 ft), and lap splices are avoided over this height.

In the design example, lap splices are excluded over the first two stories, i.e., over a height of 28.8 ft, as shown in the wall elevation of Figure 5-9. Because of potential construction difficulties in using continuous vertical bars from the foundation through Level 3, an option to use mechanical splices can be specified as shown in Figure 5-9.

It might also be a reasonable design to exclude lap splices over the first story height of 17.1 ft, and to have lap splices above the 2nd floor. In this case, however, the lap splices at the 2nd floor would be in a location where "yielding of longitudinal reinforcement is likely to occur" from seismic displacements. Thus, the 2nd floor lap splices would need to be to be designed for $1.25f_y$ in accordance with ACI 318-05 §21.9.2.3(c).

5. Design for shear

The SEAOC Blue Book Section 402.8.1 requires that "the design shear strength ϕV_n shall not be less than the shear associated with the development of the nominal flexural strength of the wall." To ensure this, the shear capacity of the wall must be sufficient to develop the flexural strength of the wall. A design for shear force based on ASCE/SEI 7-05 and ACI 318 code requirements will typically not achieve this objective. Thus, the code provisions covered in Part 5(a) should be considered as *minimum* requirements for the shear design of walls.

Designing for amplified shear forces as recommended in the Blue Book is covered in Part 5(b) below.

5a. ACI 318 requirements

Shear demand

If designing to the minimum requirements of the ACI 318 §21.9.3, the shear demand is taken directly from the design forces, factored by the load combinations discussed in Part 1 of this design example. At the base of the wall:

$$V_u = V_E = 1420 \text{ kips}$$

Shear capacity

ACI 318 §11.9 gives shear provisions for walls designed for *nonseismic* lateral forces such as wind or earth pressure. ACI 318 §21.9.4 gives shear strength provisions for walls designed for *seismic* forces.

$$V_n = A_{cv}(\alpha_c \lambda \sqrt{f_c'} + \rho_t f_y) \qquad \text{ACI 318 (Eq 21-7)}$$

Because the subject wall has a ratio of h_w/l_w greater than 2.0, $\alpha_c = 2.0$. For normal weight concrete, $\lambda = 1.0$. Therefore:

$$V_n = A_{cv}(2(1.0)\sqrt{f_c'} + \rho_t f_y)$$

As prescribed in ACI 318 §9.3.4.(a), the shear strength reduction factor, ϕ, shall be 0.6 for the design of walls if their nominal shear strength is less than the shear corresponding to development of their nominal flexural strength.

$$\phi V_n = 0.6(20\text{ in})(366\text{ in})(2(1.0)\sqrt{5000} + \rho_t(60{,}000\text{ psi})) = (621 + 263{,}000\,\rho_t)\text{ kips}$$

At each level, the amount of horizontal reinforcement provided for shear strength is given in Table 5-5. Note that for all levels above Level 2, the minimum reinforcement ratio of 0.0025 governs the amount of horizontal reinforcement. (§21.9.2.1)

Table 5-5. Horizontal reinforcement for ACI 318 shear strength requirements

Level	V_E (kips)	Horizontal Reinforcement	ρ_n	ϕV_n (kips)
Roof	93	#5@12 inches E.F.	0.00258	1300
8	253	#5@12 inches E.F.	0.00258	1300
7	423	#5@12 inches E.F.	0.00258	1300
6	604	#5@12 inches E.F.	0.00258	1300
5	794	#5@12 inches E.F.	0.00258	1300
4	996	#5@12 inches E.F.	0.00258	1300
3	1230	#5@12 inches E.F.	0.00258	1300
2	1420	#6@12 inches E.F.	0.00367	1585

ACI 318 §21.9.4.4 requires that V_n shall not be taken greater than $8A_{cv}\sqrt{f'_c}$

$$8A_{cv}\sqrt{f'_c} = 8(20\text{ in})(366\text{ in})\sqrt{5000} = 4140\text{ kips} > 1585\text{ kips} \quad \ldots \quad o.k.$$

5b. Blue Book recommendations

Shear demand

To comply with the Blue Book requirement of providing shear strength in excess of the shear corresponding to wall flexural strength, an amplified shear demand is considered.

The Blue Book (2009, Article 09.01.010) gives the following equation for the shear amplification factor, ω_v, which accounts for inelastic dynamic effects.

$$\omega_v = (1.3 + n/30), \text{ for buildings over 6 stories, where } n = \text{number of stories}$$

$$= (1.3 + 8/30) = 1.57$$

As indicated in the Blue Book, the ω_v factor is derived for analysis using inverted triangular distributions of lateral forces. If a response spectrum analysis is carried out, a somewhat lower ω_v factor can be justified in some cases, as shown in Figure 5-12.

The shear demand is based on the upper bound of flexural strength, taken as the "probable flexural strength," M_{pr}, defined in ACI 318 §2.1.

M_{pr} is calculated using a strength reduction factor ϕ of 1.0, reinforcement strength $1.25f_y$, and the load combination of ASCE/SEI 7-05 §12.3.2 that results in the greatest flexural strength (in this case corresponding to greater axial force). Table 5-6 shows the probable flexural strength along with the nominal flexural strength for comparison.

Table 5-6. Flexural strength comparison

Quantity	Axial Force Considered	Reinforcement Strength	Flexural Strength
Probable strength	$P_u = 1.41P_D + P_L = 2470$ kips	$1.25f_y = 75$ ksi	$M_{pr} = 120{,}000$ kip-ft
Nominal strength	$P_u = 0.686P_D = 1060$ kips	$f_y = 60$ ksi	$M_n = 86{,}800$ kip-ft

At the base of the wall, the magnified shear demand $V_u{*}$ is calculated as

$$V_u{*} = \omega_v(M_{pr}/M_u)(V_E) = 1.57(120{,}000 \text{ kip-ft}/73{,}000 \text{ kip-ft})(1420 \text{ kips}) = 3665 \text{ kips}$$

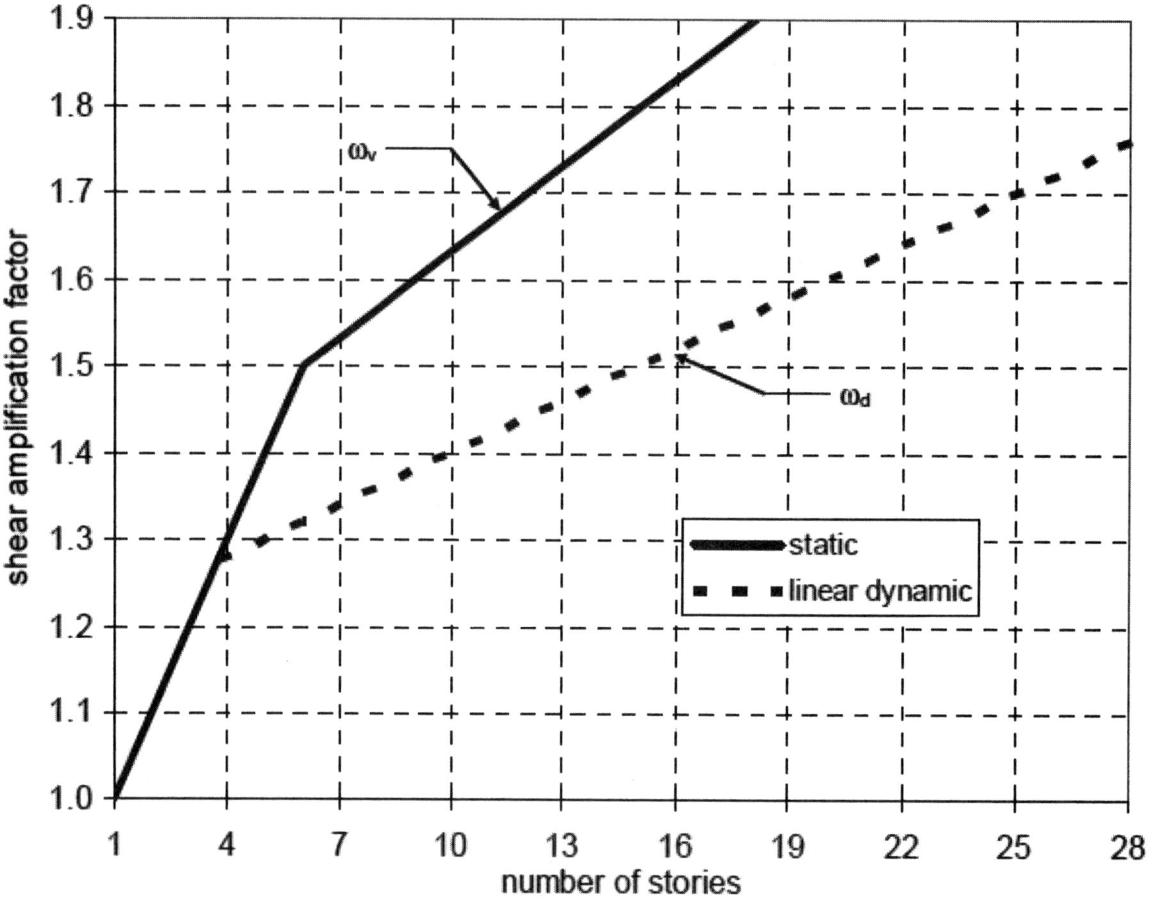

Figure 5-12. Relationship of shear amplification factor to number of stories

Shear capacity

Because this design example provides nominal shear strength to exceed the shear corresponding to flexural strength, a strength reduction factor, ϕ, of 0.75 can be used. As before, Equation (21-7) is used to calculate shear capacity:

$$\phi V_n = 0.75(20 \text{ in})(366 \text{ in})(2(1.0)\sqrt{5000} + \rho_t(60{,}000 \text{ psi})) = (776 + 329{,}000\,\rho_t) \text{ kips}$$

For the shear demand of 3665 kips, the required amount of horizontal reinforcement is calculated

$$\rho_t = (3665 \text{ kips} - 776 \text{ kips})/329{,}000 = 0.0088$$

Try #7 @ 12 inches on center each face

$$\rho_t = \frac{2(0.60 \text{ in}^2)}{6 \text{ in}(20 \text{ in})} = 0.0100 > 0.0088 \;\ldots\; o.k.$$

This amount of shear reinforcement is provided over the bottom story of the wall. For the other stories, the recommended amount of horizontal reinforcement, based on the magnified shear demand V_u*, is calculated as shown in Table 5-7. For all stories of the building, M_{pr}/M_u is that value calculated at the base plastic hinge of the wall. Likewise, the value of ω_v remains constant for all stories of the building.

Table 5-7. Horizontal reinforcement based on Blue Book shear design recommendations

Level	V_E (kips)	V_u* (kips)	Horizontal Reinforcement	ρ_t	ϕV_n (kips)
Roof	93	240	#5@12 inches E.F.	0.00258	1627
8	253	653	#5@12 inches E.F.	0.00258	1627
7	423	1092	#5@12 inches E.F.	0.00258	1627
6	604	1559	#5@12 inches E.F.	0.00258	1627
5	794	2049	#5@6 inches E.F.	0.00517	2478
4	996	2570	#6@6 inches E.F.	0.00733	3192
3	1230	3174	#6@6 inches E.F.	0.00733	3192
2	1420	3665	#7@6 inches E.F.	0.01000	4070

Paulay and Priestley (1992) recommend equations for shear strength that are somewhat different from Equation 21-7, and in which the shear strength at plastic hinge zones is taken to be less than that at other wall locations. FEMA 306 (ATC 1999) also gives equations for wall shear strength. For the wall design in this design example, the Paulay and Priestley shear strength equations result in nearly identical amounts of horizontal reinforcement as does Equation 21-7.

5c. Discussion of ACI 318 and Blue Book results for shear reinforcement

A comparison of Tables 5-5 and 5-7 shows that the Blue Book recommendation (2009, Article 09.01.010) of providing shear strength that exceeds flexural strength results in more horizontal reinforcement in the bottom four stories of the wall than that required by the code. The code approach of using $\phi = 0.6$ rather than $\phi = 0.75$ results in only a 1.25 increase over static elastic results, whereas the Blue Book approach of using ω_v and M_{pr} results in a 2.58 increase ($V_u{}^*/V_E$) for this design example. The Blue Book approach is recommended by SEAOC, as it leads to more ductile wall behavior.

In the upper four stories of the wall, the code minimum amount of horizontal steel ($\rho_n = 0.0025$) is adequate to meet both the ASCE/SEI 7-05 requirements and the Blue Book recommendations.

The wall elevation of Figure 5-9 shows the horizontal reinforcement per the Blue Book recommendation.

6. Sliding shear (shear friction)

At construction joints and flexural plastic hinge zones, walls can be vulnerable to sliding shear failure. Typically, low-rise walls are more vulnerable. If construction joint surfaces are properly prepared according to ACI 318 §11.6.9, taller walls should not be susceptible to sliding shear failure.

Sliding shear can be checked using the shear friction provisions of ACI 318 §11.6. Shear strength is computed by Equation 11-25

$$V_n = A_{vf} f_y \mu$$

μ is the coefficient of friction, which is taken as 1.0λ assuming there is a construction joint at the first story with a surface roughened to ¼-inch amplitude, where $\lambda = 1.0$ for normal weight concrete.

A_{vf} is the amount of shear-transfer reinforcement that crosses the potential sliding plane. For the wall in this design example, all vertical bars in the section are effective as shear-transfer reinforcement (ACI 318 §R11.6.7). At the base of the wall

$$A_{vf} = 30(1.56 \text{ in}^2) + 54(0.79 \text{ in}^2) = 89.5 \text{ in}^2$$

Section 11.6.7 indicates that "permanent net compression" can be taken as additive to the force $A_{vf} f_y$, thus the lower bound axial force, $0.686P_D$, can be included in Equation 11-25

$$V_n = (A_{vf} f_y + 0.686P_D)\mu = [(89.5 \text{ in}^2)(60 \text{ ksi}) + 1060 \text{ kips}](1.0) = 6430 \text{ kips}$$

Section 11.6.5 requires that the shear friction strength not be taken greater than the smallest of $0.2f_c'A_c$, $(480 + 0.08f_c')A_c$, or $1600A_c$, where A_c is the concrete area. For the example wall with $f_c' = 5000$ psi, the $(480 + 0.08f_c')A_c$ criterion governs

$$V_n \le (480 + 0.08(5000) \text{ psi})(20 \text{ in})(366 \text{ in}) = 6442 \text{ kips} > V_u{}^* = 3665 \text{ kips} \quad \ldots \quad o.k.$$

By inspection, the sliding shear capacity at higher story levels of the building is also okay.

7. Boundary zone detailing

7a. Check for special boundary elements

Boundary elements at the edge of structural walls are designed according to ACI 318 §21.9.6. For the example wall, the need for special boundary elements is determined according to ACI 318 §21.9.6.2, which applies to walls that are effectively continuous from the base of structure to top of the wall and designed to have a single critical section for flexure and axial forces. Under this section, boundary zone detailing is required if

$$c > \frac{l_w}{600(\delta_u / h_w)} \qquad \text{ACI 318 (21-8)}$$

The quantity $\frac{\delta_u}{h_w}$ shall not be taken less than 0.007.

The neutral axis depth, c, should be calculated for the axial load case that results in the largest value of c, i.e., the case with higher axial force. Therefore

$$P_u = 1.41P_D + 1.0P_L + 1.0Q_E = 2470 \text{ kips}$$

Using the same procedure as in Part 3(d) above, the depth of the compression zone is determined to be $c = 82$ in.

δ_u is the design displacement defined in ACI 318 §2.2. The commentary (ACI 318 §R2.2) indicates that the displacement is calculated considering "modification factors to account for expected inelastic response." Thus, the design displacement corresponds to C_d times δ_{xe}, according to ASCE/SEI 7-05 §12.8.6. The term δ_{xe} is the elastic displacement of the top of the wall corresponding to the design seismic force. For this building, δ_{xe} is 1.55 inches; C_d is the deflection amplification factor, equal to 5. Therefore:

$$\delta_u = \text{design displacement} = \frac{C_d \delta_{xe}}{I_E} = \frac{5(1.55 \text{ in})}{1.0} = 7.75 \text{ in}$$

$$\frac{\delta_u}{h_w} = \frac{7.75}{95.4(12)} = 0.00677 \quad \text{Use } \frac{\delta_u}{h_w} = 0.007$$

$$\frac{l_w}{600(\delta_u / h_w)} = \frac{366}{600(0.007)} = 87.1 \text{ in} > c = 82 \text{ in}$$

Therefore, special boundary elements are not required. Please refer to Example 6 for further discussion in a case where special boundary elements are required.

7b. Boundary ties per ACI §21.9.6.5

Where special boundary elements are not required, §21.9.6.5 requires the following detailing consideration.

"If the longitudinal reinforcement ratio at the wall boundary is greater than $\frac{400}{f_y}$, boundary transverse reinforcement shall satisfy §§21.6.4.2 and 21.9.6.4(a). The maximum longitudinal spacing of transverse reinforcement in the boundary shall not exceed 8 in."

In this example, the longitudinal reinforcement ratio for 15 #8 in the boundary region of the wall at the top floor is

$$\rho = \frac{15(A_b)}{t_w(1.5\text{ in} + (6)(9\text{ in}) + 1.5\text{ in})} = \frac{15(0.79\text{ in}^2)}{20\text{ in}(57\text{ in})} = 0.010$$

which is greater than $\frac{400}{f_y} = 0.00667$

Therefore, boundary ties per §21.9.6.5(a) are required at the top story, and at all other stories that have a higher reinforcement ratio.

The minimum width of the boundary element is determined based on the requirements of Section 21.9.6.4(a): "The boundary element shall extend horizontally from the extreme compression fiber a distance not less than the larger of $c - 0.1l_w$ and $\frac{c}{2}$."

$$c - 0.1l_w = 82\text{ in} - 0.1(366\text{ in}) = 45.4\text{ in} \quad \ldots \textit{ governs}$$

$$\frac{c}{2} = \frac{82\text{ in}}{2} = 41\text{ in}$$

For the wall in this design example, #4 tie sets are provided enclosing the 15 longitudinal bars at each end of the wall. The tie sets have a tie leg located at each of the longitudinal bars, as shown in Figure 5-13. The ties create a boundary element that extends 58 inches horizontally from the extreme compression fiber. (58 in > 45.4 in ... *o.k.*)

ACI 318 §21.9.6.5 requires 8-inch maximum tie spacing. The SEAOC Blue Book (1999, §402.12) requires 6-inch maximum tie spacing, but only requires boundary ties "at potential plastic hinge regions." For the example wall, the potential plastic hinge region occurs at the base and extends over the first one to two stories, as discussed in Part 4b of this example. For this example the boundary ties are spaced at 6 inches on center over the first two stories, satisfying the Blue Book and ACI requirements. Boundary ties are then spaced at 8 inches on center over the remaining height of the wall, satisfying the ACI requirement. The spacing of the boundary ties is indicated in Figure 5-9.

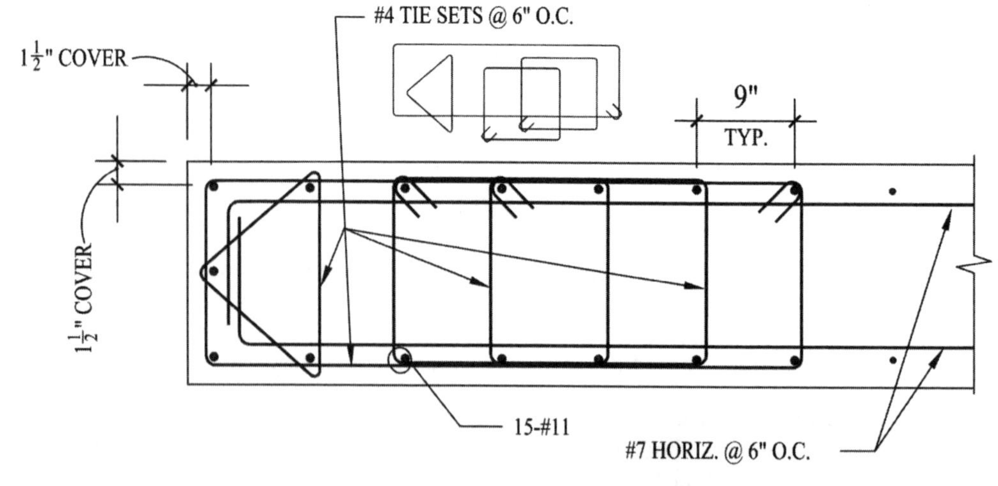

Figure 5-13. Boundary reinforcement at wall base

References

ACI-318. *Building Code Requirements for Reinforced Concrete.* American Concrete Institute, Farmington, Michigan, 2008.

ATC. *Evaluation of Earthquake Damaged Concrete and Masonry Wall Buildings*, prepared by the Applied Technology Council (ATC-43 project) for the Partnership for Response and Recovery. Federal Emergency Management Agency, Report No. FEMA-306, Washington, D.C., 1999.

CRSI. *Rebar Design and Detailing Data—ACI.* Concrete Reinforcing Steel Institute, Schaumberg, Illinois, 1996.

Jirsa, J.O., L.A. Lutz, and P. Gergely, 1979. "Rationale for Suggested Development, Splice, and Standard Hook Provisions for Deformed Bars in Tension." *Concrete International: Design & Construction*, Vol. 1, No. 7, July 1979, pp. 47-61.

Maffei, Joe. "Reinforced Concrete Structural Walls—Beyond the Code," *SEAONC Fall Seminar Proceedings.* Structural Engineers Association of Northern California, San Francisco, California, November, 1996.

Pacific Earthquake Engineering Research Center (PEER). *Tall Buildings Initiative: Guidelines for Performance-Based Seismic Design of Tall Buildings, Version 1.0*, University of California, Berkeley, November, 2010.

Paulay, T., and Priestley, M.J.N. *Reinforced Concrete and Masonry Buildings, Design for Seismic Resistance.* John Wiley & Sons, Inc., New York, 1992. (Chapter 5 covers seismic behavior and design of reinforced concrete walls, including examples. The book is not based on the ACI or UBC codes, but explains the principles that underlie several code provisions.)

Paulay, T., and Priestley, M.J.N. *Stability of Ductile Structural Walls.* ACI Structural Journal, Vol. 90, No. 4, July-August 1993.

SEAOC Seismology Committee. "Concrete Parking Structures" (Article 09.05.010). *SEAOC Blue Book Seismic Design Recommendations*, Structural Engineers Associate of California (SEAOC),Sacramento, California, www.seaoc.org/bluebook, September, 2009.

SEAOC Seismology Committee, *Recommended Lateral Force Requirements and Commentary* (Blue Book), Structural Engineers Associate of California (SEAOC), Seventh Edition, Sacramento, California, 1999.

SEAOC Seismology Committee. "Reinforced Concrete Structures" (Article 09.01.010). *SEAOC Blue Book Seismic Design Recommendations*, Structural Engineers Associate of California (SEAOC),Sacramento, California, www.seaoc.org/bluebook, September, 2009.

Standards New Zealand, 1995. *Concrete Structures Standard: Part 2—Commentary on the Design of Concrete Structures (NZS 3101: Part 2).* Standards New Zealand, May, p. 84.

Structurepoint. *spColumn Version 4.60: Design and Investigation of Reinforced Concrete Column Sections*, Structurepoint, Skokie, Illinois, 2010.

Design Example 6
Reinforced Concrete Wall with Coupling Beams

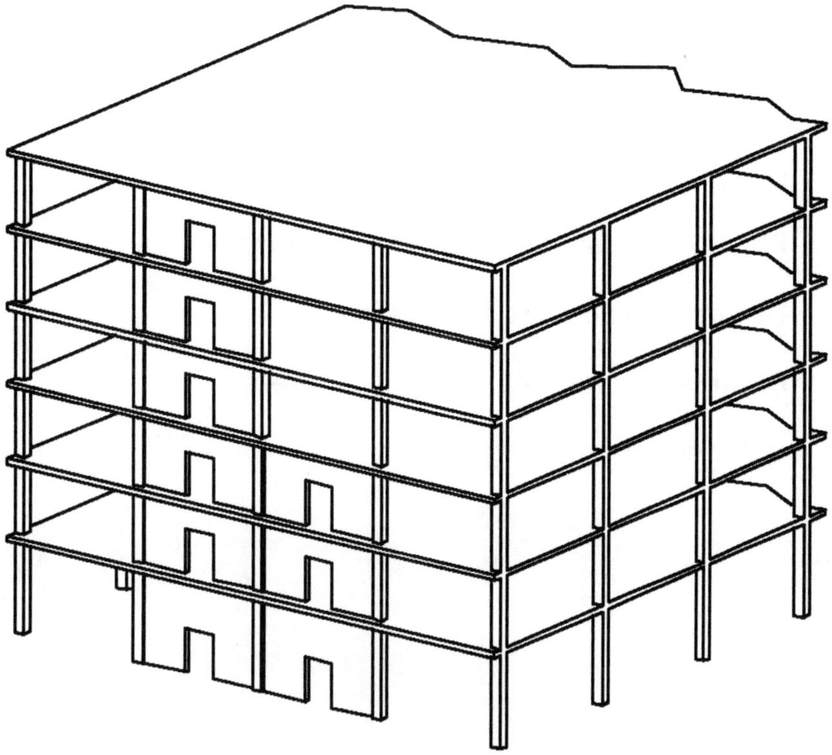

Figure 6-1. Six-story concrete office building (partial view)

Overview

The structure in this design example is a six-story office building with reinforced concrete walls as its seismic-force-resisting system. The example focuses on the design and detailing of one of the reinforced concrete walls. This is a coupled wall running in the transverse building direction and is shown in Figure 6-1. The example assumes that design lateral forces have already been determined for the building, and that the seismic moments, shears, and axial forces on each of the wall components, from the computer analysis, are given.

The purpose of this design example is to illustrate the design of coupling beams and other aspects of reinforced concrete wall design.

Outline

This design example illustrates the following parts of the design process

1. **Load combinations for design**
2. **Preliminary sizing of wall**
3. **Coupling beam design**
4. **Design of wall piers for flexure**
5. **Plastic analysis of flexural mechanism in walls**
6. **Design of wall piers for shear**
7. **Boundary zone detailing of wall piers**
8. **Detailing of coupling beams**

Given Information

$S_1 = 0.65\text{g}$

$S_S = 1.60\text{g}$

Site class D

$S_{DS} = 1.07\text{g}$

Occupancy Category I

Seismic Design Category D

Reliability/redundancy factor, $\rho = 1.0$

Seismic Importance factor, $I = 1.0$

Concrete strength, $f'_c = 4000$ psi

Steel yield strength, $f_y = 60$ ksi

The wall to be designed is one of several reinforced concrete walls in the building. The wall elevation, a plan section, and the design forces are shown in Figure 6-2. A linear static analysis of the wall for lateral forces, using a computer analysis program, gives the results shown in Figure 6-3, which shows the moments and shear for each coupling beam (i.e., wall spandrel), and the moments, shear and axial forces for each vertical wall segment (i.e., wall pier).

Lateral story displacements, corresponding to effective section properties, are also shown on the figure. In the analysis model, the member stiffness used is 30 percent of the gross member stiffness for the walls, and 10 percent of the gross member stiffness for the coupling beams. The recommendations for member stiffness assumptions are based on §5.3 of Paulay and Priestley (1992).

In this design example, the displacement output is used in Part 7b for determining the need for special boundary elements. In an actual building design, the displacements would also need to be considered for: (a) design of elements not part of the lateral-force-resisting system, (b) building separations, and (c) P-Δ analysis.

Gravity loads are not included in the computer model. Gravity effects are added separately by hand calculations.

The design and analysis of the structure is based on a response modification coefficient, R, of 5 (ASCE/SEI 7-05 Table 12.2-1) for a bearing wall system with special reinforced concrete shear walls. The deflection amplification factor, C_d, is 5. The SEAOC Blue Book (2009, Article 09.01.010) expresses the opinion that the R value for concrete bearing wall systems ($R = 5$) and that for walls in building frame systems ($R = 6$) should be the same, which may be justified based on detailing provisions. To be consistent with the current code requirements though, this design example uses $R = 5$.

This design example follows the general building code requirements of the 2009 IBC and ASCE/SEI 7-05. Unless indicated otherwise, section numbers used in this example refer to ASCE requirements. For structural concrete design, the 2009 IBC §1901.2 references the *American Concrete Institute Building Code* (ACI 318). This example follows the requirements of ACI 318-08. Discussions of the SEAOC Blue Book recommendations refer to the document *Recommended Lateral Force Recommendations and Commentary* (SEAOC, 1999) as well as the Blue Book online articles on specific topics (SEAOC, 2009) as applicable.

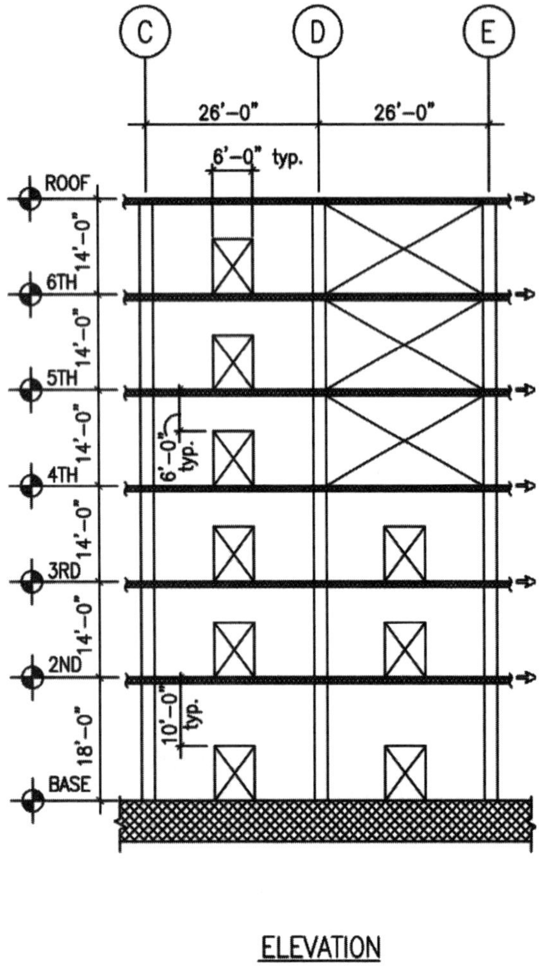

LATERAL FORCES F_{xi}	GRAVITY LOADS EACH COLUMN	
	P_D	P_L
337k	148	11
319k	309	28
259k	471	48
198k	634	69
138k	796	90
78k	962	110

P_D = EXCLUDES WALL WEIGHT

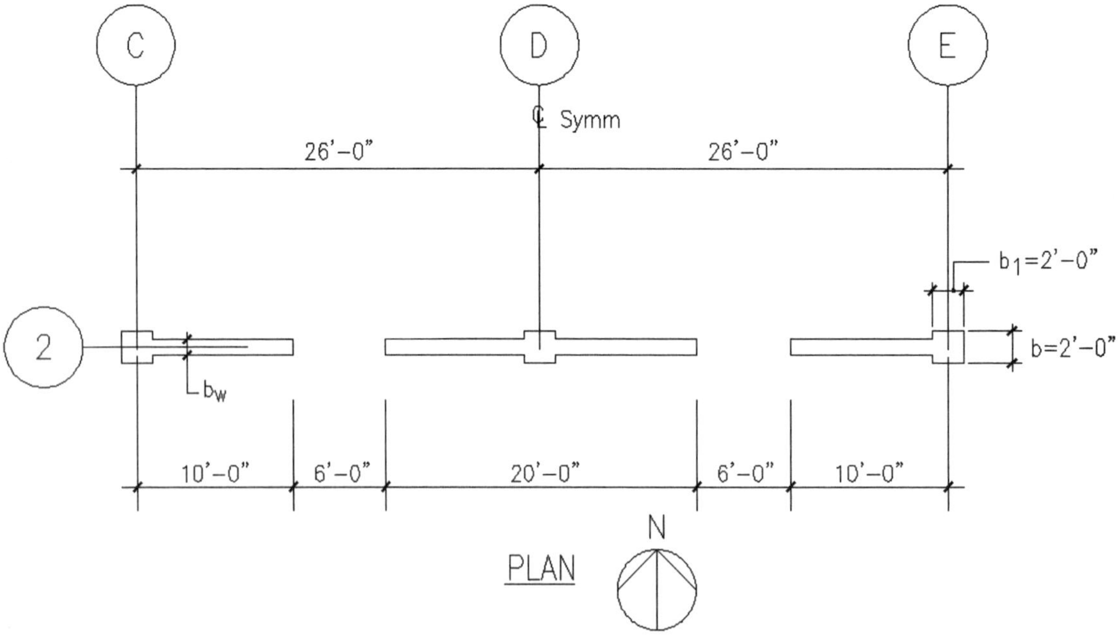

Figure 6-2. Wall elevation, plan section, and design forces

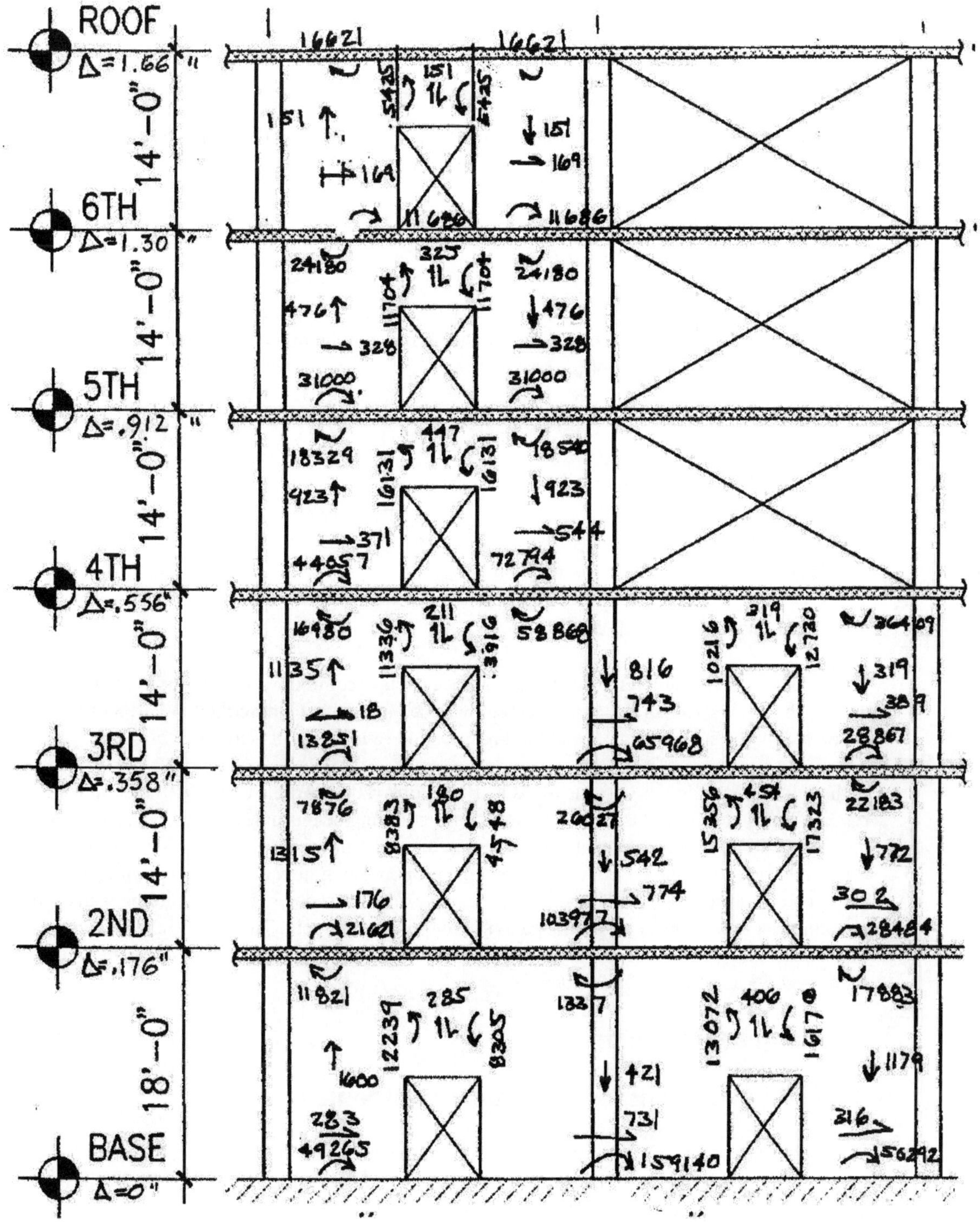

Figure 6-3. Results of ETABS computer analysis (kips, inches)

Calculations and Discussion **Code Reference**

1. Load combinations for design

Load combinations for reinforced concrete are discussed in detail in Part 1 of Design Example 5. As in that example, the governing load combinations become

$$(1.2 + 0.2S_{DS})D + \rho Q_E + L + 0.2S$$

$$(0.9 - 0.2S_{DS})D + \rho Q_E + 1.6H$$

Because there is no lateral earth pressure, $H = 0$, and since there is no snow load, $S = 0$. As indicated in §12.4.2.3 the load factor on L in the above combination is permitted to equal 0.5 since the given structure is an office building with $L_o = 50$ psf per IBC Table 1607.1.

With $S_{DS} = 1.07$ and $\rho = 1.0$, the governing load combinations for this design example are

$$1.41D + Q_E + 0.5L$$

$$0.686D + Q_E$$

The forces shown in Figure 6-3 correspond to Q_E.

2. Preliminary sizing of wall

For walls with diagonally reinforced coupling beams, the required wall thickness is often dictated by the layering of the reinforcement in the coupling beam, described in Part 8 of this example. For the subject wall, a wall thickness, b_w, of 16 inches will be tried.

Although not required by code, the SEAOC Blue Book (2009, Article 09.01.010) recommends rotation limits of 0.03 to 0.05 radians for confined coupling beams unless higher values can be justified by testing specimens that have aspect ratios and reinforcement similar to those to be used in the design. Rotation limits can affect the proportioning of walls so that coupling beams are not too short relative to wall piers. This design example assumes that the building walls and coupling beams have been proportioned to satisfy this requirement. This can be checked using the displacements δ_u from Table 6-12 and calculating the corresponding coupling beam rotation θ_{cb} as described in Part 5 of this example.

3. Coupling beam design

3a. Requirement for diagonal reinforcement

Code requirements for the diagonal reinforcement of coupling beams (ACI 318 §21.9.7.2), are based on the clear length-to-overall depth ratio for the coupling beam, l_n/h, and on the level of shear stress in the coupling beam.

For the wall in this design example $l_n/h = 72$ in/72 in $= 1.00$ for the typical coupling beam, and $l_n/h = 72$ in/120 in $= 0.60$ for the coupling beams at the second floor.

As shown in Table 6-1 (5th column), for four of the nine coupling beams the shear exceeds $4\lambda\sqrt{f'_c}A_{cw}$, where $A_{cw} = b_w h$ and $\lambda = 1.0$ for normal weight concrete. For these coupling beams, diagonal reinforcement is required.

For the five coupling beams that have lower shear stress, diagonal reinforcement is not required by ACI 318. Designing these five coupling beams without diagonal reinforcement, using horizontal reinforcement to resist flexure and vertical stirrups to resist shear, might lead to cost savings in the labor to place the reinforcing steel.

In this design example, however, diagonal reinforcement is used in all of the coupling beams of the wall because: (a) it can simplify design and construction to have all coupling beams detailed similarly, (b) research results show that diagonal reinforcement improves coupling beam performance, even at lower shear stress levels, as discussed in the SEAOC Blue Book (1999, §C407.7), and (c) uniform and consistent yielding up the height of the structure results in better overall performance.

Table 6-1. Coupling beam forces and diagonal reinforcement

Grid Line	Level	V_u (kips)	h (in)	$V_u/B_w h\sqrt{f'_c}$ (1)	Diagonal bars	A_d (in^2)	α (degrees)	ϕV_n (kips)	$\phi V_n/V_u$
C-D	Roof	151	72	2.1	4 #8	3.16	37.9	175	1.16
C-D	6th	325	72	4.5	4 #11	6.24	37.9	345	1.06
C-D	5th	447	72	6.1	6 #11	9.36	36.0	495	1.11
C-D	4th	211	72	2.9	4 #9	4.00	37.9	221	1.05
C-D	3rd	180	72	2.5	4 #9	4.00	37.9	221	1.23
C-D	2nd	285	120	2.3	4 #9	4.00	53.1	288	1.01
D-E	4th	319	72	4.4	6 #9	6.00	36.0	317	0.99
D-E	3rd	454	72	6.3	6 #11	9.36	36.0	495	1.09
D-E	2nd	406	120	3.3	4 #11	6.24	53.1	449	1.11

Note:

1. Diagonal bars are required per ACI 318 §21.9.7.2 when this ratio exceeds 4 and $l_n/h < 2$.

3b. Design of diagonal reinforcement

Diagonal reinforcement is provided in the coupling beams according to Equation 21-9 of ACI 318 §21.9.7.4

$$\phi V_n = \phi 2 A_{vd} f_y \sin\alpha \le 10\sqrt{f'_c}A_{cw} \qquad \text{(Eq 21-9)}$$

Each group of diagonal bars must consist of at least four bars per ACI 318 §21.9.7.4(b). The calculation of the required diagonal reinforcement is shown in Table 6-1. For coupling beams with higher shear stresses, six bars are needed in each group, as shown in Table 6-1.

The angle, α, of the diagonal bars is calculated based on the geometry of the reinforcement layout, as shown in Figure 6-4. The value of α depends somewhat on the overall dimension of the diagonal bar group and on the clearance between the diagonal bar group and the corner of the wall opening. This affects the dimension, x, shown in Figure 6-4 and results in a slightly different value of α for a group of six bars compared to that for a group of four bars, as shown in Table 6-1.

As discussed in Part 8 of this design example, ACI 318 provides different options for the detailing of coupling beams, and each option may require a number of preliminary design iterations to determine required bar sizes and the lateral dimensions of the diagonal bar group. Preliminary design iterations are not shown in this design example.

The provided diagonal bars are shown in Figure 6-5. The diagonal reinforcement provides both the shear and flexural resistance for the coupling beam. The vertical component of the bar forces provides resistance to shear per Equation 21-9. The horizontal component of the bar forces provides resistance to the moment, equal to $Vl_n/2$ at each end of the coupling beam. As can be derived from the geometry of the beam, Equation 21-9 automatically provides adequate shear strength in the coupling beam corresponding to the flexural forces.

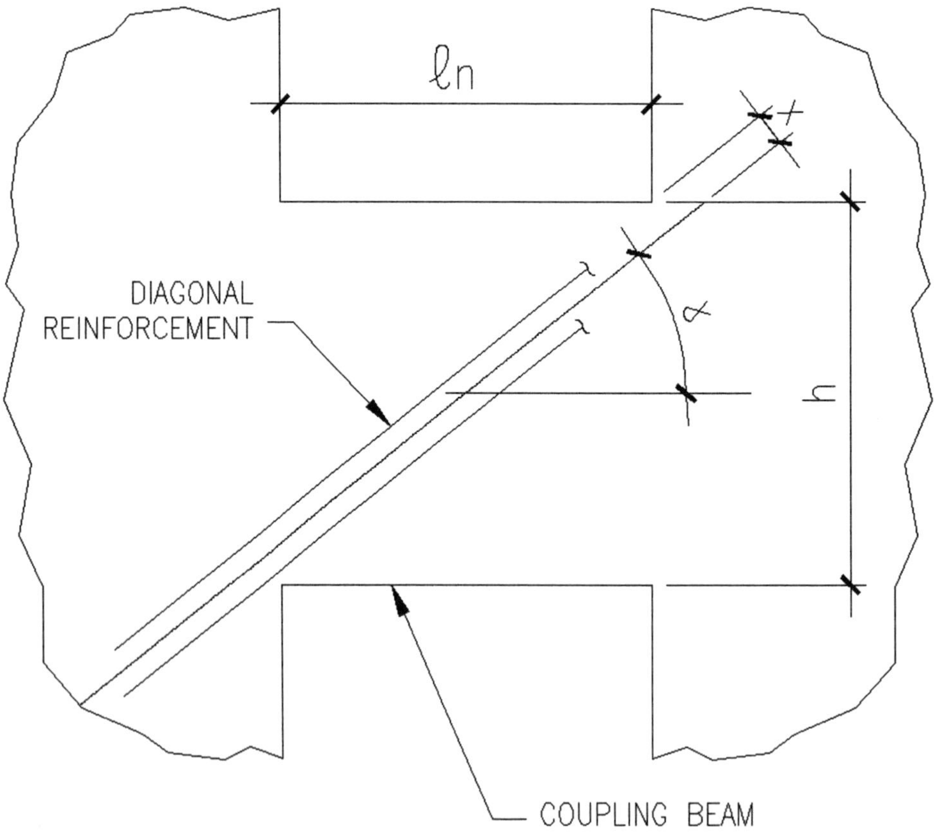

Figure 6-4. Geometry of coupling beam diagonal bars

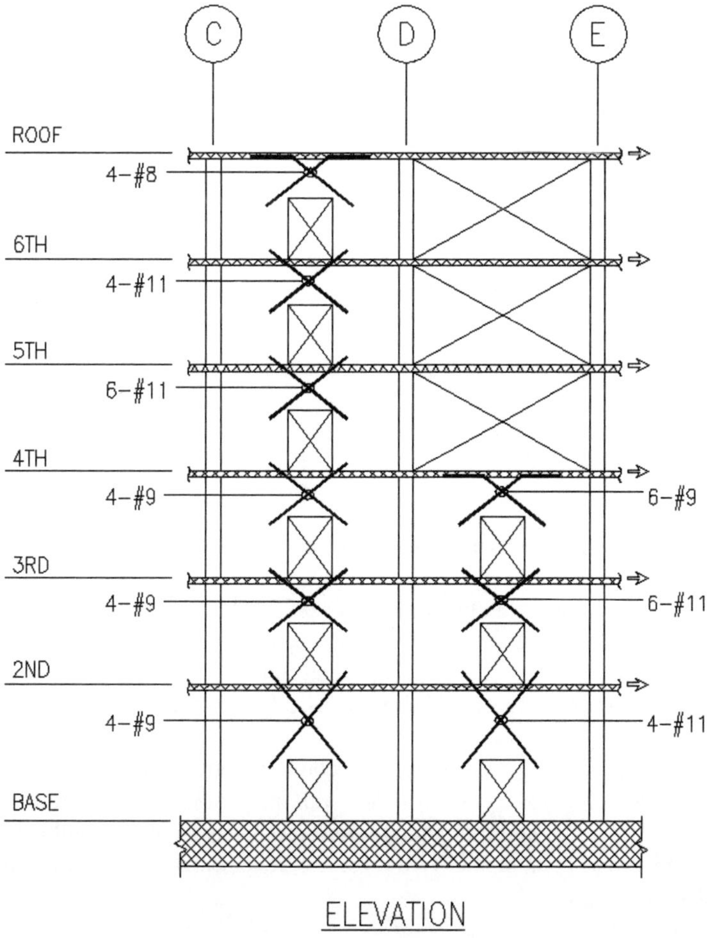

Figure 6-5. Diagonal bars provided in coupling beams

4. Design of wall piers for flexure

The design of the vertical wall segments for flexure is carried out following the procedures and recommendations given for conventional "solid" walls. This is shown in Part 3 of Design Example 5. From Figure 6-3, the critical wall segments (i.e., those with the highest moments or earthquake axial forces) include the wall pier at the fourth floor on line D, and the wall piers at the base on lines C and E. The 20-foot-long wall pier on line D at the base is also checked.

4a. Critical moments and axial forces

As can be seen from Figure 6-2, the gravity loads on each wall pier are not concentric with the centroid of the wall pier cross section. Therefore, gravity load moments must be considered in the design of flexural reinforcement. The dead and live loads (except wall self-weight) in Figure 6-2 act at the column grid lines, and have an eccentricity, e_{DF}, with respect to the section centroid, as given in Table 6-3 (Note: The calculation of weights and section centroids, e_{DF} and e_{DW}, is not shown). The wall self-weight provides additional dead load at each level, equal to the values given in Table 6-2.

Table 6-2. Dead load from wall self-weight

Level	Line C		Line D		Line E	
	Sum of Wall Weight (kips)	Eccentricity, e_{DW} (ft)[1]	Sum of Wall Weight (kips)	Eccentricity, e_{DW} (ft)[1]	Sum of Wall Weight (kips)	Eccentricity, e_{DW} (ft)[1]
Above 6th	26	2.06	26	–2.06	0	
Above 5th	53	2.06	53	–2.06	0	
Above 4th	79	2.06	79	–2.06	0	
Above 3rd	106	2.06	132	–3.71	26	–2.06
Above 2nd	132	2.06	185	–2.65	53	–2.06
At base	166	2.03	252	–1.94	86	–2.00

Note:

1. e_{DW} = distance between centroid of weight and centroid of wall section.

The calculation of the factored forces on the critical wall piers is shown in Table 6-3. In this table, gravity moments are calculated about the section centroid, using the gravity loads acting at the column centerline, P_{DF} and P_L, plus the dead load from wall self-weight, P_{DW}. Earthquake moments, M_E, are taken from Figure 6-3.

Loads are factored according to the combinations discussed in Part 1 of this design example, giving two cases for each wall pier: minimum axial force and maximum axial force. The minimum axial force case is based on the combination of E_h with 0.686D, and the maximum axial force case is based on the combination of E_h with $1.41D + 0.5L$.

Considering that larger axial compression generally increases flexural strength, potentially governing combinations are shown as shaded areas in Table 6-3.

Table 6-3. Calculation of factored axial forces and moments on critical wall piers

Level	Line	P_{DF} (kips)	e_{DF} (ft)	P_{DW} (kips)	e_{DW} (ft)	P_L (kips)	Direction of force	P_E (kips)	M_E (kip-ft)	M_D (kip-ft)	M_L (kip-ft)	Minimum Axial		Maximum Axial		See Figure
												P_u	M_u	P_u	M_u	
4th	D	471	4.13	79	–2.06	48	west	–923	–6070	1763	200	–546	–4860	–121	–3477	6-7(a)
4th	D	471	4.13	79	–2.06	48	east	923	6070	1763	200	1300	7280	1725	8663	6-7(b)
1st	C	962	–4.13	166	2.03	110	west	1600	–4105	–3595	–454	2373	–6571	3249	–9415	6-7(b)
1st	C	962	–4.13	166	2.03	110	east	–1600	4105	–3595	–454	–827	1632	49	–1205	6-7(c)
1st	E	962	4.13	86	–2.00	110	west	–1179	–4191	3776	454	–460	–1600	357	1376	6-7(c)
1st	E	962	4.13	86	–2.00	110	east	1179	4191	3776	454	1898	6782	2715	9758	6-7(b)
1st	D	962	0	252	–1.94	110	west	–421	–13,250	–538	0	411	–13,619	1350	–14,011	6-7(d)

Notes:

P_{DF} = dead load distributed over floor area, which acts at the column line.

e_{DF} = distance between P_{DF} and centroid of wall section.

P_{DW} = dead load from wall self-weight.

e_{DW} = distance between P_{DW} and centroid of wall section.

4b. Vertical reinforcement

The program spColumn (Structurepoint, 2010) is used to design the reinforcement in each wall pier. Figure 6-6 shows a wall section with the typical layout of vertical reinforcement. Typical reinforcement in the "column" portion of the wall piers is eight #9 and typical vertical reinforcement in the wall web is #7@12 inches on each face. The spColumn results of Figures 6-7a, 6-7b, and 6-7c show that this reinforcement is adequate in all locations except line C at the first floor and line D at the fourth floor where eight #11 are required instead of eight #9. Figure 6-7d shows that the typical reinforcement provides adequate flexural strength to the 20-foot-long wall pier on line D. The points in these figures indicate the (M_u/ϕ, P_u/ϕ) coordinate of each critical force. These values are obtained using the corresponding strength reduction factor ϕ per ACI 318 §9.3.2 as discussed in the previous design example. Selected results of such analysis for the most severe cases are summarized in Table 6-4. For all other load cases the wall section is tension controlled and $\phi = 0.9$ is used.

Table 6-4. Calculation of ϕ value for selected load cases

Level	Line	Direction of force	ϕP_n (kips)	ϕM_n (kip-ft)	c (in)	ε_t	ϕ
1st	C	West	3250	9,790	79.4	0.00187	0.65
1st	C	West	2380	12,200	40.0	0.00668	0.90
1st	E	East	1900	10,700	34.9	0.00809	0.90
1st	E	East	2720	11,400	55.0	0.00404	0.82

Figure 6-8 shows the vertical reinforcement provided in the wall piers to satisfy flexural strength requirements. Note that the vertical reinforcement at the fourth floor piers is increased to eight #11 for both columns and #8@12 inches at the wall webs. The reasons for this will be discussed in Part 5 of this design example.

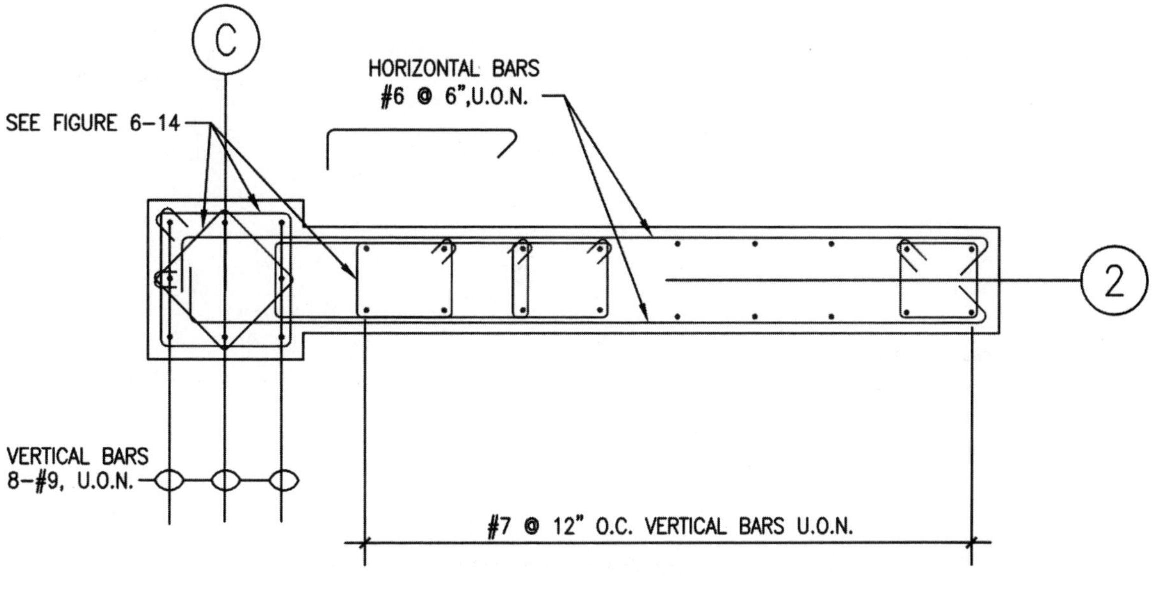

Figure 6-6. Section through wall pier in vicinity of line C

a.

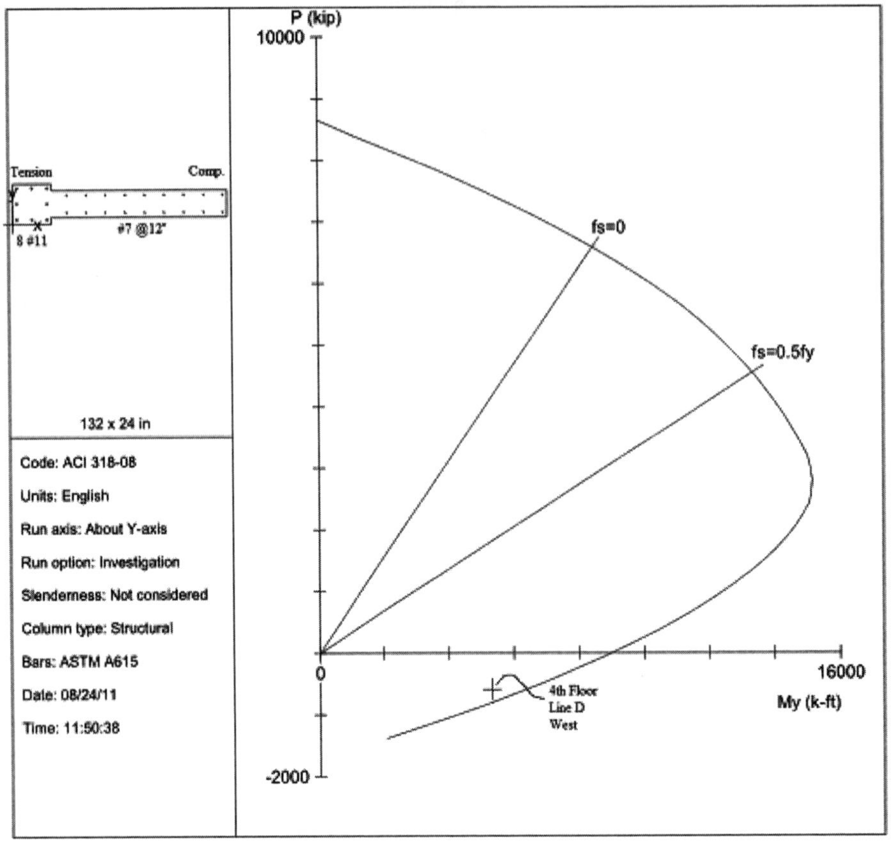

b.

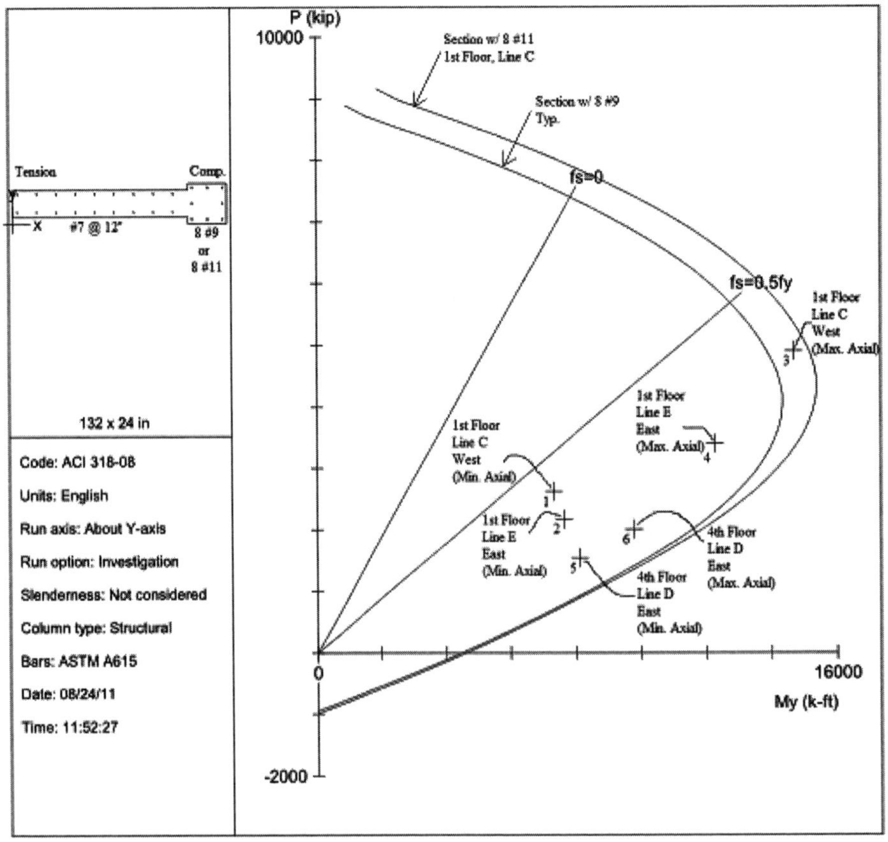

Figure 6-7. spColumn results for design of vertical reinforcement

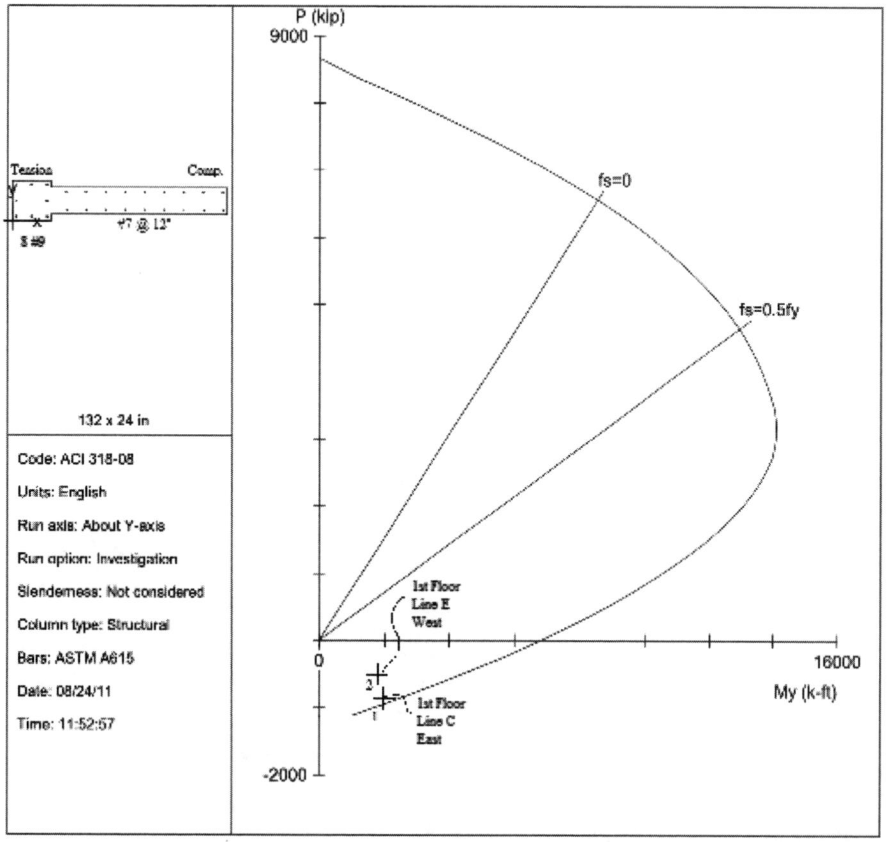

c.

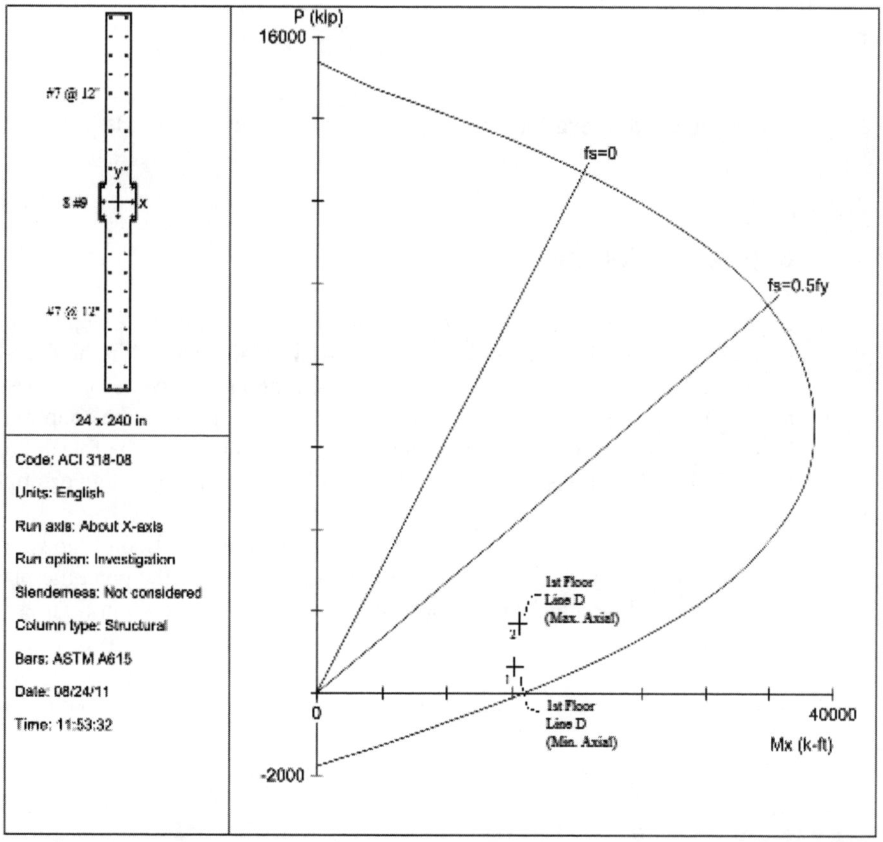

d.

Figure 6-7 (continued). spColumn results for design of vertical reinforcement

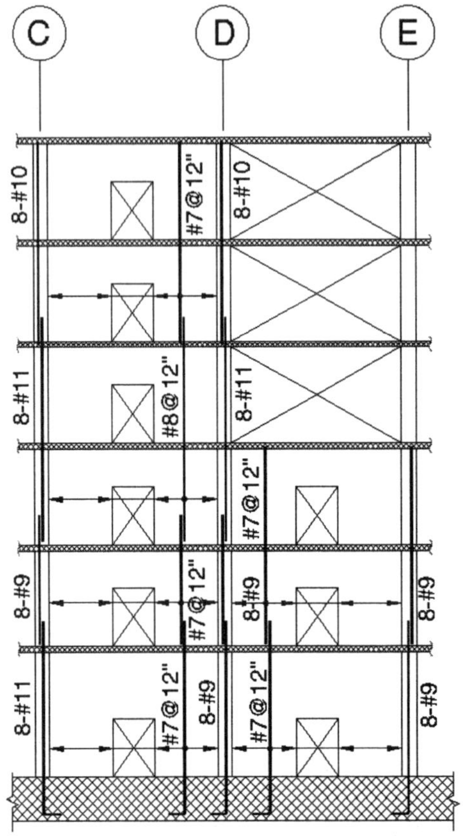

Figure 6-8. Elevation showing vertical wall reinforcement

4c. Lap splice locations

In general, lap splices should be avoided in potential plastic hinge regions of concrete structures. This is discussed in Part 4b of Design Example 5 and in the Blue Book (1999, §C404.3). For this example wall, plastic hinging is expected (and designed to occur) at the base of each wall pier at the ground floor and in the coupling beams. Plastic hinging may also be possible above the wall setback, in the fourth floor wall piers. (This will be investigated in more detail in Part 5 of this design example.)

Figure 6-8 shows the vertical wall reinforcement and the location of lap splices. The lap splices of the vertical reinforcement are located to avoid the potential plastic hinge regions in first floor and fourth floor wall piers, which are shown in Figures 6-10 and 6-11 in Parts 5b and 5c.

Any lap splices that are in potential plastic hinge locations are required by ACI 318 §21.9.2.3(c) to be designed for $1.25f_y$.

5. Plastic analysis of flexural mechanism in wall

This part of the design example presents a plastic analysis methodology that is not a code requirement. It is included to assist the reader in understanding the post-elastic behavior of coupled structural walls and how they can be analyzed for seismic forces when elements of the wall are yielding.

Plastic analyses are not required by the ASCE/SEI 7-05 or ACI 318, but they are recommended in the SEAOC Blue Book: (a) to establish shear demand corresponding to flexural strength, and (b) to identify potential plastic hinge regions where special boundary and splicing requirements may be necessary. With the trend toward nonlinear analysis methods, the ability to use plastic analyses will become increasingly important. The first three chapters of the textbook *Plastic Design in Steel* (ASCE, 1971) summarize the basic principles and methods of plastic design.

Given below is an illustration of plastic analysis for the reinforced concrete walls and coupling beams of this design example.

5a. Probable flexural strength

The "probable flexural strength," M_{pr}, will be determined in calculating shear demands, according to the Blue Book recommendations. As defined in ACI 318 §R21.5.4, M_{pr} is calculated assuming a tensile stress in the longitudinal bars of $1.25 f_y$ and a strength reduction factor, ϕ, of 1.0. For the purposes of this plastic analysis, we will neglect earthquake axial forces, in calculating M_{pr} for each wall pier and assume an axial force of $1.2P_D + 0.5P_L$. In reality, the wall pier with earthquake axial tension will have a decreased flexural strength, while the wall pier with earthquake axial compression will have an increased flexural strength. These effects tend to cancel out so that our plastic analysis will give an acceptable estimate of (a) the governing mechanism of response, and (b) the shear corresponding to the development of a mechanism at probable flexural strength. Table 6-5 shows M_{pr} values for the critical wall piers, based on the spColumn results shown in Figure 6-9.

a.

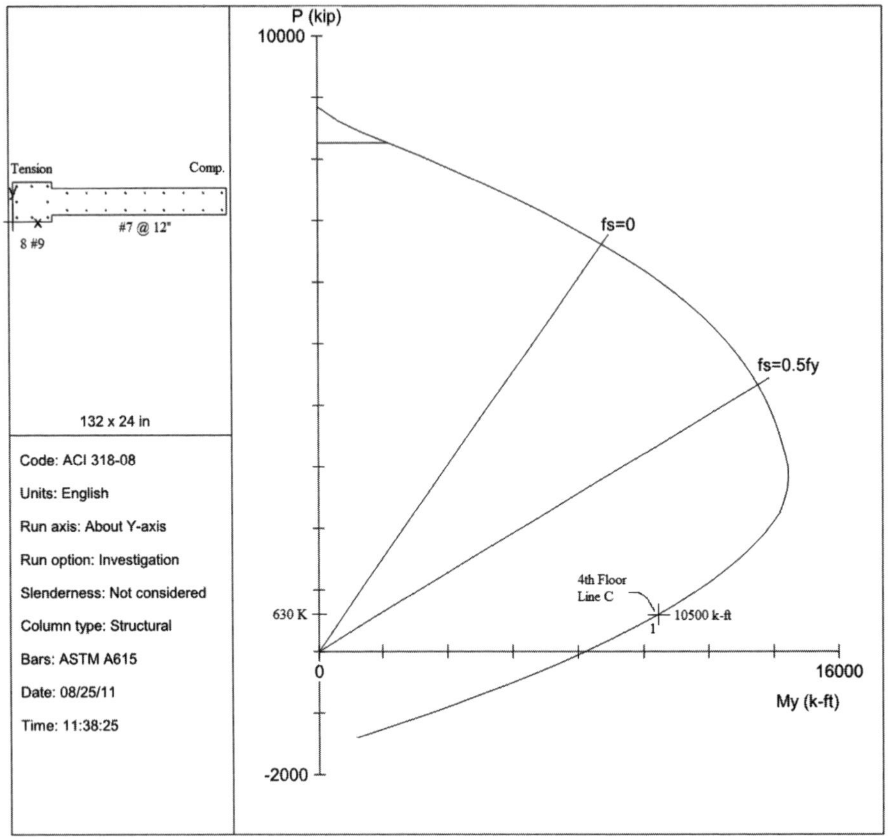

b.

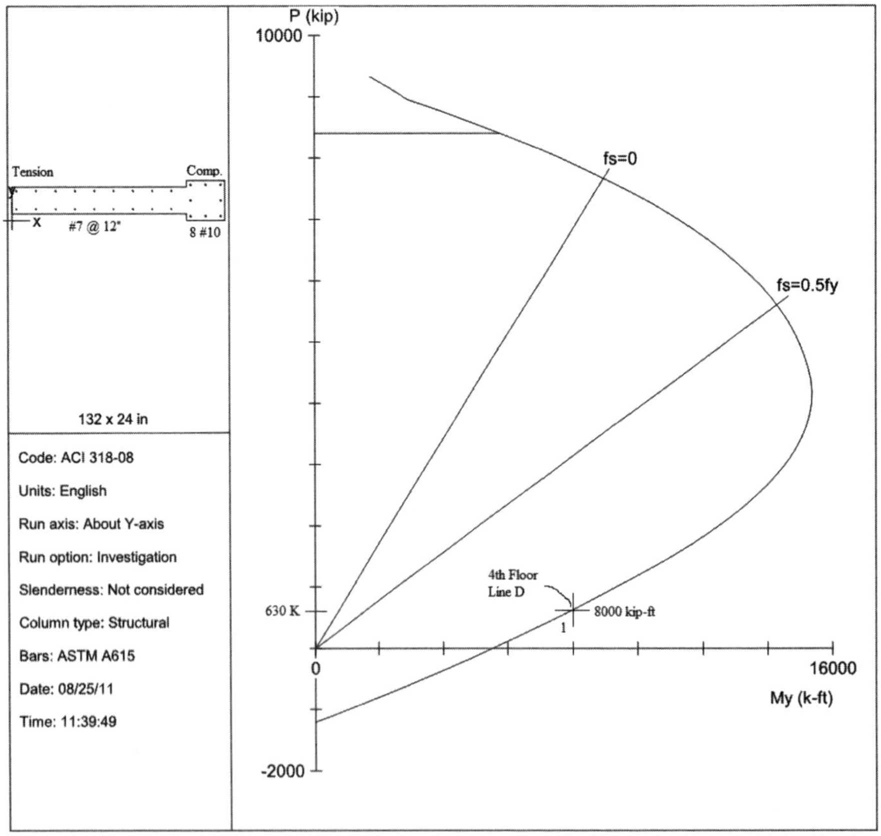

Figure 6-9. spColumn calculation of probable flexural strength $M_{pr}(f_y = 75$ ksi, $\phi = 1.0)$

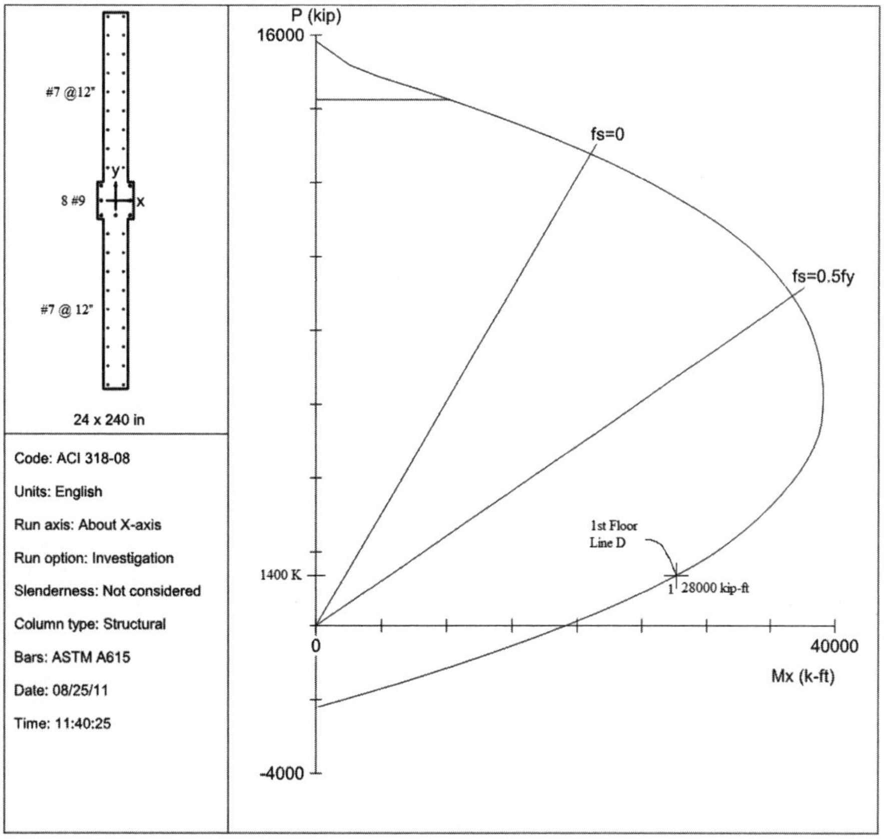

c.

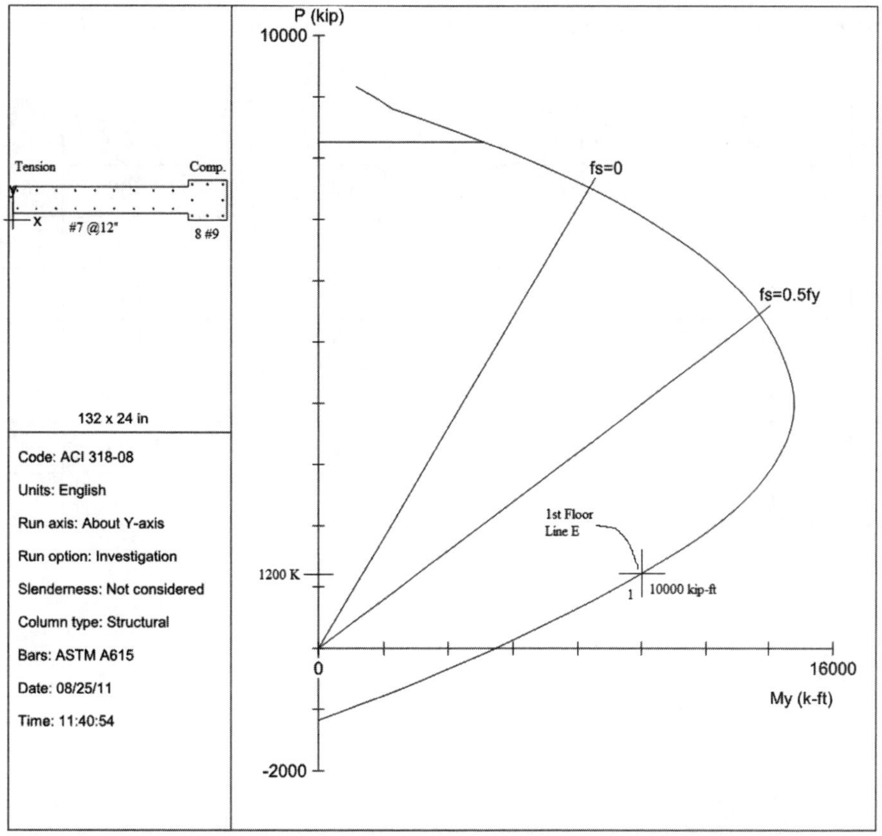

d.

Figure 6-9 (continued). spColumn calculation of probable flexural strength $M_{pr}(f_y = 75$ ksi, $\phi = 1.0)$

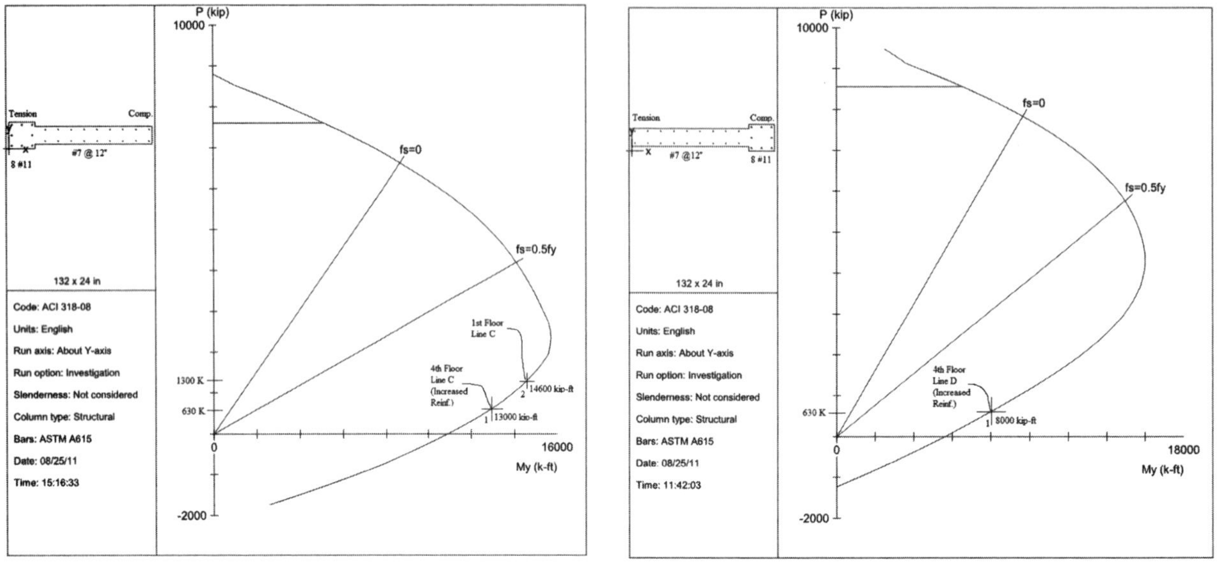

e. **f.**

Figure 6-9 (continued). spColumn calculation of probable flexural strength M_{pr}(f_y = 75 ksi, ϕ = 1.0)

Table 6-5. Approximate probable flexural strengths of wall piers for plastic analysis

Level	Grid Line	Reinforcement of Column Portion	Axial Force Considered $1.2P_D + 0.5P_L$ kips	M_{pr} (kip-ft)	See Figure
4th	C	8 #9	630	10,500	6-9(a)
4th	D	8 #10	630	8,000	6-9(b)
1st	C	8 #11	1300	14,600	6-9(e)
1st	D	8 #9	1400	28,000	6-9(c)
1st	E	8 #9	1200	10,000	6-9(d)
4th	C	8 #11	630	13,000	6-9(e)
4th	D	8 #11	630	8,000	6-9(f)

5b. Mechanism with plastic hinging at the base

The preferred behavior of the wall occurs when plastic hinges occur at the base of the wall piers and in the coupling beams. This produces the desirable situation of flexural yielding, energy dissipation, and avoidance of shear failures.

Table 6-6 shows calculations of the shear strength of the preferred plastic mechanism, which has plastic hinges forming at the base of each wall pier and in each coupling beam, as shown in Figure 6-10. The equivalent plastic hinge length at the pier base, l_p, is taken equal to 5 feet.

The plastic hinge length is used in the calculation of external work shown in Table 6-6. The calculation is not sensitive to the value of l_p assumed, since $l_p/2$ is subtracted from h_i, the height above the base. In this case, the value of 5 feet is taken as one-half the wall length of the external wall piers. Although the central pier is longer, it is assigned the same plastic hinge length.

Plastic lateral story displacements, Δ_i, increase linearly with height above the midpoint of the base plastic hinges. The value of Δ_i is arbitrarily set equal to 1.00 foot at the roof. The external work equals the sum of each lateral story force, F_{xi}, times Δ_i.

The plastic rotation angle of the wall piers, θ, equals the roof displacement divided by the roof height above the midpoint of the plastic hinge. Thus, θ = 1.00/85.5. The plastic rotation angle and internal work of the coupling beams can be calculated as

$$\theta_{cb} = \theta \frac{l_c}{l_n}$$

where

l_n = clear length of the coupling beam

l_c = distance between centroids of wall pier sections

Internal work = $\Sigma(\theta_{cb} \times M_{pr})$ for each end of each coupling beam

$= \Sigma(\theta_{cb} \times 1.25V_n l_n/2)$

$= \Sigma(\theta \times 1.25V_n l_c/2)$

$= \Sigma(\theta \times 1.25V_n l_c)$ for each coupling beam (sum of 2 ends)

The internal work of the base plastic hinges equals the sum of M_{pr} times θ for each of the three base plastic hinges. The summation of the internal work is shown in Table 6-6. Using principles of virtual work and equating internal work with external work gives the solution of $V = 2651$ kips.

Table 6-6. Plastic mechanism calculations assuming plastic hinging at base and in all coupling beams[1]

External Work					
Level	h_i (ft)	$h_i - l_p/2$ (ft)	Δ_i (ft)	$\frac{F_i}{V}$	Work/V (ft)
R	88	85.5	1.000	0.254	0.254
6th	74	71.5	0.836	0.240	0.201
5th	60	57.5	0.673	0.195	0.131
4th	46	43.5	0.509	0.149	0.076
3rd	32	29.5	0.345	0.104	0.036
2nd	18	15.5	0.181	0.058	0.011
Sum				1.000	0.708
Internal Work, Coupling Beams					
Grid Line	Level	$1.25V_n$ (kips)	l_c (ft)	Work (kip-ft)	
C-D	Roof	291	21.5	73	
C-D	6th	575	21.5	145	
C-D	5th	825	21.5	208	
C-D	4th	368	21.5	93	
C-D	3rd	368	21.5	93	
C-D	2nd	480	21.5	121	
D-E	4th	528	21.5	133	
D-E	3rd	825	21.5	208	
D-E	2nd	748	21.5	188	
				1262	
Internal Work, Wall Piers				$\theta = 1.00/85.5$	
Grid Line	Level	M_{pr} (kip-ft)		Work (kip-ft)	
C	base	14,600		171	
D	base	28,000		327	
E	base	10,000		117	
				615	
$V = (1262 + 615)/0.708 = 2651$ kips					

Note:

1. See Figure 6-10 for illustration of hinge locations.

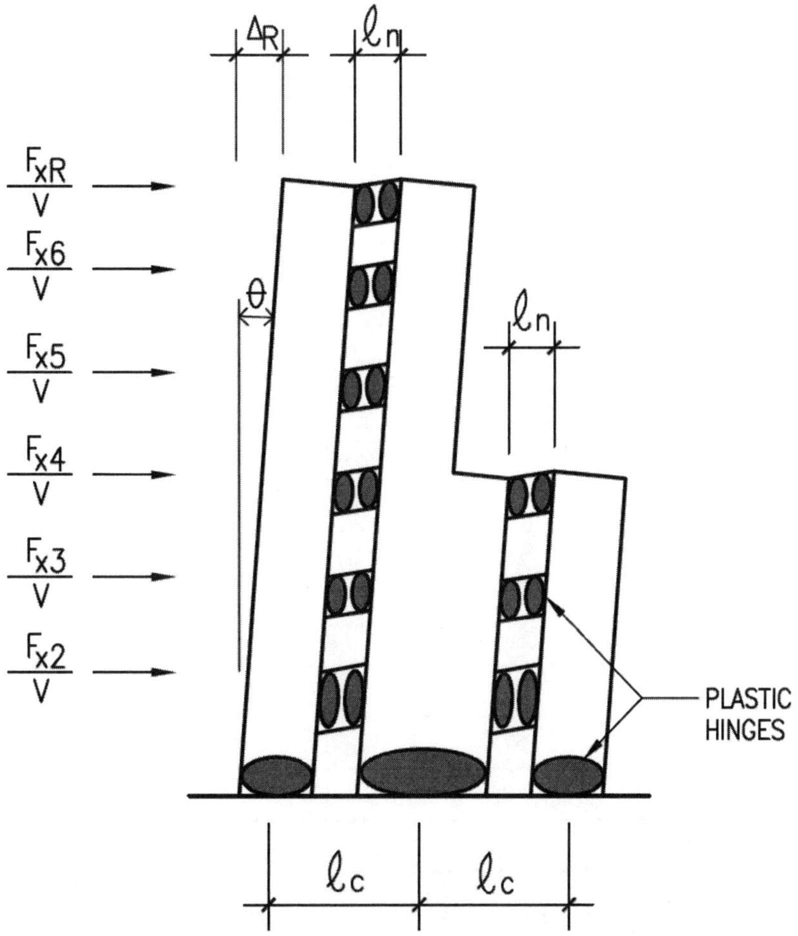

Figure 6-10. Mechanism with plastic hinges at base of wall

5c. Mechanism with plastic hinging at the fourth floor

Table 6-7 shows calculations of the shear strength of another possible plastic mechanism, shown in Figure 6-11, which has plastic hinges forming at the fourth floor wall piers and only in the coupling beams at the fifth, sixth, and roof levels. This plastic mechanism is less desirable than a mechanism with hinging at the base because energy dissipation is concentrated in fewer yielding locations, and because plastic rotations in the wall piers would need to be much greater to achieve the same roof displacement.

As in the previous calculation, plastic lateral story displacements, Δ_i, increase linearly with height above the midpoint of the base plastic hinges, and Δ_i is set equal to 1.00 foot at the roof. For this mechanism, the plastic rotation angle of the wall piers, θ, equals 1.00/39.5. The plastic analysis solution, based on equating internal and external work, gives $V = 2569$ kips. Because this is less than 2651 kips, the mechanism having plastic hinging at the fourth floor governs (i.e., is more likely to form than the preferred base mechanism shown in Figure 6-10).

Table 6-7. Plastic mechanism calculations assuming plastic hinging at fourth-floor piers[1]

External Work					
Level	h_i (ft)	$h_i - l_p/2$ (ft)	Δ_i (ft)	$\frac{F_{xi}}{V}$	Work/V (ft)
Roof	42	39.5	1.000	0.254	0.254
6th	28	25.5	0.646	0.240	0.155
5th	14	11.5	0.291	0.195	0.057
4th			0.000	0.149	0.000
3rd			0.000	0.104	0.000
2nd			0.000	0.058	0.000
Sum				1.000	0.466
Internal Work, Coupling Beams					
Grid Line	Level	$1.25V_n$ (kips)	l_c (ft)	Work (kip-ft)	
C-D	Roof	291	17	125	
C-D	6th	575	17	248	
C-D	5th	825	17	355	
Sum				728	
Internal Work, Wall Piers				θ = 1.00/39.5	
Grid Line	Level	M_{pr} (kip-ft)		Work (kip-ft)	
C	4th	10,500		266	
D	4th	8,000		203	
Sum				469	
$V = (728 + 469)/0.466 = 2569$ kips					

Note:

1. See Figure 6-11 for illustration of hinge locations.

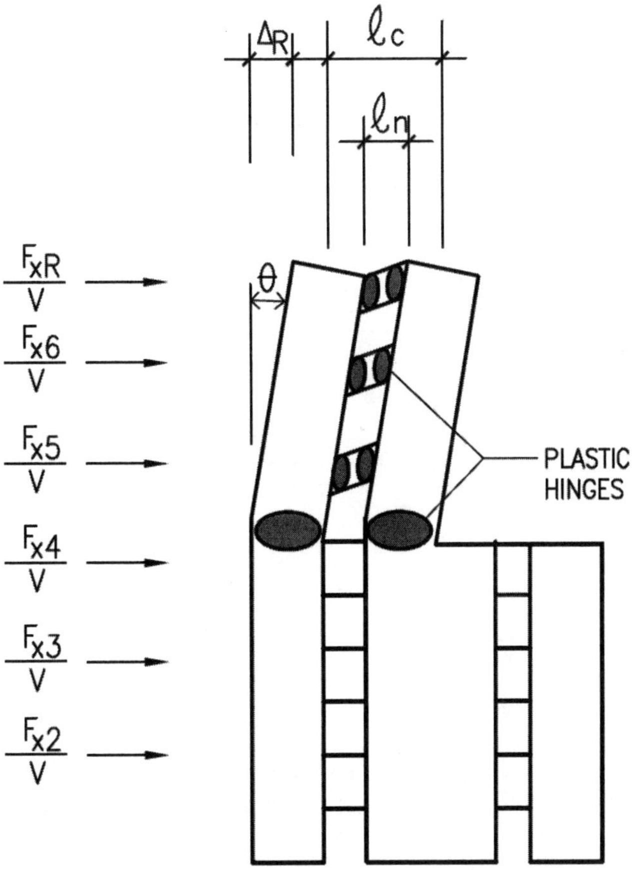

Figure 6-11. Mechanism with plastic hinges at fourth-floor wall piers

To help prevent plastic hinging in the fourth-floor piers, their flexural strength can be increased. Reinforcement of the column portions of these wall piers is increased to 8 #11. Table 6-8 shows revised internal work calculations. The solution gives $V = 2703$ kips. Because this is greater than 2651 kips, the preferred mechanism now governs.

Note that the calculation of the governing plastic limit load, V, depends on the assumed vertical distribution of lateral forces, which in actual seismic response can vary significantly from the inverted triangular pattern assumed. Thus the difference between $V = 2651$ kips and 2703 kips does not absolutely ensure against plastic hinging in the fourth-floor wall piers.

Additionally, nonlinear response-history analyses by computer generally show less predictability of yield locations than plastic analyses imply. For the wall of this design example, a response-history analysis might show some wall-pier yielding both at the base and at the fourth floor. Interaction of the wall with other walls in the structure and with gravity framing can also influence the mechanism of yielding.

Plastic analyses are simpler to carry out and understand than most other analysis methods, particularly nonlinear response-history analyses, and they offer valuable insight into the seismic performance of a structure. For this design example, the plastic analyses indicate that strengthening the fourth floor piers will protect the upper stories above the setback against high ductility demands, and make it more likely that the preferred mechanism will form.

Table 6-8. Plastic mechanism calculations assuming plastic hinging at fourth-floor piers—revised for stronger piers at fourth floor

Internal Work, Wall Piers				$\theta = 1.00/39.5$	
Grid Line	Level	M_{pr} (kip-ft)		Work (kip-ft)	
C	4th	13,000		329	
D	4th	8,000		203	
Sum				532	
$V = (728 + 532)/0.466 = 2703$ kips					

6. Design of wall piers for shear

In this part, the wall piers will be designed for shear. Both the ACI 318 and Blue Book approaches will be illustrated. Design for the minimum ACI 318 requirements is given in Part 6a below.

As discussed in Part 5 of Design Example 5, the SEAOC Blue Book contains more restrictive requirements than do the ASCE/SEI 7-05 and ACI 318 standards for the shear design of reinforced concrete walls. The SEAOC approach, in Part 6b of this design example, is recommended for the reasons given in Design Example 5.

6a. Design under ACI 318 requirements

Shear demand

If designing to the minimum requirements of the ACI 318, the shear demand is taken directly from the design forces, factored by the load combinations discussed in Part 1. For the example wall, all of the significant shear on the wall piers results from earthquake forces, thus $V_u = V_E$, where the values V_E are those shown in Figure 6-3. The highest shears are at the fourth floor, line D, with $V_E = 544$ kips in an 11-foot-long wall pier (48.5 kip-ft), and at the first floor, line D, with $V_E = 731$ kips in a 20-foot-long wall pier (36.6 kip-ft).

Shear capacity

ACI 318 §11.9 gives shear strength provisions for walls designed for *non-seismic* lateral loads such as wind or earth pressure. Section 21.9.4 gives shear strength provisions for walls designed for *seismic* forces.

In Equation 21-7, wall shear strength depends on α_c, which depends on the ratio h_w/l_w.

$$V_n = A_{cv}(\alpha_c \lambda \sqrt{f'_c} + \rho_t f_y) \qquad (21\text{-}7)$$

Per §21.9.4.2, the ratio h_w/l_w is taken as the larger of that for the individual wall pier and for the entire wall.

Overall wall h_w/l_w = 88 ft/54 ft = 1.63

11 ft long by 8 ft clear-height pier h_w/l_w = 8 ft/11 ft = 0.73

20 ft long by 8 ft clear-height pier h_w/l_w = 8 ft/20 ft = 0.40

Thus the value $h_w/l_w = 1.63$ governs for all wall piers. The coefficient, α_c, varies linearly from 3.0 for $h_w/l_w = 1.5$ to 2.0 for $h_w/l_w = 2.0$.

$$\alpha_c = 3.0 - 1.0(1.63 - 1.5)/(2.0 - 1.5) = 2.74$$

As prescribed in ACI 318 §9.3.4(a), the shear strength reduction factor, ϕ, shall be 0.6 for the design of walls if their nominal shear strength is less than the shear corresponding to development of their nominal flexural strength. For normal weight concrete $\lambda = 1.0$. For the 11-foot-long wall piers

$$\phi V_n = 0.6(16\text{ in})l_w[2.74(1.0)\sqrt{4000} + \rho_t(60{,}000\text{ psi})] = l_w(1.66\text{ kip-in} + 576\text{ kip-in }\rho_t)$$

For the wall sections with highest shear, the amount of horizontal shear reinforcement is given in Table 6-9.

Table 6-9. Design for shear by the ACI 318

Level	Grid Line	l_w (in)	V_u (kips)	Horizontal Reinforcement	ρ_t	ϕV_n (kips)	$V_u/\phi A_{cv}\sqrt{f'_c}$ (1)
4th	C	132	371	#4@10 inches E.F.	0.00250	409	4.63
4th	D	132	544	#6@10 inches E.F.	0.00550	637	6.79
1st	C	132	283	#4@10 inches E.F.	0.00250	409	3.53
1st	D	240	731	#4@10 inches E.F.	0.00250	744	5.02
1st	E	132	316	#4@10 inches E.F.	0.00250	409	3.95

Note:

1. Under ACI 318 §21.9.4.4, the value of $V_u/\phi A_{cv}\sqrt{f'_c}$ shall not exceed 10 for any wall pier, or 8 for an entire wall section.

As shown above, for all wall pier locations except the fourth floor at line D, the minimum reinforcement ratio of 0.0025 (required under ACI 318 §21.9.2.1) is sufficient to meet ACI 318 shear strength requirements.

6b. Design using Blue Book recommendations

Shear demand

To comply with the Blue Book requirement of providing shear strength in excess of the shear corresponding to wall flexural strength, an amplified shear demand must be considered. The shear corresponding to flexural strength has been calculated by the plastic analysis in Part 5 of this design example as $V = 2651$ kips at the base of the wall.

The Blue Book (2009, Article 09.01.010) gives the following equation for the shear amplification factor, ω_v, that accounts for inelastic dynamic effects.

$$\omega_v = 0.9 + n/10, \text{ for buildings up to six stories, where } n = \text{number of stories}$$

$$= 0.9 + 6/10 = 1.5$$

As indicated in the Blue Book, the ω_v, factor is derived for analysis using inverted triangular distributions of lateral forces. If a response spectrum analysis is carried out, a slightly lower ω_v factor can be justified in some cases.

At the base of the wall, the magnified shear demand, $V_u{}^*$, is calculated as

$$V_u{}^* = \omega_v V_p = \omega_v(2651 \text{ kips}) = 1.5(2651 \text{ kips}) = 3977 \text{ kips}$$

In the plastic analysis, the amplification effect considered by ω_v can instead be considered by using a different vertical distribution of the lateral forces, f_{xi}. Rather than using the inverted triangular distribution, a vertical distribution with a resultant located lower in the building, such as a uniform distribution pattern, could be used in the plastic analysis to give shear forces.

Shear capacity

Because we are designing for the nominal shear strength to exceed the shear corresponding to flexural strength, a strength reduction factor, ϕ, of 0.75 can be used (see ACI 318 §9.3.4). As before, ACI 318 Equation 21-7 is used to calculate shear capacity

$$V_n = A_{cv}(\alpha_c \lambda \sqrt{f_c'} + \rho_t f_y) \qquad (21\text{-}7)$$

$$\phi V_n = 0.75(16 \text{ in})l_w[2.74(1.0)\sqrt{4000} + \rho_t(60{,}000 \text{ psi})] = l_w[2.08 \text{ kip-in} + (720 \text{ kip-in})\rho_t]$$

For the shear demand of 3977 kips over the net wall length of 42 feet (504 inches) at the first floor, the required amount of horizontal reinforcement is calculated

$$\phi V_n = 504(2.08 + 720\,\rho_t) = 1050 + 363{,}000\,\rho_t \geq 3977 \text{ kips}$$

$$\rho_t = (3977 \text{ kips} - 1050 \text{ kips})/363{,}000 = 0.00806$$

Try #6 @ 6 inches o.c. each face

$$\rho_t = 2(0.44 \text{ in}^2)/(6 \text{ in})(16 \text{ in}) = 0.00917 \; \ldots \; o.k.$$

For the other stories of the building, the shear demands are magnified from the analysis results by the same proportion as for the first floor. The recommended amount of horizontal reinforcement can be calculated as shown in Table 6-10.

Table 6-10. Design for shear by the Blue Book recommendations

Level	V_E (kips)	V_u* (kips)[1]	l_w net (in)	Horizontal Reinforcement	ρ_n	ϕV_n (kips)
6th	338	1026	264	#5@12 inches E.F.	0.00323	1163
5th	656	1991	264	#6@6 inches E.F.	0.00917	2291
4th	915	2777	264	#7@6 inches E.F.	0.01250	2925
3rd	1150	3491	504	#6@6 inches E.F.	0.00917	4374
2nd	1250	3795	504	#6@6 inches E.F.	0.00917	4374
1st	1310	3977	504	#6@6 inches E.F.	0.00917	4374

Note:

1. V_u* = magnified shear demand.

6c. Recommended shear reinforcement

A comparison of Tables 6-9 and 6-10 shows that the Blue Book recommendations for ensuring that shear strength exceeds flexural capacity results in increased horizontal reinforcement compared to that required by the ASCE/SEI 7-05 and ACI 318 codes. The Blue Book approach is recommended, as it leads to more ductile wall behavior.

7. Boundary zone detailing of wall piers

Boundary elements at the edges of structural walls are designed according to ACI 318 §21.9.6. This section provides two alternative criteria for determining where special boundary elements are required, each of which could be applied to the example wall, depending on whether the wall is considered as a whole or if each wall pier is considered independently.

If considered as a whole, the example wall is not "designed to have a single critical section for flexure and axial loads." It is, instead, designed to have plastic hinging in the coupling beams and at the base of each wall pier, as shown in Part 5. Considering the wall as a whole, the need for special boundary elements should be determined according to ACI 318 §21.9.6.3 rather than §21.9.6.2. In accordance with this interpretation, the evaluation of special boundary elements by the gross-section-stress criteria is shown in Part 7a.

Alternatively, if the wall piers of the example wall are considered independently, then ACI 318 §21.9.6.2 can be interpreted to apply because each wall pier is continuous and is designed to have a single critical section (i.e., the plastic hinge) at the base of the wall pier for flexure and axial forces. Using this interpretation, the evaluation of special boundary elements based on neutral axis depth is shown in Part 7b. Detailing of the boundary elements is covered in Parts 7c and 7d.

7a. Evaluation per ACI 318 §21.9.6.3 (based on $0.2f_c'$ stress limit)

ACI 318 §21.9.6.3 requires structural walls to have special boundary elements at boundaries and edges around openings where the maximum extreme fiber compressive stress, corresponding to factored forces including earthquake effects, exceeds $0.2f_c'$. The special boundary element is permitted to be discontinued where the calculated compressive stress is less than $0.15f_c'$. Stresses are calculated for the factored forces using a linearly elastic model and gross section properties.

The stress analysis for this example is performed as shown in Table 6-11. The results are shown in Figure 6-12. The shaded area in the figure indicates where special boundary elements are required.

Table 6-11. Stress analysis for special boundary element requirement by ACI 318

Line D:

Level	Minimum axial		Maximum axial						Minimum axial		Maximum axial	
	P_u (kips)	M_u (kip-ft)	P_u (kips)	M_u (kip-ft)	A_g (in²)	I_g (in⁴)	X_o (to the left) (in)	Wall length (in)	f_L (ksi)	f_R (ksi)	f_L (ksi)	f_R (ksi)
6th	−33	−598	98	−175	2304	3,590,000	70.5	120	0.13	−0.11	0.08	0.01
	269	1350	400	1773	2304	3,590,000	70.5	120	−0.20	0.34	−0.24	0.47
5th	−228	−1782	49	−874	2304	3,590,000	70.5	120	0.32	−0.39	0.23	−0.12
	724	3384	1001	4292	2304	3,590,000	70.5	120	−0.48	0.87	−0.58	1.14
4th	−546	−4843	−121	−3446	2304	3,590,000	70.5	120	0.90	−1.04	0.76	−0.62
	1300	7289	1725	8686	2304	3,590,000	70.5	120	−1.15	1.77	−1.30	2.19
3rd	−287	−5846	309	−6216	4032	18,440,000	120	240	0.39	−0.53	0.56	−0.41
	1345	5148	1941	4778	4032	18,440,000	120	240	−0.07	0.74	0.11	0.85
2nd	137	−9018	903	−9392	4032	18,440,000	120	240	0.74	−0.67	0.96	−0.51
	1221	8312	1987	7938	4032	18,440,000	120	240	−0.35	0.95	−0.13	1.11
1st	412	−13585	1351	−13941	4032	18,440,000	120	240	1.16	−0.96	1.42	−0.75
	1254	12915	2193	12559	4032	18,440,000	120	240	−0.70	1.32	−0.44	1.52

Table 6-11. Stress analysis for special boundary element requirement by ACI 318 (continued)

Line C:

Level	Minimum axial		Maximum axial						Minimum axial		Maximum axial	
	P_u (kips)	M_u (kip-ft)	P_u (kips)	M_u (kip-ft)	A_g (in^2)	I_g (in^4)	X_o (to the left) (in)	Wall length (in)	f_L (ksi)	f_R (ksi)	f_L (ksi)	f_R (ksi)
6th	269	-1350	400	-1773	2304	3,590,000	61.5	120	0.93	-0.15	0.54	0.17
	-33	598	98	175	2304	3,590,000	61.5	120	-0.14	0.10	0.01	0.08
5th	724	-3384	1001	-4292	2304	3,590,000	61.5	120	1.01	-0.35	1.32	-0.40
	-228	1782	49	874	2304	3,590,000	61.5	120	-0.47	0.25	-0.16	0.19
4th	1300	-4894	1725	-6291	2304	3,590,000	61.5	120	1.57	-0.39	2.04	-0.48
	-546	2448	-121	1051	2304	3,590,000	61.5	120	-0.74	0.24	-0.27	0.15
3rd	1643	-2750	2216	-4640	2304	3,590,000	61.5	120	1.28	0.18	1.92	0.05
	-627	-542	-54	-2432	2304	3,590,000	61.5	120	-0.16	-0.38	0.48	-0.50
2nd	1952	-3871	2672	-6252	2304	3,590,000	61.5	120	1.64	0.09	2.45	-0.06
	-678	-267	42	-2648	2304	3,590,000	61.5	120	-0.24	-0.35	0.56	-0.50
1st	2374	-6599	3250	-9474	2304	3,590,000	61.5	120	2.39	-0.26	3.36	-0.44
	-826	1611	50	-1264	2304	3,590,000	61.5	120	-0.69	-0.04	0.28	-0.23

Line E:

Level	Minimum axial		Maximum axial						Minimum axial		Maximum axial	
	P_u (kips)	M_u (kip-ft)	P_u (kips)	M_u (kip-ft)	A_g (in^2)	I_g (in^4)	X_o (to the left) (in)	Wall length (in)	f_L (ksi)	f_R (ksi)	f_L (ksi)	f_R (ksi)
3rd	134	-647	649	1363	2304	3,590,000	70.5	120	0.21	-0.05	-0.04	0.51
	772	4165	1287	6175	2304	3,590,000	70.5	120	-0.65	1.02	-0.90	1.58
2nd	-190	-194	473	2306	2304	3,590,000	70.5	120	-0.04	-0.11	-0.34	0.59
	1354	4554	2017	7054	2304	3,590,000	70.5	120	-0.49	1.34	-0.79	2.04
1st	-460	-1583	358	1411	2304	3,590,000	70.5	120	0.17	-0.46	-0.18	0.39
	1898	6799	2716	9793	2304	3,590,000	70.5	120	-0.78	1.95	-1.13	2.80

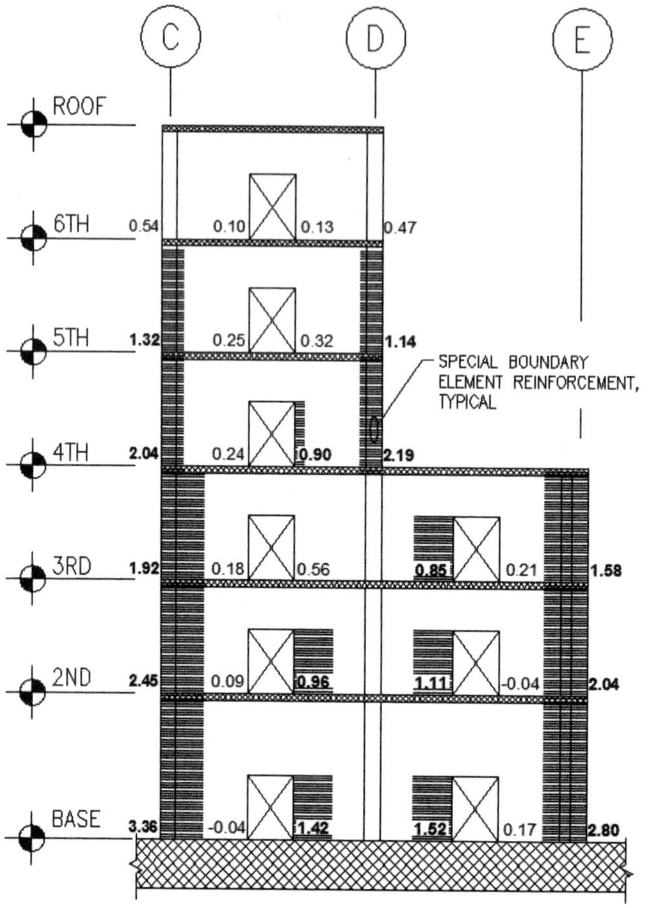

Figure 6-12. Compressive stresses (ksi) and required locations of special boundary elements according to ACI 318 §21.9.6.3

7b. Evaluation per ACI 318 §21.9.6.2 (based on neutral axis depth)

Considering each wall pier independently, and assuming that plastic hinging occurs at the base as shown in Figure 6-10, there are three wall piers to be evaluated, shown in the first three rows of Table 6-12.

Given the possibility of plastic hinging at the fourth floor, as shown in Figure 6-11, wall piers at lines C and D are also evaluated using wall height and relative displacement between the fourth floor and the roof. This is shown in the last two rows of Table 6-12.

Table 6-12. Requirement for special boundary element by ACI318 §21.9.6.2.

Wall Pier	Load Direction	P_u (kips)	M_u (kip-ft)	V_u (kips)	c (in)	ϕM_n (kip-ft)	h_w (ft)	δ_{xe} (in)	δ_u (in)	$\frac{l_w}{600(\delta_u/h_w)}$ (in)
Assuming plastic hinging at the base, per Figure 6-10.										
Line C 1st-Roof	W	3250	−9,470	283	79.4	9,778	88	1.66	8.30	28.0
	E	50	−1,260	283	24.5	8,326				28.0
Line D 1st-Roof	W	1350	−13,900	731	56.0	23,300	88	1.66	8.30	50.9
	E	2190	12,600	731	71.6	27,500				50.9
Line E 1st-4th	W	358	1,410	316	25.7	7,610	46	0.566	2.78	31.4
	E	2720	9,790	316	55.0	11,400				31.4
Assuming plastic hinging at the 4th floor, per Figure 6-11.										
Line C 4th-Roof	W	1720	−6,290	371	31.0	11,000	42	1.10	5.50	20.2
	E	−121	1,050	371	23.2	8,150				20.2
Line D 4th-Roof	W	−121	−3,450	544	23.2	8,150	42	1.10	5.50	20.2
	E	1720	8,690	544	31.0	11,100				20.2

According to ACI 318 §21.9.6.2, boundary zone detailing is required if

$$c > \frac{l_w}{600(\delta_u/h_w)} \qquad (21\text{-}8)$$

Where the quantity $\frac{\delta_u}{h_w}$ shall not be taken less than 0.007.

The compression depth, c, should be calculated for the axial force case that results in the largest value of c. Therefore

$$P_u = 1.41P_D + 0.5P_L + 1.0Q_E$$

Using the same procedure as in Part 4b, the depth of the compression zone is determined for each wall pier.

The term δ_u is the design displacement defined in ACI 318 §2.2, taken at the top of the wall (the roof level). The commentary (ACI 318 §R2.2) indicates that the displacement is calculated considering "modification factors to account for expected inelastic response." Thus, the design displacement corresponds to C_d times δ_{xe} according to ASCE/SEI 7-05 §12.8.6. The term δ_{xe} is the elastic displacement corresponding to the design loads, using effective section properties. For the example wall, δ_{xe}, at each level is shown in Figure 6-3 and the deflection amplification factor C_d is equal to 5.

The calculations in Table 6-12 show that special boundary elements are required at the locations shown shaded in Figure 6-13.

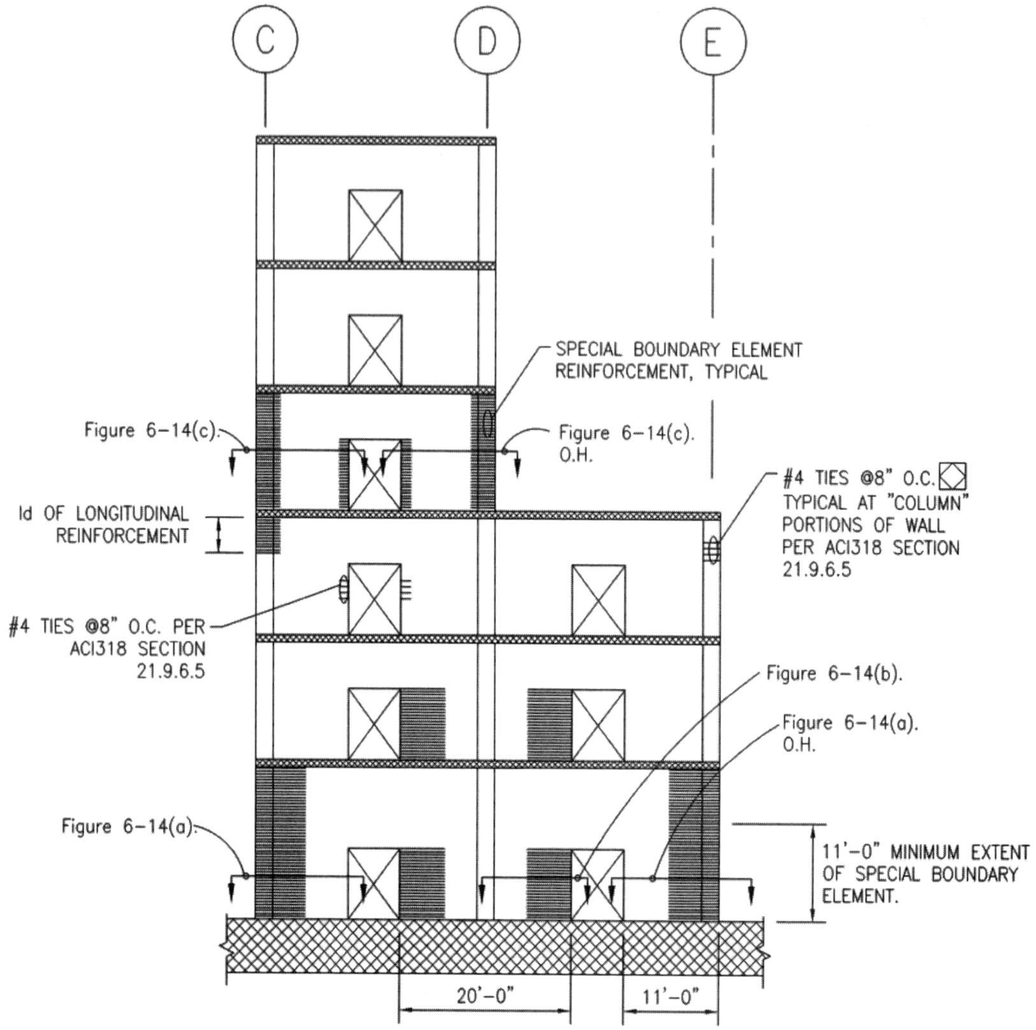

Figure 6-13. Required locations of special boundary elements according to ACI 318 §21.9.6.2.

7c. Detailing special boundary elements

Whether the approach of ACI 318 §21.9.6.2 or §21.9.6.3 is used, boundary elements are detailed according to §21.9.6.4. In this section we will use the results of Part 7b and detail the boundary elements required.

The minimum width of the boundary element is determined based on the requirements of § 21.9.6.4(a): "The boundary element shall extend horizontally from the extreme compression fiber a distance not less than the larger of $c - 0.1l_w$ and $\frac{c}{2}$."

As defined in ACI 318 §21.0, c is the distance from the extreme compression fiber to the neutral axis calculated for the factored axial force and nominal flexural strength resulting in the largest neutral axis depth. In applying these requirements to the example wall, the wall piers with the largest neutral axis depth govern the design.

Where special boundary elements are required, they are extended vertically from the critical section a distance not less than the larger of l_w or $M_u/4V_u$ per ACI 318 §21.9.6.2(b) as shown in Figure 6-13.

At doorways, the extent of boundary reinforcement is limited to the height of the doorway. At the base of line D, where the required vertical extent of boundary element reinforcement is greater than the first-story height, boundary element reinforcement is provided over the height of the doorways on both the first and second stories.

Per ACI 318 §21.9.6.4(d), the boundary element transverse reinforcement at the wall base shall extend into the support at least l_d of the largest longitudinal reinforcement in the special boundary element, or at least 12 inches when the special boundary element terminates on a footing or mat. In this example, boundary element transverse reinforcement is extended below the fourth floor a distance l_d at line C.

The calculation of required width and vertical extent of the special boundary elements is shown in Table 6-13.

Table 6-13. Required width and vertical extent of special boundary elements

			Minimum Width of Boundary Element				Minimum Extent of Boundary Element		
Wall Pier	Boundary	Governing c (in)	$c - 0.1l_w$ (in)	$c/2$ (in)	Provide Width (in)	Figure	l_w (ft)	$\frac{M_u}{4V_u}$ (ft)	Provided extent (ft)
Line C, 1st-Roof Line E, 1st-4th	Column	79.4	66.2	39.0	69.5	6-14a	11	8.37	18
Line D, 1st-Roof	Doorway	71.6	47.6	41.2	50.5	6-14b	20	4.75	16*
Line C, 4th-Roof Line D, 4th-Roof	Column	31.0	17.8	18.7	33.5	6-14c	11	4.24	14
Line C, 4th-Roof Line D, 4th-Roof	Doorway	23.2	10.0	10.6	13.5	6-14c	11	1.59	8*

* Extent limited to height of wall boundary, equal to height of door opening.

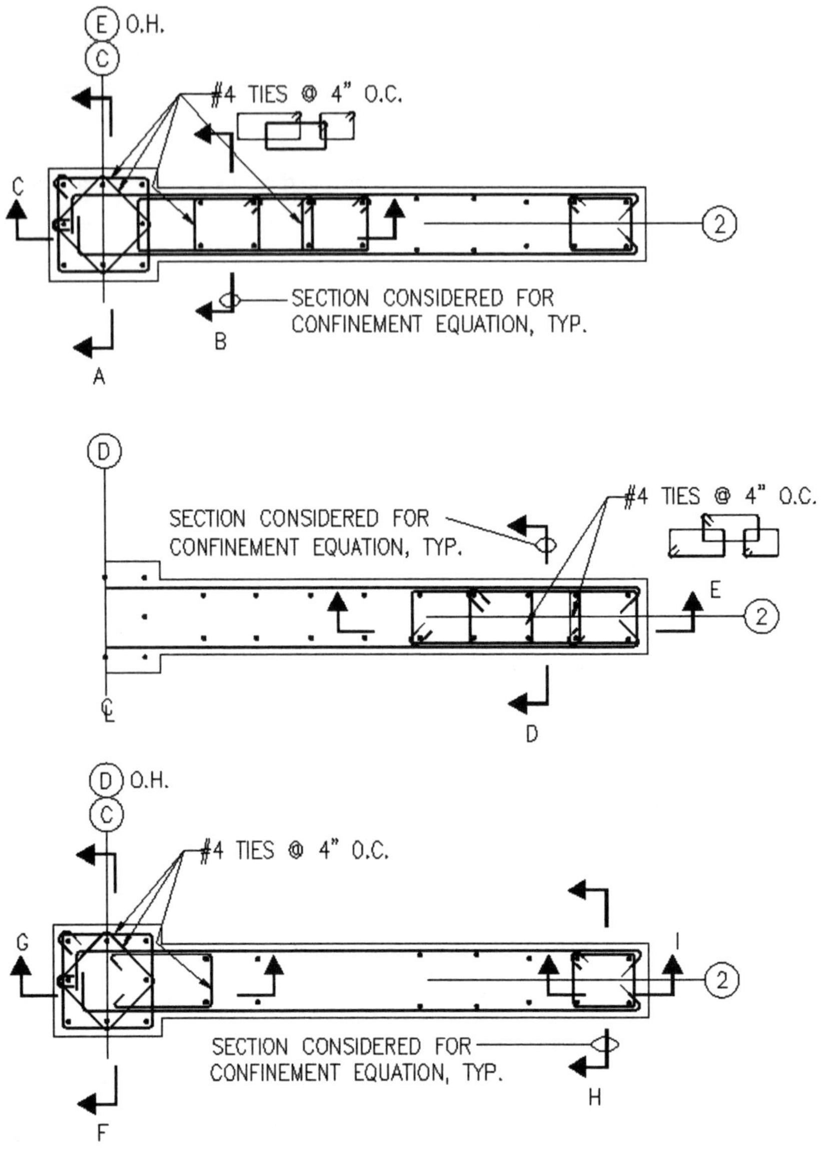

Figure 6-14. Boundary ties required by the ACI 318 procedure

The required area of boundary ties is calculated according to ACI 318 Equation (21-5).

$$A_{sh} = 0.09sb_c f'_c/f_{yt} \qquad (21\text{-}5)$$

Calculations of A_{sh} are given in Table 6-14, corresponding to section cuts A, B, C, D, E, F, G, H, and I through the boundary zones as shown in Figure 6-14.

Table 6-14. Required boundary zone ties by the ACI 318 procedure

Section Cut	h_c (in)	S (in)	A_{sh} Required (in^2)	Tie legs	A_{sh} Provided (in^2)
A	21.5	4	0.52	3 #4	0.60
B	13.5	4	0.32	2 #4	0.40
C	68.0	4	1.63	9 #4	1.80
D	13.5	4	0.32	2 #4	0.40
E	59.0	4	1.42	8 #4	1.60
F	21.5	4	0.52	3 #4	0.60
G	32.0	4	0.77	4 #4	0.80
H	13.5	4	0.32	2 #4	0.40
I	12	4	0.29	2 #4	0.40

Note:

1. See Figure 6-14.

Blue Book recommendations

ACI Equation (21-5), which specifies the required area of confining reinforcement, comes from the requirements for special moment frame columns and is referenced by §21.9.6.4(c) for special boundary elements of walls. For special moment frame columns, the SEAOC Blue Book (2009, Article 09.05.010) recommends an alternative equation that accounts for axial load and effectiveness of confining reinforcement, based on detailing:

$$A_{sh} = 0.2k_n k_p (A_g / A_{ch})(f'_c / f_{yt}) s b_c$$

Where $k_n = [0.6 + 0.4(n/n_{ls})][(h_x + 12)/20]$

$(h_x + 12)/20 \geq 1.0$

n = total number of longitudinal bars in the column cross section

n_{ls} = number of longitudinal bars that are laterally supported by the corner of hoops or by seismic hoops or crossties that are ≥ 135 degrees

h_x = center-to-center horizontal spacing of crossties or hoop legs

$k_p = P_u / A_g f'_c \geq 0.2$

The Blue Book does not provide a comparable recommendation for confinement of special wall boundaries because it is difficult to quantify the axial load on a wall boundary. However, Elwood et al (2009) recommend an alternative formulation of the k_p term for use with wall boundary elements, based on the expected strain demand:

$$k_p = 120(c/l_w)(\delta_u/h_w)$$

Elwood et al suggest that an upper limit on k_p may be appropriate to avoid congestion of ties but do not provide a recommendation for the upper limit.

This recommendation for boundary confinement could be checked as well. For this design example, the requirements of ACI 318 are followed.

7d. Additional boundary ties

Where special boundary elements are not required, ACI 318 §21.9.6.5(a) requires the following detailing consideration.

"If the longitudinal reinforcement ratio at the wall boundary is greater than $\frac{400}{f_y}$, boundary transverse reinforcement shall satisfy §§21.6.4.2 and 21.9.6.4(a). The maximum longitudinal spacing of transverse reinforcement in the boundary shall not exceed 8 in."

At the doorway end of walls where vertical reinforcement is #7@ 12 inches o.c.

$$\rho = \frac{2(0.60 \text{ in})}{12 \text{ in}(16 \text{ in})} = 0.00625 < \frac{400}{f_y} = 0.00667$$

Therefore, no boundary ties are required.

At the doorway end of walls where vertical reinforcement is #8@ 12 inches o.c.

$$\rho = \frac{2(0.79 \text{ in})}{12 \text{ in}(16 \text{ in})} = 0.00823 > \frac{400}{f_y} = 0.00667$$

Therefore, #4@ 8 inches o.c. ties are provided as a minimum.

At column end of walls with 8 #9 longitudinal bars,

$$\rho = \frac{8(1.0 \text{ in})}{24 \text{ in}(24 \text{ in})} = 0.0139 > \frac{400}{f_y} = 0.00667$$

Therefore #4@ 8 inches o.c. ties are used as a minimum tie throughout the columns. The same ties are used for the column portion of the wall pier D at the first three floors, although ties are not strictly required by the code since this portion is not a wall boundary.

Figure 6-13 shows the location of the boundary ties provided to satisfy ACI 318 §21.9.6.5.

8. Detailing of coupling beams

ACI 318 §21.9.7.4(c) and (d) provide two options for confinement of diagonal bars in coupling beams. The first option (c) provides transverse reinforcement around each of the two groups of diagonal bars individually, while the second option (d) provides transverse reinforcement confining the entire beam cross section. The second option was introduced in ACI 318-08 and tends to result in more easily constructible reinforcement layouts. Both options are discussed for comparison in this design example.

8a. Option 1: Ties around diagonal bars per ACI 318 §21.9.7.4(c)

In the first option, per ACI 318 §21.9.7.4(c), "diagonal bars shall be enclosed by transverse reinforcement having out-to-out dimensions not smaller than $b_w/2$ in the direction parallel to b_w and $b_w/5$ along the other sides." This design example assumes that these ties are #4, and the recommended layering of reinforcement in the coupling beams is shown in Figure 6-15. The proposed layering corresponds to a clear cover of 1 inch in the coupling beam and 1⅜ inches in the wall pier. The layering shown in Figure 6-15 results in a diagonal bar cage with lateral "core" dimensions of 9.0 inches by 14.8 inches, measured outside-to-outside of the ties. Ties extend over the portion of the diagonal bars within the coupling beam length, as shown in Figure 6-16.

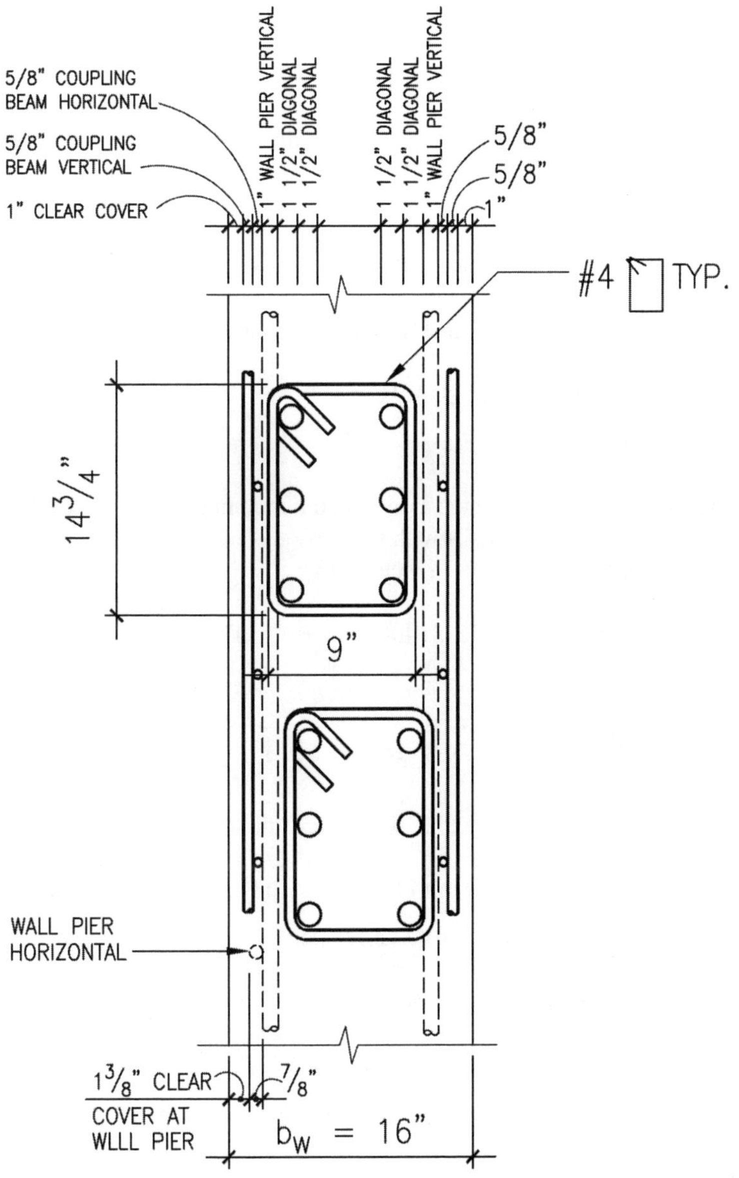

Figure 6-15. Section through coupling beam showing layering of reinforcement

Under the requirements of ACI 318 §21.9.7.4(c), the required transverse reinforcement around diagonal bars "shall satisfy ACI 318 §21.6.4.2 and 21.6.4.4, shall have spacing measured parallel to the diagonal bars satisfying 21.6.4.3(c) and not exceeding six times the diameter of the diagonal bars, and shall have spacing of crossties of legs of hoops measured perpendicular to the diagonal bars not exceeding 14 in." §21.6.4.3(c) requires a maximum tie spacing not exceeding

$$s_o = 4 + \left(\frac{14 - h_x}{3}\right) \quad \text{where } 4 \text{ in} \le s_o \le 6 \text{ in} \qquad \text{(Eq 21-2)}$$

h_x is the maximum horizontal spacing of hoop legs on all faces of the member.

In this example we have

$$4 + \left(\frac{14 - h_x}{3}\right) = 4 + \left(\frac{14 - 14}{3}\right) = 4.0 \text{ in} \quad \textit{governs}$$

$$6d_b = 6(1.00 \text{ in}) = 6.0 \text{ in. (for \#8 diagonal bars)}$$

Thus

$s = 4.0$ in is chosen as the required spacing.

Equations 21-4 and 21-5 must be checked in each direction.

$$A_{sh} = 0.3(sb_c f'_c / f_{yt})[(A_g / A_{ch}) - 1] \qquad \text{(Eq 21-4)}$$

$$A_{sh} = 0.09 sb_c f'_c / f_{yt} \qquad \text{(Eq 21-5)}$$

The quantity A_g is calculated assuming the minimum cover per §7.7 around each diagonal bar core. For walls with #11 bars and smaller, without exposure to weather, this minimum cover equals ¾ inch. Thus

$$A_g = [9.0 + 2(0.75)][14.8 + 2(0.75)] = 171 \text{ in}^2$$

and

$$A_{ch} = 9.0(14.8) = 133 \text{ in}^2$$

For the required area of steel in the vertical direction, $b_c = 9$ in

$$A_{sh} = 0.3[(4 \text{ in})(9 \text{ in})(4 \text{ ksi}) / 60 \text{ ksi}](171/133 - 1) = 0.206 \text{ in}^2 \qquad \text{(Eq 21-4)}$$

$$A_{sh} = 0.09(4 \text{ in})(9 \text{ in})(4 \text{ ksi}) / 60 \text{ ksi} = 0.216 \text{ in}^2 \quad \textit{governs} \qquad \text{(Eq 21-5)}$$

For steel in the horizontal direction, $b_c = 14.8$ in

$$A_{sh} = 0.3[(4 \text{ in})(14.8 \text{ in})(4 \text{ ksi}) / 60 \text{ ksi}](171/133 - 1) = 0.338 \text{ in}^2 \qquad \text{(Eq 21-4)}$$

$$A_{sh} = 0.09(4 \text{ in})(14.8 \text{ in})(4 \text{ ksi}) / 60 \text{ ksi} = 0.355 \text{ in}^2 \quad \textit{governs} \qquad \text{(Eq 21-5)}$$

A single #4 tie around the six diagonal bars provides two tie legs in each direction, thus $A_{sh} = 0.40$ in^2. Using #3 perimeter ties with one #3 crosstie provides $A_{sh} = 0.22$ in^2 across the shorter core direction and $A_{sh} = 0.33$ in^2 across the longer core direction, which would not quite meet the A_{sh} requirement of 0.355 in^2.

Per §21.9.7.4(c), cross-ties shall not be spaced more than 14 inches on center. For the heaviest diagonal reinforcement of 6 #11 bars, the center-to-center dimension of the #11 bars is given as 12 inches in Figure 6-16. The center-to-center hoop dimension in this direction thus equals 12 inches plus one diameter of a #11 bar plus one diameter of a #4 tie, equal to 12.0 + 1.40 + 0.5 = 13.9 inches. Since this is less than 14 inches, a crosstie is not needed.

The diagonal bars must be developed for tension into the wall piers. By ACI code, bars are required to have a development length of l_d. For coupling beams, however, ACI 318 §21.9.7.4(b) requires an increase of 25 percent in the development length. Paulay and Priestley (1992) recommend increasing the development length by 50 percent instead of 25 percent. Following the ACI 318 requirement, the bars are extended a distance of $1.25l_d$ beyond the face of the supporting wall pier, as shown in Figure 6-16, where l_d is the development length of a straight bar as determined under ACI 318 §12.2.

Per ACI 318 §21.9.7.4(c), confinement around the diagonal bars is required at the intersection of the diagonals, either by continuing the ties through this intersection or by providing "alternatively configured reinforcement satisfying the spacing and volume ratio requirements of the transverse reinforcement along the diagonals." This is provided by a combination of hoops and crossties at the intersection.

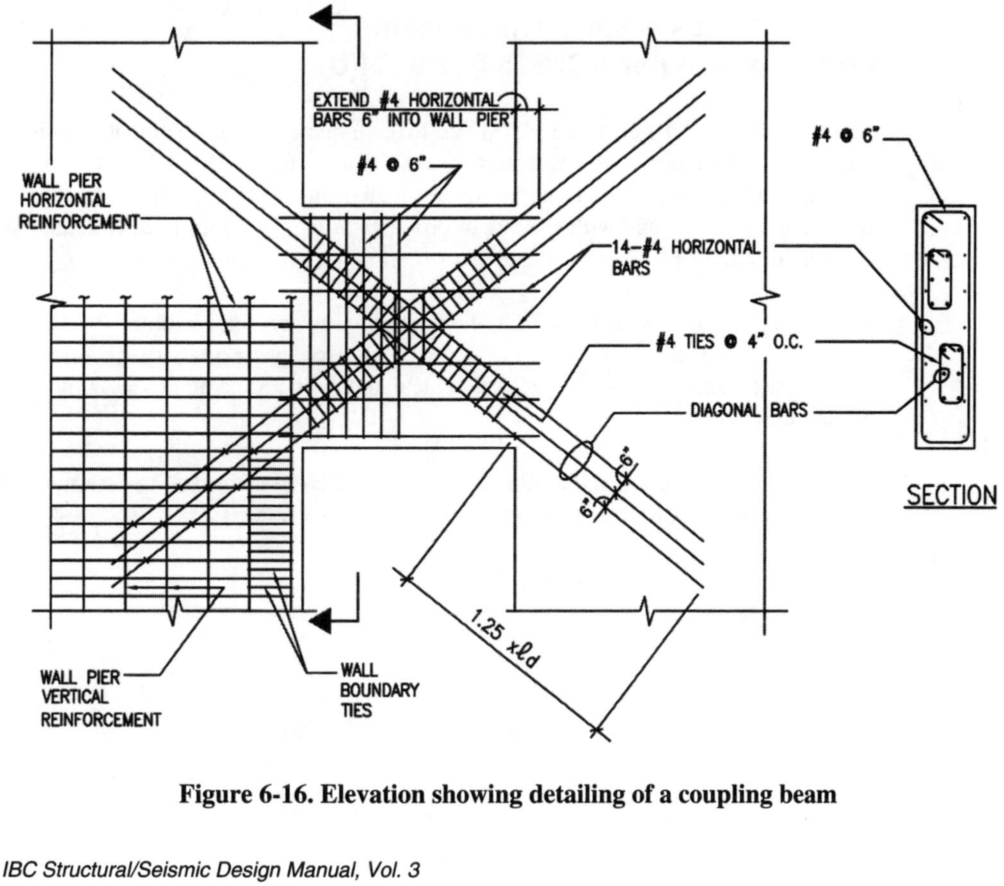

Figure 6-16. Elevation showing detailing of a coupling beam

Per ACI 318 §21.9.7.4(c), "additional longitudinal and transverse reinforcement shall be distributed around the beam perimeter with total area in each direction not less than $0.002b_w s$ and spacing not exceeding 12 in."

For #4@6 inches transverse (vertical) bars

$$0.002b_w s = 0.002(16 \text{ in})(6 \text{ in}) = 0.192 \text{ in}^2 \leq 0.40 \text{ in}^2 \ \ldots \ o.k.$$

For 14 #4 longitudinal (horizontal) bars:

$$0.002b_w s = 0.002(16 \text{ in})(72 \text{ in} / 7) = 0.329 \text{ in}^2 \leq 0.40 \text{ in}^2 \ \ldots \ o.k.$$

The Blue Book (2009, Section §402.13) requires that the reinforcement parallel to the longitudinal axis of the beam be at least #3 in size, spaced at not more than 12 inches on center. The reinforcement transverse to the longitudinal axis of the beam must be at least #3 in size, spaced at not more than 6 inches on center.

Figure 6-16 shows the recommended parallel and transverse reinforcement: 14 #4 bars longitudinally and #4 ties @ 6 inches transversely, which also meets the ACI 318 requirements.

Per the Blue Book recommendations (2009, Article 09.01.010), the longitudinal reinforcement is extended 6 inches into the wall pier, as shown in Figure 6-16. Longitudinal reinforcement is *not* developed for tension because it is undesirable to increase the probable moment strength of the coupling beam beyond that provided by the diagonal bars.

8b. Option 2: Ties around entire beam cross section per ACI 318 §21.9.7.4(d)

In the second option, per ACI 318 §21.9.7.4(d), "transverse reinforcement shall be provided for the entire beam cross section satisfying 21.6.4.2, 21.6.4.4, and 21.6.4.7, with longitudinal spacing not exceeding the smaller of 6 in. and six times the diameter of the diagonal bars, and with spacing of crossties or legs of hoops both vertically and horizontally in the plane of the beam cross section not exceeding 8 in."

In this example, for #8 diagonal bars, the maximum tie spacing permitted by ACI is the smaller of 6 inches or $6d_b = 6(1.00 \text{ in}) = 6.0$ in. However, to satisfy the volume ratio requirements of ACI 318 §21.6.4.4 calculated next, a spacing of $s = 4.0$ in is chosen for design.

ACI 318 §21.6.4.4 requires that Equations 21-4 and 21-5 be checked to determine the minimum area of transverse reinforcement in each direction.

$$A_{sh} = 0.3(sb_c f'_c / f_{yt})[(A_g / A_{ch}) - 1] \qquad \text{(Eq 21-4)}$$

$$A_{sh} = 0.09 s b_c f'_c / f_{yt} \qquad \text{(Eq 21-5)}$$

For the coupling beams at the second floor in this example,

$$A_g = (16 \text{ in})(120 \text{ in}) = 1920 \text{ in}^2$$

and

$$A_{ch} = (14 \text{ in})(118 \text{ in}) = 1652 \text{ in}^2$$

For the required area of steel in the vertical direction, $b_c = 14$ in

$$A_{sh} = 0.3[(4 \text{ in})(14 \text{ in})(4 \text{ ksi}) / 60 \text{ ksi}](1920/1652 - 1) = 0.181 \text{ in}^2 \qquad \text{(Eq 21-4)}$$

$$A_{sh} = 0.09(4 \text{ in})(14 \text{ in})(4 \text{ ksi}) / 60 \text{ ksi} = 0.336 \text{ in}^2 \quad \textit{governs} \qquad \text{(Eq 21-5)}$$

For steel in the horizontal direction, $b_c = 118$ in

$$A_{sh} = 0.3[(4 \text{ in})(118 \text{ in})(4 \text{ ksi}) / 60 \text{ ksi}](1920/1652 - 1) = 1.53 \text{ in}^2 \qquad \text{(Eq 21-4)}$$

$$A_{sh} = 0.09(4 \text{ in})(118 \text{ in})(4 \text{ ksi}) / 60 \text{ ksi} = 2.83 \text{ in}^2 \quad \textit{governs} \qquad \text{(Eq 21-5)}$$

In the vertical direction, a single #4 tie around the beam cross section provides two tie legs thus $A_{sh} = 0.40$ in^2, which is greater than the required 0.336 in^2. However, the spacing of crossties or legs of hoops in the plane of the beam cross section must be limited to 8 inches, so a vertical crosstie is provided as well.

In the horizontal direction, the two legs of the #4 tie plus 14 #4 horizontal crossties provides a total of $A_{sh} = 3.2$ in^2, which satisfies the required 2.83 in^2. This arrangement also satisfies the 8-inch maximum spacing between crossties or legs of hoops.

For the coupling beams at other floors, performing these same calculations with the beam depth of 72 inches results in the same design as above, except that the shallower beam requires fewer horizontal crossties (8 rather than 14) to maintain a similar spacing and satisfy transverse reinforcement area requirements.

Figures 6-17 and 6-18 show the resulting reinforcement layout for this design example. ACI 318 §21.9.7.4(d) permits hoops configured per ACI 318 §21.5.3.6, consisting of a stirrup having seismic hooks at both ends and closed by a crosstie.

Per ACI 318 §21.5.3.6 and §21.6.4.2, consecutive crossties are alternated end for end along the longitudinal reinforcement. Per ACI 318 §21.9.7.4(d) each crosstie and each hoop leg must engage a longitudinal bar of equal or larger diameter, so #4 longitudinal bars are placed around the perimeter of the beam at the ends of each crosstie and hoop leg.

Per the Blue Book recommendations (2009, Article 09.01.010), the longitudinal reinforcement is extended 6 inches into the wall pier, as shown in Figure 6-18. Longitudinal reinforcement is *not* developed for tension because it is undesirable to increase the probable moment strength of the coupling beam beyond that provided by the diagonal bars.

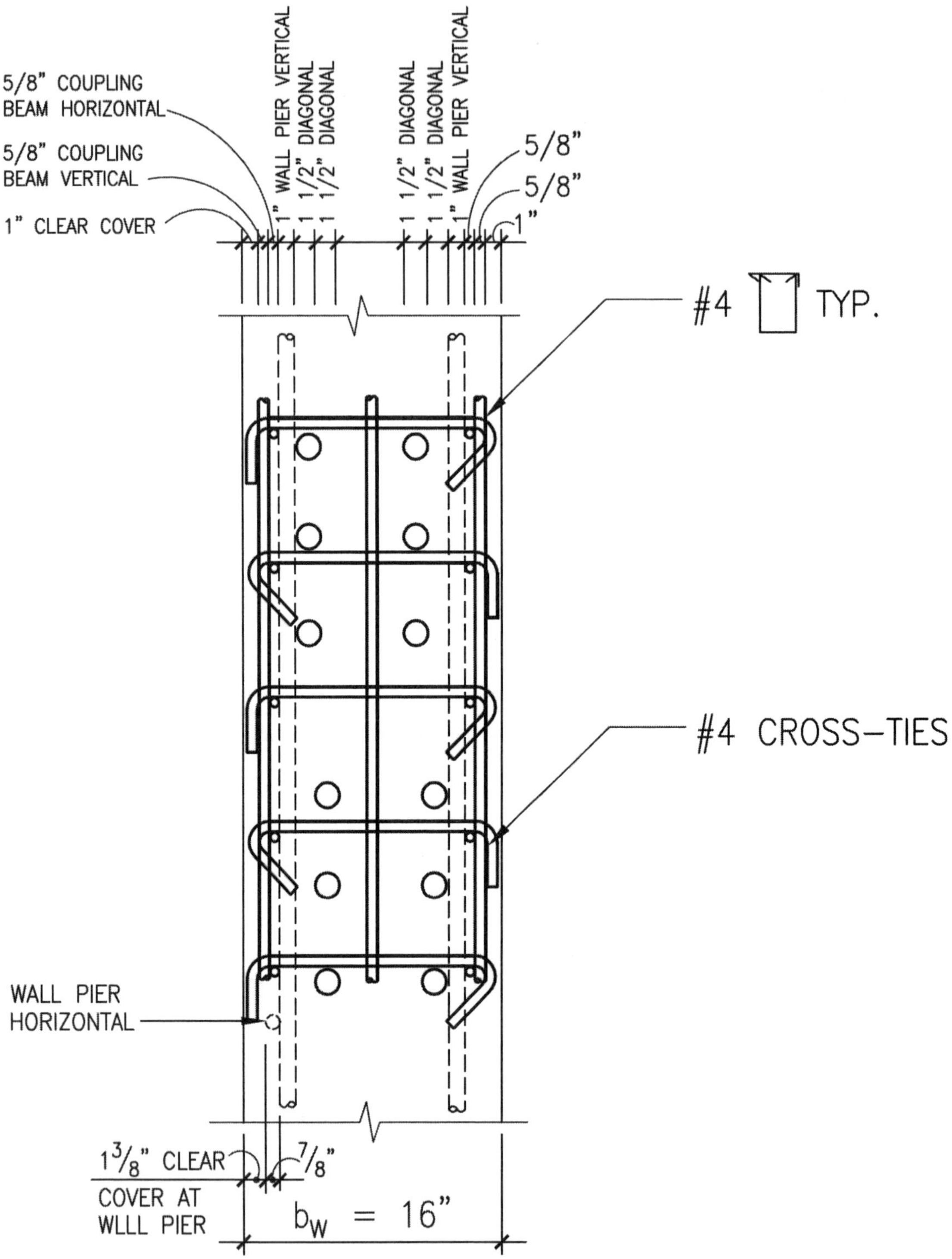

Figure 6-17. Section through coupling beam showing layering of reinforcement

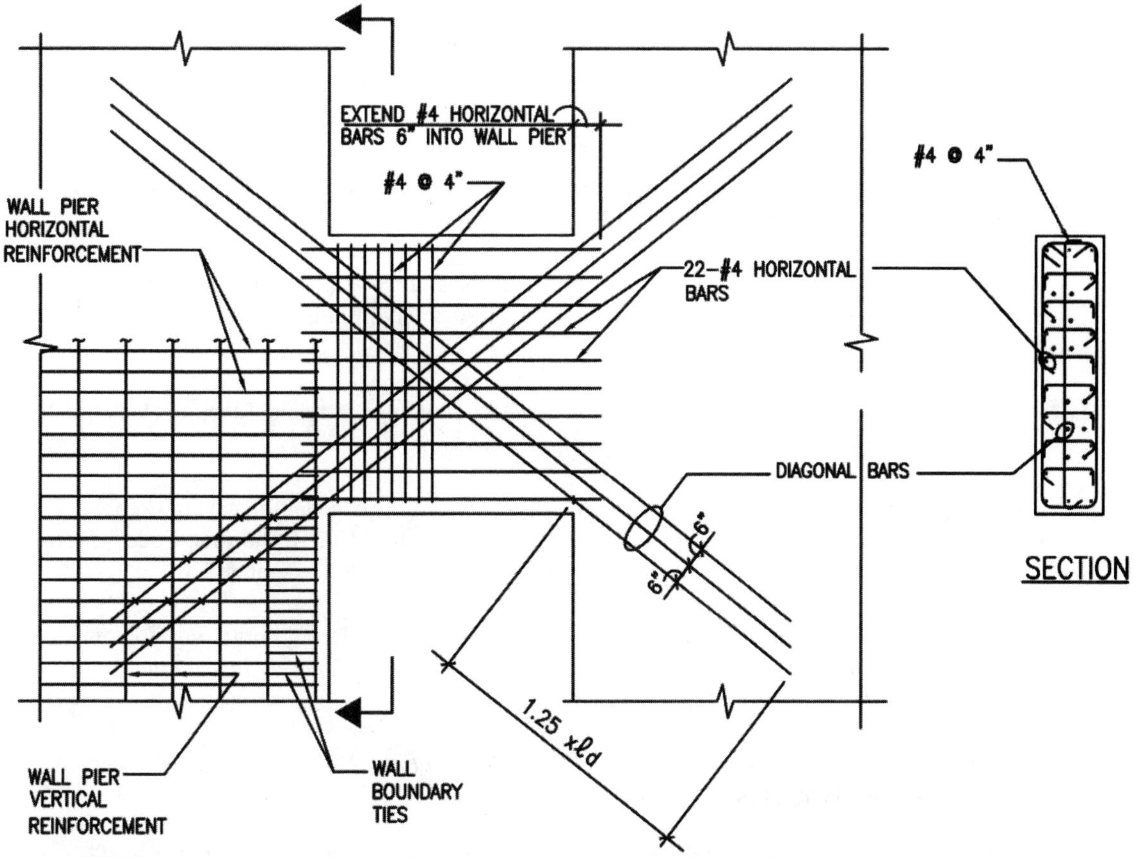

Figure 6-18. Elevation showing detailing of a coupling beam

References

ACI 318. *Building Code Requirements for Reinforced Concrete.* American Concrete Institute, Farmington, Michigan, 2008.

ASCE. *Plastic Design in Steel, A Guide and Commentary*, American Society of Civil Engineers, New York, 1971.

ASCE/SEI 7-05. *Minimum Design Loads for Buildings and Other Structures*, Structural Engineering Institute of the American Society of Civil Engineers, Reston, Virginia, 2005.

ATC-40. *Seismic Evaluation and Retrofit of Concrete Buildings*, Applied Technology Council, Redwood City, California, 1996.

Elwood, Kenneth J., Joe Maffei, Kevin A. Riederer, and Karl Telleen. "Improving Column Confinement Part 2: Proposed new provisions for the ACI 318 Building Code," *Concrete International*, Volume 31, No. 12, pages 41-48, December 2009.

Elwood, Kenneth J., Joe Maffei, Kevin A. Riederer, and Karl Telleen. "Improving Column Confinement Part 1: Assessment of design provisions," *Concrete International*, Volume 31, No. 11, pages 32-48, November 2009.

FEMA 273. *NEHRP Guidelines for the Seismic Rehabilitation of Buildings*, Federal Emergency Management Agency, Washington, D.C., 1997.

FEMA 306/307. *Evaluation of Earthquake Damaged Concrete and Masonry Wall Buildings,* Federal Emergency Management Agency, Washington, D.C., 1998.

Ghosh, S.K. "Design of Reinforced Concrete Buildings under the 1997 UBC," *Building Standards,* May-June, pp. 20-24. International Conference of Building Officials, Washington, D.C., 1998.

Maffei, J. "Reinforced Concrete Structural Walls: Beyond the Code" SEAONC 1996 Fall Seminar Notes, Structural Engineers of Northern California, San Francisco, California, 1996.

Paulay, T. and Priestley, M.J.N. *Reinforced Concrete and Masonry Buildings, Design for Seismic Resistance*, John Wiley & Sons, New York, N.Y., 1992.

SEAOC Seismology Committee, *Recommended Lateral Force Requirements and Commentary* (Blue Book), Structural Engineers Associate of California (SEAOC), Seventh Edition, Sacramento, California, 1999.

SEAOC Seismology Committee. "Reinforced Concrete Structures (Article 09.01.010)." *SEAOC Blue Book Seismic Design Recommendations*, Structural Engineers Associate of California (SEAOC),Sacramento, California, www.seaoc.org/bluebook, First Printing, September, 2009.

Structurepoint. *spColumn Version 4.60: Design and Investigation of Reinforced Concrete Column Sections*, Structurepoint, Skokie, Illinois, 2010.

Design Example 7
Reinforced Concrete Special Moment-Resisting Frame

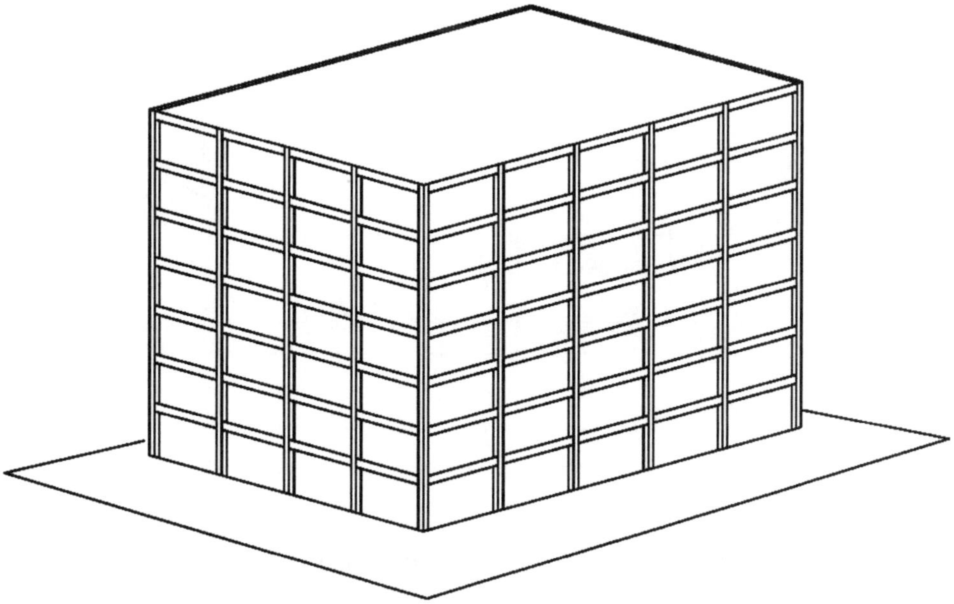

Figure 7-1. Seven-story concrete SMRF building

Overview

Concrete frame buildings, especially older, non-ductile frames, have frequently experienced significant structural damage in earthquakes. A number have collapsed. Following the 1971 San Fernando earthquake, special requirements for ductile concrete frames were introduced in the code.

Today these ductile frames are designated as SMRF (Special Moment-Resisting Frames). All reinforced concrete moment-frame structures built in Seismic Design Category D, E, or F locations must be SMRF as required by Table 12.2-1. Ordinary Moment-Resisting Frames (OMRF) are prohibited in Seismic Design Category locations C, D, E or F, and Intermediate Moment-Resisting Frames (IMRF) are prohibited in Seismic Design Category locations D, E, or F.

This example illustrates the seismic design of a seven-story concrete SMRF. A conceptual elevation of the building is shown in Figure 7-1. This is a reinforced concrete office building with the typical floor plan shown in Figure 7-2. The building has seven stories and has a SMRF on each perimeter wall. A typical building elevation is shown in Figure 7-3.

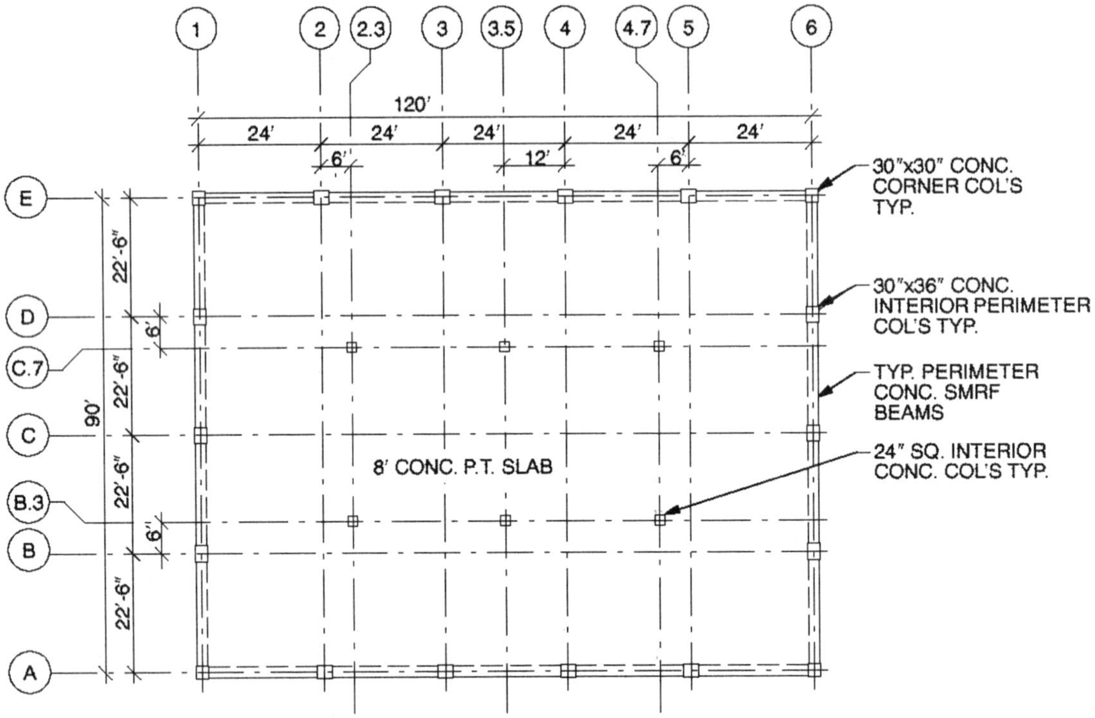

Figure 7-2. Typical floor plan

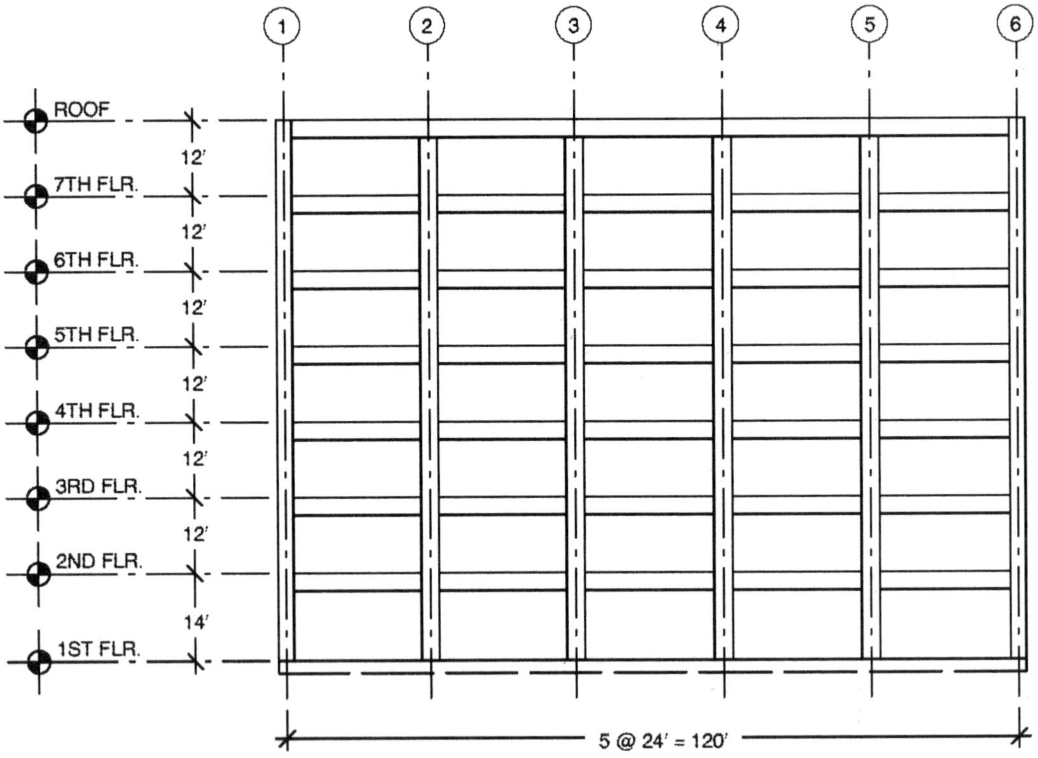

Figure 7-3. Typical frame elevation, line A

Outline

Determine the controlling seismic parameters and/or design the following structural elements.

1. **Site ground motion**
2. **Design base shear coefficient**
3. **Reliability/redundancy factor**
4. **Combined effect of horizontal and vertical earthquake induced forces**
5. **Vertical distribution of seismic forces**
6. **Frame nodal and member forces**
7. **Analysis and evaluation of frame drifts**
8. **Beam design**
9. **Column design**
10. **Joint shear analysis**
11. **Detailing of beams and columns**
12. **Foundation considerations**

Given Information

The building has a floor system consisting of post-tensioned slabs and girders. Vertical loads are carried by a frame system. Use of perimeter SMRF and the interior frames is designed to allow freedom for tenant improvements.

Seismic and site data:

Mapped spectral response accelerations (5% critical damping, Site Class B)

$S_S = 1.5$ at 0.2-second response §22.0

$S_1 = 0.60$ at 1.0-second response

$I = 1.0$ (Occupancy Category II) §11.5.1

Site Class D

Average story weights (for seismic design):

Roof:

Roofing	9.0 psf
Concrete slab (8-inch)	100.0
Girders	15.0
Columns	4.0
Partitions	5.0
Curtain wall	5.0
Ceiling + Elec. + Mech.	5.0
Misc.	3.0
Total	146.0 psf

Typical floor:	($6^{th} - 7^{th}$)	($3^{rd} - 5^{th}$)	(2nd floor)
Covering	2.0 psf	2.0 psf	2.0 psf
Concrete slab (8-inch)	100.0	100.0	100.0
Girders	15.0	24.0	24.0
Columns	8.0	8.0	10.0
Partitions*	10.0	10.0	10.0
Curtain wall	10.0	10.0	10.0
Ceiling + Elec. + Mech.	5.0	5.0	5.0
Misc.	3.0	3.0	3.0
Total	150.0 psf	159.0 psf	161.0 psf

* Partitions are 20 psf for gravity calculations and 10 psf for seismic calculations.

Structural Materials:

Concrete $f'_c = 4000$ psi (normal weight)

Reinforcing A706, ($f_y = 60$ ksi) for longitudinal bars

Reinforcing A615, Grade 60 ($f_y = 60$ ksi) for stirrups

Calculations and Discussion	Code Reference

1. Site ground motion — §11.4

The maximum considered earthquake spectral response acceleration for short periods, S_{MS}, and at 1.0-second period, S_{M1}, adjusted for site class effects, shall be determined by Equations 11.4-1 and 11.4-2.

$F_a = 1.0$ — T 11.4-1

$F_v = 1.5$ — T 11.4-2

$S_{MS} = F_a S_S = (1.0)(1.5g) = 1.5$ — Eq 11.4-1

$S_{M1} = F_v S_1 = (1.5)(0.60g) = 0.90$ — Eq 11.4-2

Five-percent-damped design spectral response acceleration at short periods, S_{DS}, and at 1.0-second period, S_{D1}, adjusted for site class effects, shall be determined by Equations 11.4-3 and 11.4-4.

$S_{DS} = 2/3\ S_{MS} = (2/3)(1.5) = 1.0$ — Eq 11.4-3

$S_{D1} = 2/3\ S_{M1} = (2/3)(0.90) = 0.60$ — Eq 11.4-4

2. Design base shear coefficient — §12.2-1

The coefficients for a reinforced concrete special moment-resisting frame building are

Response modification factor, $R = 8$ — T 12.2-1

System overstrength factor, Ω_o, = 3

Deflection amplification factor, $C_d = 5.5$

The approximate fundamental building period is

$T_a = C_t h_n^x = (0.016)(86\text{ ft})^{0.9} = 0.88\text{ sec}$ — Eq 12.8-7

Alternately,

$T_a = 0.1N = 0.1(7) = 0.7\text{ sec}$ — Eq 12.8-8

Therefore, use $T_a = 0.88$ sec.

The seismic base shear, V, in a given direction shall be determined in accordance with the following equation

$V = C_s W$ — Eq 12.8-1

The calculation of seismic response coefficient, C_s, shall be

$$C_s = \frac{S_{DS}}{\frac{R}{I}} = \frac{1.0}{\left(\frac{8.0}{1.0}\right)} = 0.125$$ — Eq 12.8-2

but need not exceed the following

$$C_s = \frac{S_{D1}}{\left(\frac{R}{I}\right)T} = \frac{0.60}{\left(\frac{8.0}{1.0}\right)0.88\text{ sec}} = 0.085 \text{ for } T \le T_L$$ Eq 12.8-3

or

$$C_s = \frac{S_{D1}T_L}{T^2\left(\frac{R}{I}\right)} \text{ for } T \ge T_L$$ Eq 12.8-4

The value of seismic response coefficient, Cs, shall not less than

$$C_s = 0.044S_{DSI} = 0.044(1.0)(1.0) = 0.044$$ Eq 12.8-5, Supplement No. 2

In addition, for buildings located where S_1 is equal to or greater than 0.6, the value of the seismic response coefficient, C_s, shall not be less than

$$C_s = \frac{0.5S_1}{\left(\frac{R}{I}\right)} = \frac{0.50(0.60)}{\left(\frac{8.0}{1.0}\right)} = 0.038$$ Eq 12.8-6

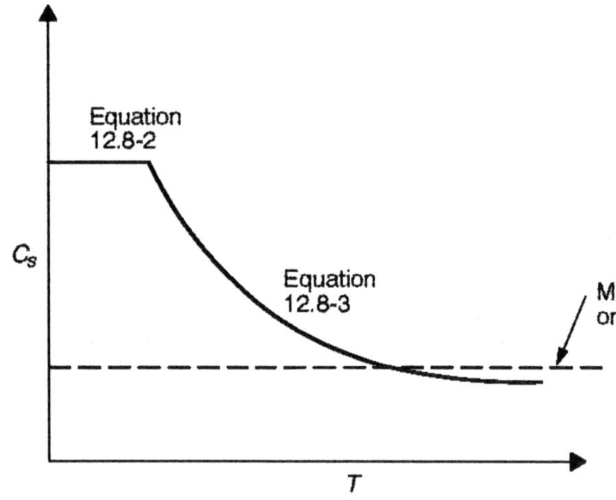

Figure 7-4. Typical response spectrum

Therefore, Equation 12.8-3 controls the base shear calculation for this building and the seismic coefficient is thus

$$V = 0.085W$$

3. Redundancy factor §12.3.4.2

In structures assigned to Seismic Design Catetory D, E, or F, ρ shall be 1.3 unless one of the following two conditions is met, whereby ρ is permitted to be taken as 1.0

a. Each story resisting more than 35 percent of the base shear in the direction of interest shall comply with Table 12.3-3, or

b. Structures that are regular in plan at all levels, provided that the seismic force resisting systems consist of at least two bays of seismic force-resisting perimeter framing on each side of the structure in each story resisting more than 35 percent of the base shear.

Based on item b above, this building has a ρ = 1.0.

4. Combined effect of horizontal and vertical earthquake induced forces §12.4.2

$$E = E_h + E_v \qquad \text{Eq 12.4-1}$$

$$E = E_h - E_v \qquad \text{Eq 12.4-2}$$

$$E_h = \rho Q_E \qquad \text{Eq 12.4-3}$$

$$E_v = 0.2 S_{DS} D \qquad \text{Eq 12.4-4}$$

$$E = E_h + E_v$$

$$= \rho Q_E + 0.2 S_{DS} D$$

$$= 1.0 Q_E + 0.2(1.0) D$$

$$= Q_E + 0.2D$$

5. Vertical distribution of seismic forces §12.8.3

In this part, the seismic forces on the concrete frame are determined.

5a. Story masses (weights) are calculated in Table 7-1

Table 7-1. Calculation of building and story weights

Level	Area (sf)	w_i (psf)	W_i (kips)
Roof	10,800	146.0	1,577
7th	10,800	150.0	1,620
6th	10,800	150.0	1,620
5th	10,800	159.0	1,717
4th	10,800	159.0	1,717
3rd	10,800	159.0	1,717
2nd	10,800	161.0	1,739
Totals	75,600		11,707

5b. Base shear and vertical distribution of shear

Using the results of Part 2, the base shear is

$$V = 0.085W = 0.085(11{,}707 \text{ kips}) = 995 \text{ kips}$$

The vertical distribution of shear is determined from Equations 12.8-11, 12.8-12 and 12.8.13

$$F_x = C_{vx}V \qquad \text{Eq 12.8-11}$$

$$C_{vx} = \frac{w_x h_x^k}{\sum_{i=1}^{n} w_i h_x^k} \qquad \text{Eq 12.8-12}$$

k = a distribution exponent related to the building period as follows

k = 1.0 for 0.5-sec period or less

k = 2.0 for 2.5-sec period or more

For T_a between 1.0 and 2.5 seconds, use linear interpolation to determine k.

Therefore, $k = 1.0 + \left(\frac{(0.88 \text{ sec} - 0.5 \text{ sec})}{(2.5 \text{ sec} - 0.5 \text{ sec})}\right)(1.0) = 1.19$

$$V_x = \sum_{i=1}^{n} F_i \qquad \text{Eq 12.8-13}$$

The calculation of story forces and story shears is shown in Table 7-2 below.

Table 7-2. Vertical distribution of shear

Level	(2) W_x (kips)	(3) ΣW_x (kips)	(4) h_x (ft)	(5) h (ft)	(6) $W_x h_x^k$ (k-ft)	(7) $\frac{W_x h_x^k}{\Sigma W_i h_i^k}$ %	(8) F_i (kips)	(9) ΣF_i (kips)
R	1,577	1,577	86	12	316,144	25.5	253	253
7	1,620	3,197	74	12	271,582	21.9	218	471
6	1,620	4,817	62	12	220,019	17.7	176	647
5	1,717	6,534	50	12	180,528	14.5	145	792
4	1,717	8,251	38	12	130,231	10.5	104	896
3	1,717	9,968	26	12	82,907	6.7	66	963
2	1,739	11,707	14	14	40,197	3.2	32	995
Totals	11,707				1,241,608	100.0%	995	
$V = 995$								

6. Frame nodal and member forces

The longitudinal frame along Line A is designed in this example. First, dead and live loads on the beams are determined using a tributary width of 15 feet. The gravity loads applied to the beams in the frame analysis are summarized below in Table 7-3.

Table 7-3. Beam gravity loads for analysis

Framing Level	Dead Load (plf)	Live Load (plf)
Roof	2190	300
7	2400	750
6	2400	750
5	2535	750
4	2535	750
3	2535	750
2	2565	750

A torsional analysis of the building using a 5 percent accidental torsion (using an eccentricity equivalent to 5 percent of the perpendicular building dimension) gives results such that all frames on the four faces of the building resist torsional shears of approximately 2 percent of the base shear. Thus the seismic forces in the frame analysis were increased by 2 percent to account for accidental torsion (per §12.8.4.2). Each line of perimeter frames should be designed to resist a base shear of 52 percent of the total building design base shear, V.

A two dimensional frame analysis is performed for the frame along Line A. The frame forces are determined from story forces above. Forces are distributed to frame nodes in proportion to their location along line A. Thus, at longitudinal frames with six columns, 10 percent of the story force is applied to end column nodes and 20 percent of the story force is applied to the interior column nodes. The force distribution at transverse frames with five columns is 12.5 percent to exterior column nodes and 25 percent to interior column nodes. The frame nodal loads for longitudinal and transverse frames are summarized in Table 7-4. Frame joint and member numbers are shown in Figures 7-5 and 7-6.

Table 7-4. Column nodal forces for analysis

Level	Story Forces (kips)	Long. Frame End Column Node Forces (kips)	Long. Frame Interior Col. Node Forces (kips)	Trans. Frame End Column Node Forces (kips)	Trans. Frame Interior Col. Node Forces (kips)
Roof	263	13.1	26.3	16.4	32.9
7	227	11.3	22.7	14.2	28.3
6	183	9.2	18.3	11.4	22.9
5	151	7.5	15.1	9.4	18.9
4	108	5.4	10.8	6.8	13.5
3	69	3.4	6.9	4.3	8.6
2	33	1.7	3.3	2.1	4.2
Total	1034	51.6	103.4	64.6	129.3

The loads shown in columns 3 and 4 of Table 7-4 add to 52 percent of the design base shear. The loads in columns 5 and 6 also add to 52 percent of the design base shear. To account for torsion, a load factor of 1.04 × 50 percent = 52 percent was used to define the frame nodal loads. This problem was solved on a two-dimensional frame program. Any elastic finite element analysis program could be used, including those with three-dimensional capability.

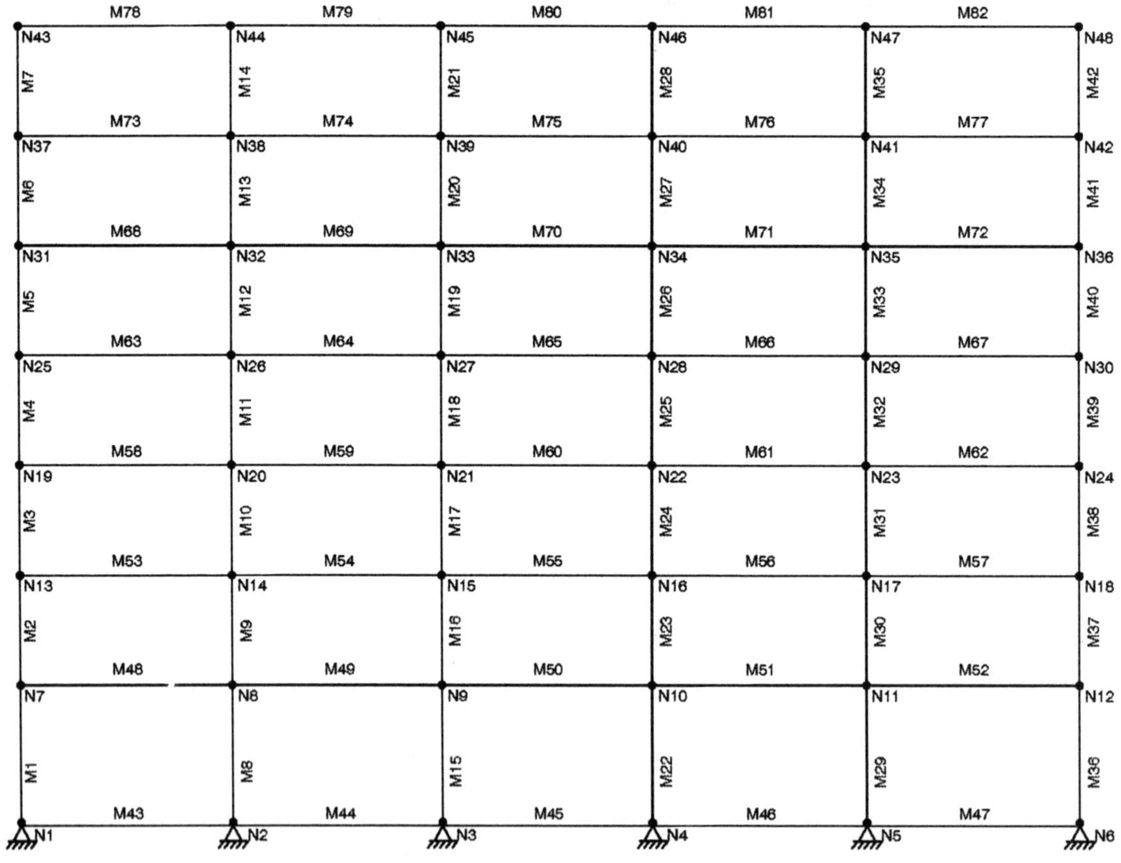

Figure 7-5. Computer model of the frame on Line A

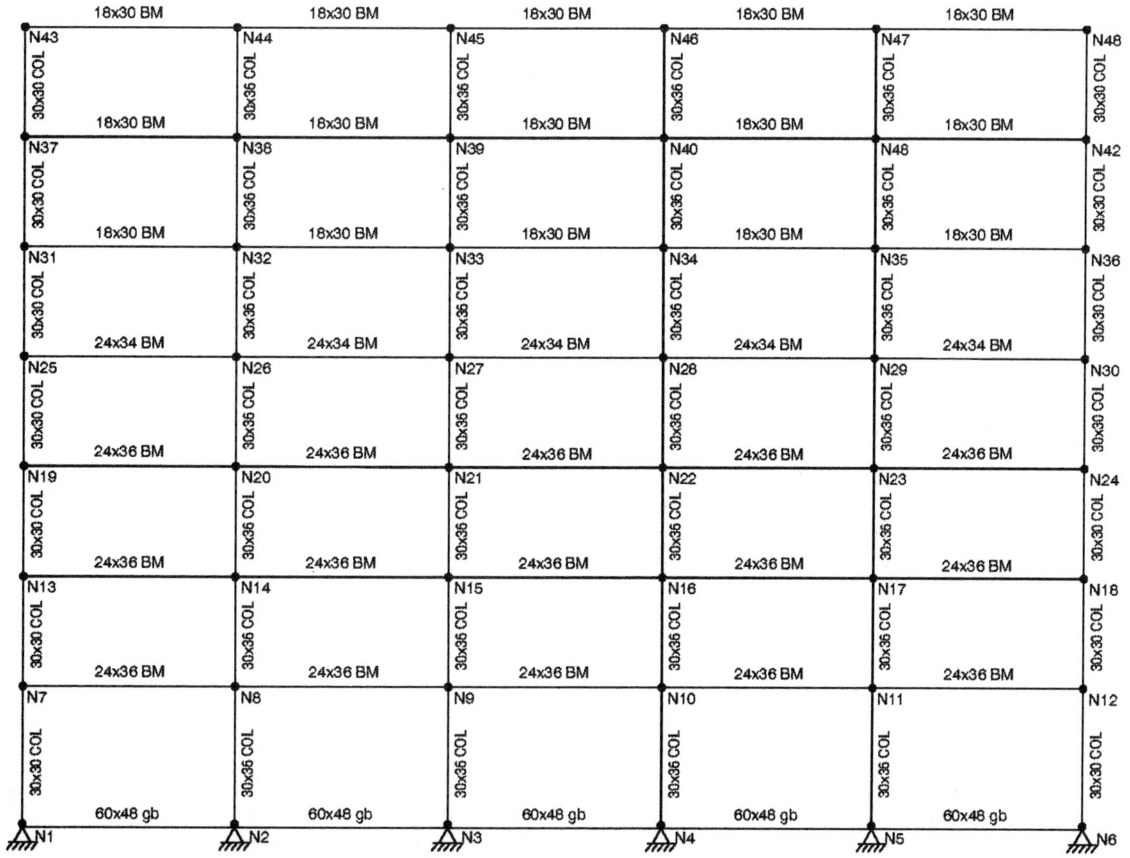

Figure 7-6. Computer model of the frame on Line A with beam & column sizes

7. Analysis and evaluation of frame drifts

Under §12.12.1, allowable story drifts, Δ_a, are limited to $0.020h_{sx}$ for drifts corresponding to the maximum inelastic response displacement δ_x. Under §12.8.6, $\delta_x = \frac{C_d \delta_{xe}}{I}$ or $\delta_x = \frac{5.5\delta_{xe}}{1.0} = 5.5\delta_{xe}$. Table 6-5 summarizes the calculation of the allowable frame drifts.

Table 7-5. Allowable story deformations and displacements

Story	Total Height (ft)	Story Height (ft)	Allowable δ_x (in)	Sum $\Sigma\delta_x$ (in)	Allowable δ_{xe} (in)	$\Sigma\delta_{xe}$ (in)
Roof	86	12	2.88	20.64	0.524	3.755
7	74	12	2.88	17.76	0.524	3.231
6	62	12	2.88	14.88	0.524	2.707
5	50	12	2.88	12.00	0.524	2.183
4	38	12	2.88	9.12	0.524	1.659
3	26	12	2.88	6.24	0.524	1.135
2	14	14	3.36	3.36	0.611	0.611

The frame analysis is thus performed using a standard frame analysis program. Columns, beams, and grade beams were sized to meet allowable drift limits. Member section properties were chosen to represent the cracked structure. In accordance with ACI 318-08 §10.10.4.1, 70 percent of the gross section properties are used for columns and 35 percent of gross section properties are used for beams to estimate the contribution of cracked sections on the frame behavior.

Selected sections were 30×30 corner columns, 30×36 interior perimeter columns, 24×36 beams and 60×48 foundation grade beams. The designer must size a frame that meets drift limitations and also meets strength criteria. For the design of this frame, the controlling parameters were frame stiffness and strength of beams. Using the member sizes chosen, frame analysis gives the lateral story displacements, shown below in Table 7-6. Note that the frame analysis gives δ_{xe} deflections, thus the comparison is made using δ_{xe} deflections, and that the ρ factor is not used in the deflection analysis.

Table 7-6. Displacements determined from analysis

Story	Total Height (ft)	Story Height (ft)	From Analysis δ_{xe} Story Drifts (in)	Maximum Allowable δ_{xe} Story Drifts (in)	From Analysis $\Sigma\delta_{xe}$ (in)	Maximum Allowable $\Sigma\delta_{xe}$ (in)
Roof	86	12	0.33	0.52	3.16	3.75
7	74	12	0.46	0.52	2.83	3.23
6	62	12	0.50	0.52	2.37	2.71
5	50	12	0.48	0.52	1.87	2.18
4	38	12	0.49	0.52	1.39	1.66
3	26	12	0.49	0.52	0.90	1.13
2	14	14	0.41	0.61	0.41	0.61

As shown in Table 7-6, story drifts are determined to be within allowable limits. The iteration between frame stiffness and member strengths has resulted in a frame design with conservative drifts. The designer must iterate between frame analysis and member section design.

8. Beam design

8a. Load combinations

The next procedure is frame member design. Frame beams are designed to support gravity loads and resist seismic forces. Beams are sized to limit frame drift and to resist the corresponding moment with a nominal strength ϕM_n. The ϕ factor for bending analysis is 0.90. The controlling load combinations are given in §12.4.2.3 and are summarized below.

$$(1.2 + 0.2S_{DS}) + \rho Q_E + 0.5L$$

§12.4.2.3
Load Comb. 5

Thus,

$$1.2D + 0.2(1.0g)D + 0.5L + \rho Q_E = 1.4D + 0.5L + \rho Q_E$$

Alternately,

$$(0.9 - 0.2S_{DS})D + \rho Q_E \qquad \text{Load Comb. 7}$$

$$0.9D - 0.2(1.0g)D - \rho Q_E = 0.7D - \rho Q_E$$

8b. Design requirements for frame beams

The nominal beam strength is calculated using the following formula and ignoring compression steel for simplicity

$$\phi M_n = \phi A_s f_y \left(d - \frac{a}{2} \right) \geq M_u$$

Note that standard practice has been to consider the frame beam to have a combined section including consideration of the contribution of the adjacent slab for both compression and tension stresses (ACI 318-08 §21.5.1.4). Chapter 21 of ACI 318-08 is the required code for design of concrete SMRF elements under the 2009 IBC.

The probable flexural strength, M_{pr}, is calculated per ACI 318-08 §21.5.4.1 using $1.25f_y$ for the reinforcing steel stress. Recalculating the beam strength using $\phi = 1.0$ gives

$$M_{pr} = 1.25 A_s f_y \left(d - \frac{a_{pr}}{2} \right)$$

The shear strength of the beam, ϕV_n, must be designed to be greater than required to resist V_e due to M_{pr} at both ends of the beam plus shear from gravity loads. L is the distance from column face to column face. For this example, the distance is L = 24 ft minus 36 in = 21 ft, 0 in center of column to center of column. The ϕ factor for shear analysis is 0.75 per ACI 318-08 §9.3.2.3. Thus, the ultimate shear load is calculated as

$$V_e = \frac{+M_{pr} - (-M_{pr})}{L} + \frac{w_{\text{factored gravity}} L}{2} \leq \phi V_n$$

$$\phi V_n = \phi V_c + \phi V_s$$

$$\phi V_c = 0; \quad \phi V_s = 0.75 A_v f_y \frac{d}{s}$$

Under ACI 318-08 §21.5.4.2, the shear contribution from concrete, V_c, is considered to be zero when both of the following conditions occur: 1) the earthquake induced shear force represents more than one half of the total shear force; and 2) factored axial compressive force is less than $A_g f'_c/20$.

In the region of plastic hinges, transverse ties are required to resist shear forces.

Maximum spacing of ties cannot exceed — ACI 318-08 §21.5.3.2

1. $d/4$.
2. eight times the diameter of the smallest longitudinal reinforcement
3. 24 times the diameter of the hoop bars
4. 12 inches

An example beam design for beam 48 (Figure 7-7) is shown. The controlling load combinations including seismic forces are Formulas 5 and 7. Depending on the direction of seismic inertial force, seismic moments add with gravity moments at one beam end and subtract at the other end.

Beyond regions of potential plastic hinges, stirrups with seismic ties are required at a maximum spacing of $d/2$ throughout the length of the beam under ACI 318-08 §21.5.3.4.

Diagrammatic shear and moment diagrams are shown in Figure 7-7.

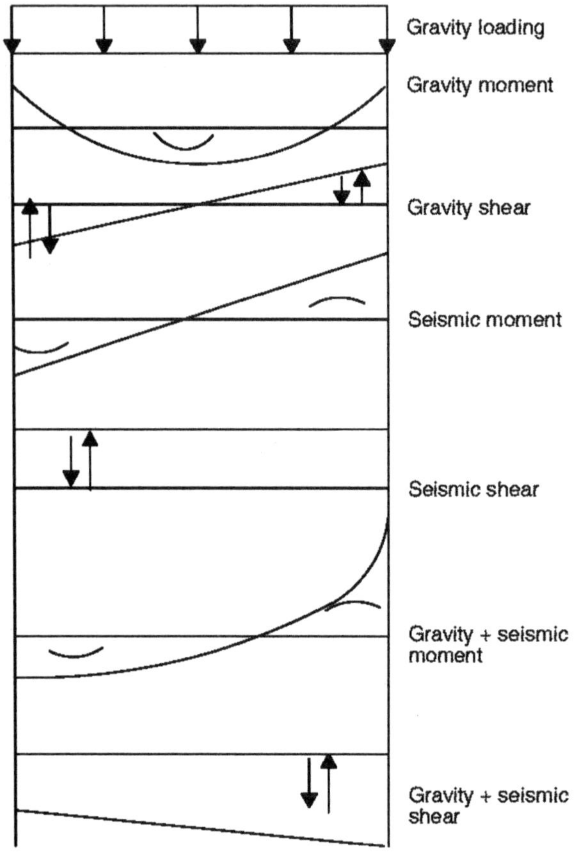

Figure 7-7. Moment and shear diagrams for beams

A review of the moment and shear diagrams for gravity loads and seismic loads (Figure 7-7) will help the designer realize that seismic moment and negative gravity moment at beam ends will be additive for top reinforcement design and subtractive for bottom reinforcement design. Because seismic moment is usually considerably greater than gravity moment, the reinforcement design will be controlled by load combinations including seismic loads, however, greater amounts of top reinforcement will be required than bottom reinforcement. Because the frame behavior produces beam moments as depicted in Figure 7-7, load combination 5 will maximize negative moments for top reinforcement design and load combination 7 will maximize positive moments for bottom reinforcement design.

An example calculation for beam 48 follows.

From the frame analysis, load combination 5, negative moment is –706 kip-ft. For a beam with $b = 24$ in and $h = 36$, $d = 33$

Try 5 #10 top bars, $A_s = 6.35\ \text{in}^2$

Per ACI 318-08 §§10.2.7 and 10.5.1

$$A_{s,\min} = \frac{200 b_w d}{f_y} = \frac{200(24\ \text{in})(33\ \text{in})}{60{,}000\ \text{psi}} = 2.64\ \text{in}^2 \le 6.35\ \text{in}^2 \quad \therefore\ o.k.$$

$$a = \frac{(6.35\ \text{in}^2)(60{,}000\ \text{psi})}{0.85(4000\ \text{psi})(24\ \text{in})} = 4.67\ \text{in}$$

$$\phi M_n = (0.90)(6.35\ \text{in}^2)(60{,}000\ \text{psi})\left(33\ \text{in} - \frac{4.67\ \text{in}}{2}\right)\left(\frac{1}{12\ \text{in}}\right)\left(\frac{1\ \text{kip}}{1000\ \text{lb}}\right)$$

$$= 876\ \text{kip-ft} \ge 707\ \text{kip-ft} \quad \therefore\ o.k.$$

From the frame analysis, Equation 7, positive moment is 442 kip-ft.

Try 5 #7 bottom bars, $A_s = 3.0\ \text{in}^2$

Per ACI 318-08 §13.2.4

$$b = 52\ \text{in};\ [24\ \text{in} + \text{minimum of } 36\ \text{in} - 8\ \text{in} = 28\ \text{in or } 4(8\ \text{in}) = 32\ \text{in}]$$

$$a = \frac{(3.0\ \text{in}^2)(60{,}000\ \text{psi})}{0.85(4000\ \text{psi}(52\ \text{in})} = 1.02\ \text{in}$$

$$\phi M_n = (0.90)(3.0\ \text{in}^2)(60{,}000\ \text{psi})\left(33\ \text{in} - \frac{1.021\ \text{in}}{2}\right)\left(\frac{1}{12\ \text{in}}\right)\left(\frac{1\ \text{kip}}{1000\ \text{lb}}\right)$$

$$= 439\ \text{kip-ft} \ge 360\ \text{kip-ft} \quad \therefore\ o.k.$$

Thus the beam 48 design will have 5 #10 top bars and 5 #7 bottom bars. Note that ACI 318-05 §21.5.2.2 requires that positive moment strength (bottom reinforcement) be a minimum 50 percent of negative moment strength at the joints and that neither the positive or negative moment strength along the beam be less than one fourth of the strength at either joint (end). In this example, 439 kip-ft is 50 percent of 876 kip-ft, therefore, OK.

8c. Beam skin reinforcement

If the overall depth of a beam exceeds 36 inches, longitudinal skin reinforcement shall be distributed along both side faces of a beam for a distance $h/2$ from the tension face per ACI 318-08 §10.6.7. The skin reinforcement shall be spaced a distance s per ACI 318-08 §10.6.4.

$$s = 15\left(\frac{40{,}000}{f_s}\right) - 2.5c_c \quad f_s = 2/3 f_y \qquad \text{ACI 318-08 (Eq 10-4)}$$

$$s = 15\left(\frac{40{,}000}{\frac{2}{3}(60{,}000)}\right) - 2.5(3\ \text{in})$$

$$s = 7\tfrac{1}{2}\ \text{in}$$

But not greater than

$$12\,(40{,}000/f_s)$$

$$s = 12\left(\frac{40{,}000}{\frac{2}{3}(60{,}000)}\right) = 12\text{ in}$$

For this example, skin reinforcement is not required.

8d. Beam shear design

The design shear force, V_e, is determined from consideration of the statical forces on the portion of the member between faces of the joints per ACI 318-08 §§21.3.4.1 and 21.5.4.1. For shear design, the designer allows for plastic hinge formation, which will produce shear forces greater than those from frame analysis.

$$V_e = \frac{+M_{pr} - (-M_{pr})}{L} + \frac{w_{\text{gravity}} L}{2}$$

$$-a = \frac{(1.25)(6.35)(60{,}000\text{ psi})}{0.85(4000\text{ psi})(24\text{ in})} = 5.84\text{ in}$$

$$-M_{pr} = [(1.25)(6.35\text{ in}^2)(60{,}000\text{ psi})]\left(33\text{ in} - \frac{5.84\text{ in}}{2}\right)\left(\frac{1}{12{,}000}\right) = 1194\text{ kip-ft}$$

$$+a_{pr} = \frac{(1.25)(3.0)(60{,}000\text{ psi})}{0.85(4000\text{ psi})(52\text{ in})} = 1.27\text{ in}$$

$$+M_{pr} = [(1.25)(3.0\text{ in}^2)(60{,}000\text{ psi})]\left(33\text{ in} - \frac{1.27\text{ in}}{2}\right)\left(\frac{1}{12{,}000}\right) = 607\text{ kip-ft}$$

Shear from dead load is calculated from load combination 5.

Beam length $L = 24\text{ ft} - 3\text{ ft}/2 - 3\text{ ft}/2 = 21\text{ ft}$
$L = 21\text{ ft, 3 in}$ at end bays.

$$V_{\text{gravity}} = [(1.40)(2565\text{ plf}) + (0.50)(750\text{ plf})]\left(\frac{21.25\text{ ft}}{2}\right) = 42\text{ kips}$$

$$\therefore V_u = \frac{(1194\text{ kip-ft} + 607\text{ kip-ft})}{21.25\text{ ft}} + 42\text{ kips} = 127\text{ kips}$$

The design shear, V_e, is thus the sum of the shear from the plastic end moments plus the gravity shear.

Seismic stirrups at the plastic hinge regions are calculated as shown below. Note that the plastic hinge region is a distance of $2d$ from the column face.

Try #5 ties with four vertical legs at 6-inch spacing over the $2d$ length (62 inches).

Transverse reinforcement should be proportioned to resist shear, assuming $V_c = 0$ when both of the following conditions occur (ACI 318-08 §21.5.4.2)

a. The earthquake induced shear force calculated in accordance with §21.5.4.1 represents one half or more of the maximum required shear strength within those lengths.

b. The factored axial compressive force including earthquake effects is less than $A_g f_c'/20$.

$$\phi V_n = \phi V_c + \phi V_s$$

$$\phi V_c = 0$$

$$\phi V_s = \frac{\phi A_v f_y d}{s}$$

$$\phi V_n = 0 + \frac{0.75(2)(0.31\text{ in}^2)(60{,}000\text{ psi})(33\text{ in})}{6\text{ in}} = 153\text{ kips} \geq 127\text{ kips} \quad \therefore \textit{ o.k.}$$

Therefore, use 2 legs, #5 stirrup ties at 6-inch spacing at plastic hinge regions at beam ends.

Seismic stirrups in the beam between plastic hinge regions are calculated as follows. Try #5 ties at 7-inch spacing

$$V_u = 85\text{ kips} + 42\text{ kips}\left(\frac{10\text{ ft, }9\text{ in} - 2(33\text{ in})}{10\text{ ft, }9\text{ in}}\right) = 106\text{ kips}$$

$$\phi V_s = 0.75(2)(0.31\text{ in}^2)(60{,}000\text{ psi})\ (33\text{ in})/8\text{ in} = 115\text{ kips} \geq 104\text{ kips} \quad \therefore \textit{ o.k.}$$

Therefore, the final design for Beam 48 is a 24-inch-wide by 36-inch-deep beam with 5 #10 top bars, 5 #7 bottom bars, 2 legs-#5 stirrup ties at 6-inch spacing each end with 2 legs, #5 stirrup ties at 8 inches between.

8e. Design of all Frame A beams

Following these same procedures and using the forces from the frame analysis, the Frame A beam designs for flexural strength are shown in Table 7-7.

Table 7-7. Beam member longitudinal steel design

Member	$M_{u,i}$ Eq. 12-6	$M_{u,j}$ Eq. 12-5	B (in)	h (in)	d (in)	Bar Location	No. bars	Bar Size	Bar Area (in^2)	A_s (in^2)	a (in)	ϕM_n (k-ft)	Bending Results	DCR
Level 2														
48	–442	706	24	36	33	Top	5	#10	1.27	6.35	4.67	876	OK	0.81
			52	36	33	Bottom	5	#7	0.60	3.00	1.02	439	OK	1.01
49	–430	706	24	36	33	Top	5	#10	1.27	6.35	4.67	876	OK	0.81
			52	36	33	Bottom	5	#7	0.60	3.00	1.02	439	OK	0.98
50	–430	707	24	36	33	Top	5	#10	1.27	6.35	4.67	876	OK	0.81
			52	36	33	Bottom	5	#7	0.60	3.00	1.02	439	OK	0.98
51	–430	708	24	36	33	Top	5	#10	1.27	6.35	4.67	876	OK	0.81
			52	36	33	Bottom	5	#7	0.60	3.00	1.02	439	OK	0.98
52	–442	692	24	36	33	Top	5	#10	1.27	6.35	4.67	876	OK	0.79
			52	36	33	Bottom	5	#7	0.60	3.00	1.02	439	OK	1.01
Level 3														
53	–472	730	24	36	33	Top	5	#10	1.27	6.35	4.67	876	OK	0.83
			52	36	33	Bottom	5	#7	0.60	3.00	1.02	439	OK	1.08
54	–474	747	24	36	33	Top	5	#10	1.27	6.35	4.67	876	OK	0.85
			52	36	33	Bottom	5	#7	0.60	3.00	1.02	439	OK	1.08
55	–473	746	24	36	33	Top	5	#10	1.27	6.35	4.67	876	OK	0.85
			52	36	33	Bottom	5	#7	0.60	3.00	1.02	439	OK	1.08
56	–474	745	24	36	33	Top	5	#10	1.27	6.35	4.67	876	OK	0.85
			52	36	33	Bottom	5	#7	0.60	3.00	1.02	439	OK	1.08
57	–472	737	24	36	33	Top	5	#10	1.27	6.35	4.67	876	OK	0.84
			52	36	33	Bottom	5	#7	0.60	3.00	1.02	439	OK	1.08
Level 4														
58	–450	705	24	36	33	Top	5	#10	1.27	6.35	4.67	876	OK	0.81
			52	36	33	Bottom	5	#7	0.60	3.00	1.02	439	OK	1.03
59	–460	733	24	36	33	Top	5	#10	1.27	6.35	4.67	876	OK	0.84
			52	36	33	Bottom	5	#7	0.60	3.00	1.02	439	OK	1.05
60	–459	732	24	36	33	Top	5	#10	1.27	6.35	4.67	876	OK	0.84
			52	36	33	Bottom	5	#7	0.60	3.00	1.02	439	OK	1.05
61	–460	732	24	36	33	Top	5	#10	1.27	6.35	4.67	876	OK	0.84
			52	36	33	Bottom	5	#7	0.60	3.00	1.02	439	OK	1.05
62	–450	724	24	36	33	Top	5	#10	1.27	6.35	4.67	876	OK	0.83
			52	36	33	Bottom	5	#7	0.60	3.00	1.02	439	OK	1.03

Table 7-7. Beam member longitudinal steel design (continued)

Member	$M_{u,i}$ Eq. 12-6	$M_{u,j}$ Eq. 12-5	B (in)	h (in)	d (in)	Bar Location	No. bars	Bar Size	Bar Area (in^2)	A_s (in^2)	a (in)	ϕM_n (k-ft)	Bending Results	DCR
Level 5														
63	−378	631	24	34	31	Top	5	#10	1.27	6.35	4.67	819	OK	0.77
			50	34	31	Bottom	5	#7	0.60	3.00	1.06	411	OK	0.92
64	−389	662	24	34	31	Top	5	#10	1.27	6.35	4.67	819	OK	0.81
			50	34	31	Bottom	5	#7	0.60	3.00	1.06	411	OK	0.95
65	−388	662	24	34	31	Top	5	#10	1.27	6.35	4.67	819	OK	0.81
			50	34	31	Bottom	5	#7	0.60	3.00	1.06	411	OK	0.95
66	−389	662	24	34	31	Top	5	#10	1.27	6.35	4.67	819	OK	0.81
			50	34	31	Bottom	5	#7	0.60	3.00	1.06	411	OK	0.95
67	−378	655	24	34	31	Top	5	#10	1.27	6.35	4.67	819	OK	0.80
			50	34	31	Bottom	5	#7	0.60	3.00	1.06	411	OK	0.92
Level 6														
68	−185	430	18	30	27	Top	4	#9	1.00	4.00	3.92	451	OK	0.95
			40	30	27	Bottom	2	#6	0.44	0.88	0.39	97	OK	1.75
69	−191	451	18	30	27	Top	4	#9	1.00	4.00	3.92	451	OK	1.00
			40	30	27	Bottom	2	#6	0.44	0.88	0.39	106	OK	1.80
70	−192	452	18	30	27	Top	4	#9	1.00	4.00	3.92	451	OK	1.00
			40	30	27	Bottom	2	#6	0.44	0.88	0.39	106	OK	1.81
71	−191	451	18	30	27	Top	4	#9	1.00	4.00	3.92	451	OK	1.00
			40	30	27	Bottom	2	#6	0.44	0.88	0.39	106	OK	1.81
72	−185	446	18	30	27	Top	4	#9	1.00	4.00	3.92	451	OK	0.99
			40	30	27	Bottom	2	#6	0.44	0.88	0.39	106	OK	1.75
Level 7														
73	−129	378	18	30	27	Top	4	#9	1.00	4.00	3.92	451	OK	0.84
			40	30	27	Bottom	2	#5	0.31	0.62	0.27	75	OK	1.72
74	−138	399	18	30	27	Top	4	#9	1.00	4.00	3.92	451	OK	0.88
			40	30	27	Bottom	2	#5	0.31	0.62	0.27	75	OK	1.84
75	−137	397	18	30	27	Top	4	#9	1.00	4.00	3.92	451	OK	0.88
			40	30	27	Bottom	2	#5	0.31	0.62	0.27	75	OK	1.84
76	−138	394	18	30	27	Top	4	#9	1.00	4.00	3.92	451	OK	0.87
			40	30	27	Bottom	2	#5	0.31	0.62	0.27	75	OK	1.84
77	−130	389	18	30	27	Top	4	#9	1.00	4.00	3.92	451	OK	0.86
			40	30	27	Bottom	2	#5	0.31	0.62	0.27	75	OK	1.72

Table 7-7. Beam member longitudinal steel design (continued)

Member	$M_{u,i}$ Eq. 12-6	$M_{u,j}$ Eq. 12-5	B (in)	h (in)	d (in)	Bar Location	No. bars	Bar Size	Bar Area (in^2)	A_s (in^2)	a (in)	ϕM_n (k-ft)	Bending Results	DCR
Roof														
78	–99	288	18	30	27	Top	4	#8	0.79	3.16	3.10	362	OK	0.80
			40	30	27	Bottom	2	#5	0.31	0.62	0.27	75	OK	1.32
79	–96	300	18	30	27	Top	4	#8	0.79	3.16	3.10	362	OK	0.83
			40	30	27	Bottom	2	#5	0.31	0.62	0.27	75	OK	1.28
80	–94	296	18	30	27	Top	4	#8	0.79	3.16	3.10	362	OK	0.82
			40	30	27	Bottom	2	#5	0.31	0.62	0.27	75	OK	1.25
81	–92	294	18	30	27	Top	4	#8	0.79	3.16	3.10	362	OK	0.81
			40	30	27	Bottom	2	#5	0.31	0.62	0.27	75	OK	1.27
82	–90	277	18	30	27	Top	4	#8	0.79	3.16	3.10	362	OK	0.77
			40	30	27	Bottom	2	#5	0.31	0.62	0.27	75	OK	1.20

Notes for Table 7-7:

1. DCR = demand to capacity ratio.

With longitudinal beam reinforcement proportioned as indicated in Table 7-7 above, the plastic moment M_{pr} and shear design is as follows. Note that M_{pr} is calculated including contribution of perimeter reinforcement. $V_{U,\,(gravity)}$ is calculated as the factored combination of $D + L$ loads: $V_{U,\,(gravity)} = 1.40D + 0.50L$ from the seismic load combination.

Table 7-8. Beam member shear reinforcement design

Mem ID	Bar Loc.	A_s T&B	a_{pr} (in^2)	M_{pr} (k-ft)	V_{pr} (k)	$V_{U,GR}$ (k)	V_u (k)	ϕV_c (k)	Ties # legs	Tie Size	A_{vs} (in^2)	s (in)	ϕV_s (k)	ϕV_n (k)	Re-sult	DCR
Level 2																
48	Top	6.35	5.84	1,194	84	48	132	0.0	2	0.31	0.62	6.0	153	153	OK	0.86
	Bot-tom	3.00	1.27	607												
49	Top	6.35	5.84	1,194	84	48	132	0.0	2	0.31	0.62	6.0	153	153	OK	0.86
	Bot-tom	3.00	1.27	607												
50	Top	6.35	5.84	1,194	84	48	132	0.0	2	0.31	0.62	6.0	153	153	OK	0.86
	Bot-tom	3.00	1.27	607												
51	Top	6.35	5.84	1,194	84	48	132	0.0	2	0.31	0.62	6.0	153	153	OK	0.86
	Bot-tom	3.00	1.27	607												
52	Top	6.35	5.84	1,194	84	48	132	0.0	2	0.31	0.62	6.0	153	153	OK	0.86
	Bot-tom	3.00	1.27	607												

Table 7-8. Beam member shear reinforcement design (continued)

Mem ID	Bar Loc.	A_s T&B	a_{pr} (in²)	M_{pr} (k-ft)	V_{pr} (k)	$V_{U,GR}$ (k)	V_u (k)	ϕV_c (k)	Ties # legs	Tie Size	A_{vs} (in²)	s (in)	ϕV_s (k)	ϕV_n (k)	Re-sult	DCR
Level 3																
53	Top	6.35	5.84	1,194	84	47	131	0.0	2	0.31	0.62	6.0	153	153	OK	0.86
	Bot-tom	3.00	1.27	607												
54	Top	6.35	5.84	1,194	84	47	131	0.0	2	0.31	0.62	6.0	153	153	OK	0.86
	Bot-tom	3.00	1.27	607												
55	Top	6.35	5.84	1,194	84	47	131	0.0	2	0.31	0.62	6.0	153	153	OK	0.86
	Bot-tom	3.00	1.27	607												
56	Top	6.35	5.84	1,194	84	47	131	0.0	2	0.31	0.62	6.0	153	153	OK	0.86
	Bot-tom	3.00	1.27	607												
57	Top	6.35	5.84	1,194	84	47	131	0.0	2	0.31	0.62	6.0	153	153	OK	0.86
	Bot-tom	3.00	1.27	607												
Level 4																
58	Top	6.35	5.84	1,194	84	48	132	0.0	2	0.31	0.62	6.0	153	153	OK	0.86
	Bot-tom	3.00	1.27	607												
59	Top	6.35	5.84	1,194	84	47	132	0.0	2	0.31	0.62	6.0	153	153	OK	0.86
	Bot-tom	3.00	1.27	607												
60	Top	6.35	5.84	1,194	84	47	132	0.0	2	0.31	0.62	6.0	153	153	OK	0.86
	Bot-tom	3.00	1.27	607												
61	Top	6.35	5.84	1,194	84	47	132	0.0	2	0.31	0.62	6.0	153	153	OK	0.86
	Bot-tom	3.00	1.27	607												
62	Top	6.35	5.84	1,194	84	48	132	0.0	2	0.31	0.62	6.0	153	153	OK	0.86
	Bot-tom	3.00	1.27	607												

Table 7-8. Beam member shear reinforcement design (continued)

Mem ID	Bar Loc.	A_s T&B	a_{pr} (in²)	M_{pr} (k-ft)	V_{pr} (k)	$V_{U,GR}$ (k)	V_u (k)	ϕV_c (k)	Ties # legs	Tie Size	A_{vs} (in²)	s (in)	ϕV_s (k)	ϕV_n (k)	Re-sult	DCR
Level 5																
63	Top	6.35	5.84	1,114	78	48	126	0.0	2	0.31	0.62	6.0	144	144	OK	0.88
	Bot-tom	2.20	1.32	569												
64	Top	6.35	5.84	1,114	78	47	125	0.0	2	0.31	0.62	6.0	144	144	OK	0.88
	Bot-tom	2.20	1.32	569												
65	Top	6.35	5.84	1,114	78	47	125	0.0	2	0.31	0.62	6.0	144	144	OK	0.88
	Bot-tom	2.20	1.32	569												
66	Top	6.35	5.84	1,114	78	47	125	0.0	2	0.31	0.62	6.0	144	144	OK	0.88
	Bot-tom	2.20	1.32	569												
67	Top	6.35	5.84	1,114	78	48	126	0.0	2	0.31	0.62	6.0	144	144	OK	0.88
	Bot-tom	2.20	1.32	569												
Level 6																
68	Top	4.00	4.90	614	35	45	80	0.0	2	0.20	0.40	6.0	81	81	OK	0.99
	Bot-tom	0.88	0.49	147												
69	Top	4.00	4.90	614	35	45	80	0.0	2	0.20	0.40	6.0	81	81	OK	0.99
	Bot-tom	0.88	0.49	147												
70	Top	4.00	4.90	614	35	45	80	0.0	2	0.20	0.40	6.0	81	81	OK	0.99
	Bot-tom	0.88	0.49	147												
71	Top	4.00	4.90	614	35	45	80	0.0	2	0.20	0.40	6.0	81	81	OK	0.99
	Bot-tom	0.88	0.49	147												
72	Top	4.00	4.90	614	35	45	80	0.0	2	0.20	0.40	6.0	81	81	OK	0.99
	Bot-tom	0.88	0.49	147												

Table 7-8. Beam member shear reinforcement design (continued)

Mem ID	Bar Loc.	A_s T&B	a_{pr} (in^2)	M_{pr} (k-ft)	V_{pr} (k)	$V_{U,GR}$ (k)	V_u (k)	ϕV_c (k)	Ties # legs	Tie Size	A_{vs} (in^2)	s (in)	ϕV_s (k)	ϕV_n (k)	Re-sult	DCR
Level 7																
73	Top	4.00	4.90	614	33	45	78	0.0	2	0.20	0.40	6.0	81	81	OK	0.96
	Bot-tom	0.62	0.34	104												
74	Top	4.00	4.90	614	33	45	78	0.0	2	0.20	0.40	6.0	81	81	OK	0.96
	Bot-tom	0.62	0.34	104												
75	Top	4.00	4.90	614	33	45	78	0.0	2	0.20	0.40	6.0	81	81	OK	0.96
	Bot-tom	0.62	0.34	104												
76	Top	4.00	4.90	614	33	45	78	0.0	2	0.20	0.40	6.0	81	81	OK	0.96
	Bot-tom	0.62	0.34	104												
77	Top	4.00	4.90	614	33	45	78	0.0	2	0.20	0.40	6.0	81	81	OK	0.96
	Bot-tom	0.62	0.34	104												
Roof																
78	Top	3.16	3.87	495	28	39	67	0.0	2	0.20	0.40	6.0	81	81	OK	0.83
	Bot-tom	0.62	0.34	104												
79	Top	3.16	3.87	495	28	39	67	0.0	2	0.20	0.40	6.0	81	81	OK	0.83
	Bot-tom	0.62	0.34	104												
80	Top	3.16	3.87	495	28	39	67	0.0	2	0.20	0.40	6.0	81	81	OK	0.83
	Bot-tom	0.62	0.34	104												
81	Top	3.16	3.87	495	28	39	67	0.0	2	0.20	0.40	6.0	81	81	OK	0.83
	Bot-tom	0.62	0.34	104												
82	Top	3.16	3.87	495	28	39	67	0.0	2	0.20	0.40	6.0	81	81	OK	0.83
	Bot-tom	0.62	0.34	104												

Notes for Table 7-8:

1. DCR = demand to capacity ratio.

Having satisfied the design for both bending and shear, the final beam designs are thus chosen as shown in Table 7-9.

Table 7-9. Final beam designs

Level	Width (in)	Depth (in)	Long. Reinf. Top	Long. Reinf. Bottom	Skin Reinf.	Shear Reinf. in Hinge Regions	Shear Reinf. between Hinge Regions
Roof	18	30	4 #9	2 #6	None Req'd	2 legs #4 Ties@ 6 in	2 legs #4 Ties@ 8 in
7	18	30	4 #9	2 #5	None Req'd	2 legs #4 Ties@ 6 in	2 legs #4 Ties@ 8 in
6	18	30	4 #9	3 #7	None Req'd	2 legs #4 Ties@ 6 in	2 legs #4 Ties@ 8 in
5	24	34	5 #10	5 #7	None Req'd	2 legs #5 Ties@ 6 in	2 legs #5 Ties@ 8 in
4	24	36	5 #10	5 #8	None Req'd	2 legs #5 Ties@ 6 in	2 legs #5 Ties@ 8 in
3	24	36	5 #10	5 #8	None Req'd	2 legs #5 Ties@ 6 in	2 legs #5 Ties@ 8 in
2	24	36	5 #10	5 #7	None Req'd	2 legs #5 Ties@ 6 in	2 legs #5 Ties@ 8 in

Figure 7-8 represents a beam cross-section showing dimensions and reinforcement.

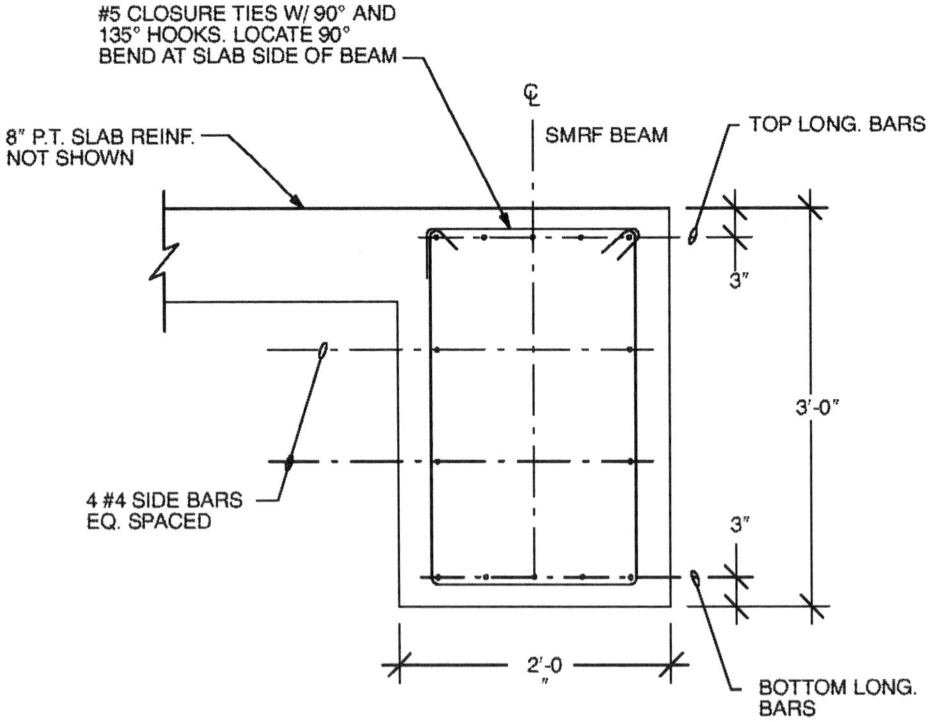

Figure 7-8. 24×36 beam at level 2

9. Column design

9a. Strong column calculation

Columns should be designed to ensure that the plastic hinges are located in the beams (i.e., strong column-weak beam behavior) and to resist column shears. To assure strong column-weak beam behavior, columns must be designed to have nominal bending strengths 120 percent stronger than beams per ACI 318-08 §21.6.2.2. This is achieved by summing the M_e of columns above and below a joint and comparing that with the sum of M_g for beams on both sides of a joint.

$$\Sigma M_{nc} \geq (6/5)\Sigma M_{nb} \qquad \text{ACI 318-08 (21-1)}$$

The controlling girder location occurs at levels 2, 3, 4, and 5. The girders are 24 inches by 34 inches with 5 #10s top, 5 #7s bottom.

Calculation of $-M_{nb}$ (negative, at beam tops)

$$a = \frac{5(1.27\text{ in}^2)(60{,}000\text{ psi})}{0.85(4000\text{ psi})(24\text{ in})} = 4.67\text{ in}$$

$$-M_{nb} = (0.90)(6.35\text{ in}^2)(60{,}000)\left(33\text{ in} - \frac{4.67\text{ in}}{2}\right)\frac{1}{12{,}000} = 876\text{ kip-ft}$$

Calculation of M_{nb} (positive, at beam bottoms)

$$a = \frac{5(0.60\text{ in}^2)(60{,}000\text{ psi})}{0.85(4000\text{ psi})(24\text{ in})} = 1.02\text{ in}$$

$$M_{nb} = (0.90)(3.0\text{ in}^2)(60{,}000)\left(33\text{ in} - \frac{2.21\text{ in}}{2}\right)\frac{1}{12{,}000} = 430\text{ kip-ft}$$

Therefore, at interior columns

$$6/5\,\Sigma\, M_{nb} = \frac{6}{5}(876\text{ kip-ft} + 430\text{ kip-ft}) = 1567\text{ kip-ft}$$

Therefore, at end columns

$$6/5\,\Sigma\, M_{nb} = \frac{6}{5}(876\text{ kip-ft}) = 1051\text{ kip-ft}$$

The girder moments are resisted by two column sections, the column above the joint and the column below the joint. The required column strengths, M_{nc} for interior and end columns are given below.

$$M_{nc} = \frac{1}{2}(1567 \text{ kip-ft}) = 784 \text{ kip-ft}$$

or

$$M_{nc} = \frac{1}{2}(1051 \text{ kip-ft}) = 526 \text{ kip-ft}$$

9b. Forces on columns due to factored load combinations

For column design, load combinations 5 and 7 are used. Also, because strength design is used, the effect of the vertical seismic component, E_v, must be included. Load combinations 5 and 7 are given below. Tables 7-10 and 7-11 provide axial forces and moments on the columns of Frame A for load combinations 5 and 7, respectively.

$$1.2D + 0.5L + 1.0E = 1.2D + 0.2D + 0.5L + 1.0Q_E = 1.4D + 0.5L + 1.0Q_E \quad (5)$$

$$0.9D - 1.0E = 0.9D - 0.2D - 1.0Q_E = 0.70D - 1.0Q_E \quad (7)$$

Table 7-10. Column loads for Load Combination 5

Member	P_u (kips)	V_u (kips)	M_u bottom (kip-ft)	M_u top (kip-ft)
1	94	47	535	–119
2	91	35	228	–194
3	89	32	179	–192
4	85	26	155	–140
5	76	19	130	–84
6	53	9	–6	–111
7	25	–7	–8	–86
8	624	101	1024	–389
9	529	101	642	–563
10	434	92	539	–563
11	342	81	500	–473
12	249	67	444	–343
13	161	47	181	–378
14	73	28	41	–296
15	631	102	1027	–394
16	535	102	638	–565
17	442	95	551	–575
18	348	84	514	–487
19	253	69	459	–354
20	164	50	190	–400
21	74	29	37	–304
22	631	102	1030	–399
23	535	100	634	–563
24	443	95	551	–574
25	348	84	514	–486
26	253	68	460	–355
27	164	50	189	–407
28	74	27	26	–296
29	635	104	1045	–404
30	538	98	623	–558
31	444	94	551	–572
32	349	84	512	–484
33	255	69	460	–355
34	165	50	187	–414
35	75	26	14	–285
36	538	62	597	–270
37	449	67	422	–372
38	356	62	365	–379
39	265	57	345	–332
40	178	49	323	–269
41	110	39	177	–309
42	48	30	80	–277

Table 7-11. Column loads for Load Combination 7

Member	P_u (kips)	V_u (kips)	M_u bottom (kip-ft)	M_u top (kip-ft)
1	–77	51	551	–160
2	–56	44	281	–242
3	–31	39	230	–243
4	–9	34	207	–192
5	9	27	183	–135
6	9	17	44	–166
7	8	6	–41	–76
8	282	103	1030	–393
9	239	100	637	–562
10	197	92	542	–565
11	156	83	503	–476
12	114	66	448	–347
13	75	48	181	–387
14	36	27	35	–294
15	289	103	1028	–396
16	245	100	637	–565
17	204	94	551	–575
18	160	84	514	–487
19	118	69	459	–354
20	77	50	190	–402
21	37	28	36	–301
22	289	102	1029	–398
23	246	100	635	–564
24	203	94	551	–574
25	160	83	514	–486
26	118	69	460	–355
27	77	50	190	–406
28	37	27	29	–298
29	294	104	1040	–400
30	250	99	628	–559
31	206	93	548	–570
32	163	82	509	–481
33	120	68	456	–352
34	79	50	186	–406
35	38	26	20	–287
36	367	59	580	–229
37	303	58	369	–323
38	235	54	313	–328
39	170	48	293	–279
40	110	41	270	–218
41	67	32	129	–253
42	29	21	35	–209

9c. Design of column for bending strength

ACI 318-08 §21.6.3.1 requires the longitudinal reinforcement ratio of columns to be between 1 and 6 percent. The design of columns is usually performed by calculating a column axial force-moment capacity (P-M) interaction diagram. The major points used to construct such a diagram are ϕP_n for compression, (ϕP_b, ϕM_b) at the balanced point, ϕM_n for pure moment, and ϕT_n for pure tension. The ϕ factor for column calculations is 0.65 for tied columns and 0.70 for spiral tied columns meeting requirements in ACI 318-08 §10.9.3. In accordance with ACI 318-08 §9.3.2.2, the ϕ factor may be increased linearly to 0.9 for columns or other axial load-carrying members as ϕP_n decreases from $0.10 f'_c A_g$ (or ϕP_b, whichever is less) to zero.

The equation for ϕP_n is given in ACI 318-08 §10.3.6.

$$\phi P_n = 0.85\phi[0.85 f'_c (A_g - A_{st}) + f_y A_{st}] \qquad \text{ACI 318-08 (10-1)}$$

Note that $\phi = 0.65$ for members with axial compression and flexure (not with spiral shear reinforcement) per ACI 318-08 §9.3.2.1.

Calculation of the balanced point is determined by using 0.002 strain for reinforcing steel at yield and 0.003 for concrete strain at crushing ACI 318-08 §10.3.2. By summing forces and moments, the balanced axial load and moment (ϕP_b, ϕM_b) can be determined. The nominal moment strength is determined by using 0.002 strain for steel yielding and by calculating tension forces and compression forces such that they add up to 0. The resulting moment is thus ϕM_n, where $\phi = 0.9$

The equation for tension members is

$$\phi T_n = \phi F_y A_{st}$$

Note that $\phi = 0.90$ for members with axial tension and axial tension with flexure per ACI 318-08 §9.3.2.1.

The designer may use a commercial program to develop a P-M diagram for the column axial load-moment interaction including effects for slenderness of columns. From the frame analysis for Frame A, the controlling load cases are summarized in Table 7-12.

Table 7-12. Critical column loads for Frame A

Column	Level	Location	Size (in)	Load Comb. Equation	P_u (kips)	V_u (kips)	M_u (kip-ft)
29	1	interior	30×36	16-5	635	105	1045
1	1	end	30×30	16-6	−77	51	551

Column 29 represents the controlling load combination for a column in compression and Column 1 represents the controlling load combination for a column in tension.

Using the program, check 30×36 interior column 29 with 20 #10 bars around perimeter. The resulting P-M diagram is shown in Figure 7-9.

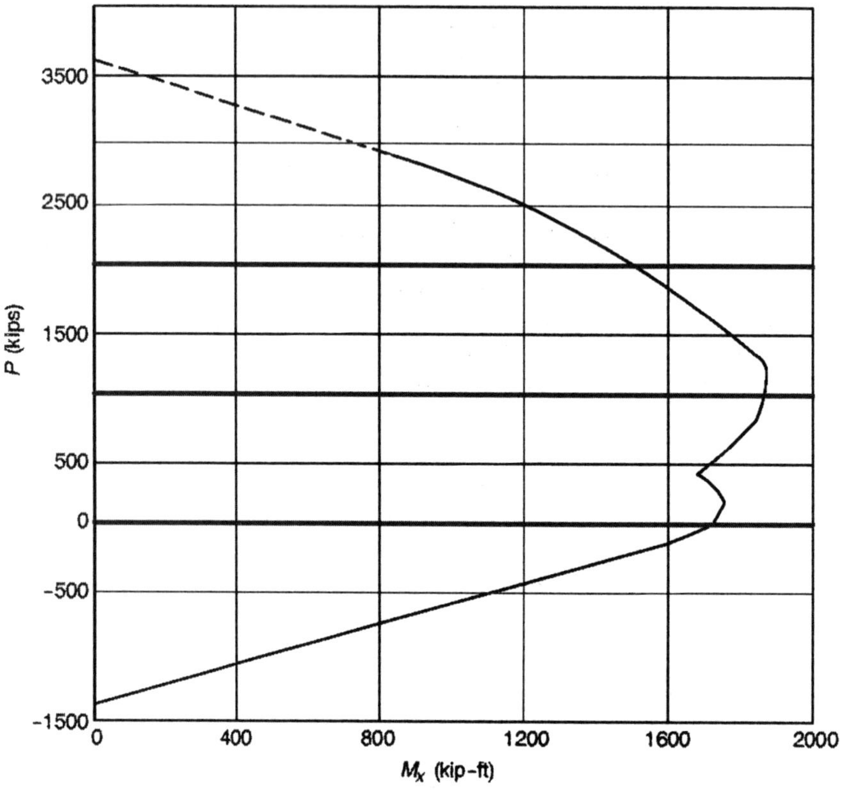

Figure 7-9. Column P-M diagram for 30-inch x 36-inch interior column

Check 30×30 corner Column 1 with 20 #10 bars around the perimeter. The resulting P-M diagram is shown in Figure 7-10.

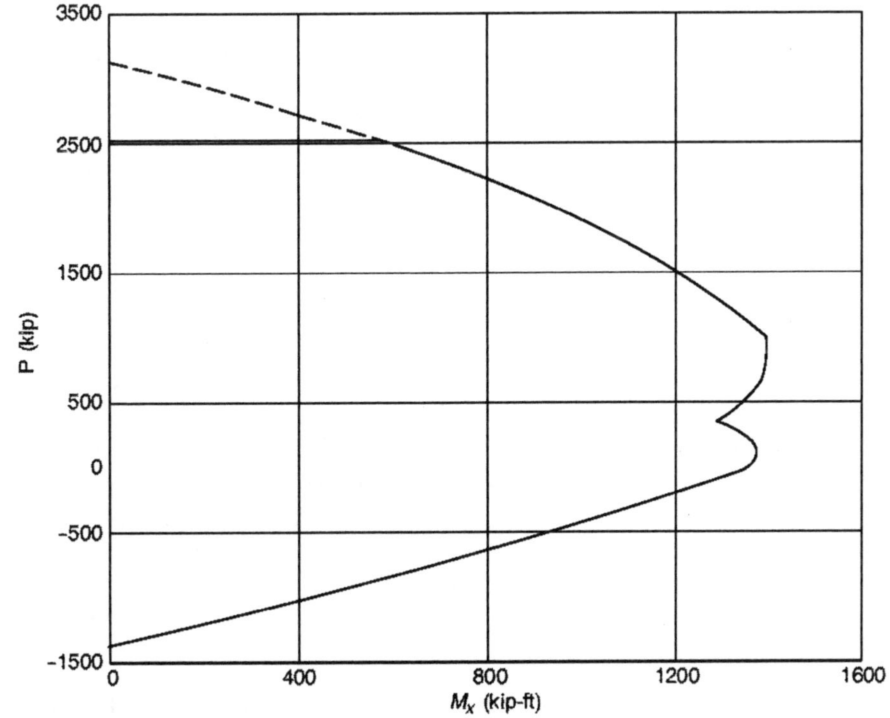

Figure 7-10. Column P-M diagram for 30-inch-square corner column

By comparing the design loads against the column P-M diagrams of Figures 7-9 and 7-10, it can be seen that both columns have adequate strength; see Table 7-13. Both column sections achieve 120 percent of beam moment strength and thus have adequate strength to develop the plastic moments of beams. ϕM_n for interior columns is approximately 2100 kip-ft and for end columns is approximately 1520 kip-ft at the axial load of approximately 600 kips.

$$\Sigma M_{nc} = \frac{6}{5} \Sigma M_{nb}$$

$$\Sigma M_{nc,\text{ interior}} = 2(2100 \text{ kip-ft}) = 4200 \text{ kip-ft} \geq 1567 \text{ kip-ft} \quad \therefore \ o.k.$$

$$\Sigma M_{nc,\text{ end}} = 2(1520 \text{ kip-ft}) = 3040 \text{ kip-ft} \geq 1051 \text{ kip-ft} \quad \therefore \ o.k.$$

It is assumed by the code that the design of columns to be 120 percent greater in flexural strength than girders will assure plastic hinge formation in the beams. This is probably true in most cases. Since that is what is required in ACI 318-08, it is what is shown in this example.

Some engineers believe that they should design the columns to develop the strength of the beam plastic moments, M_{pr}. While this is not explicitly required by ACI 318-08, it is probably a good idea. The reasoning is that the yielding elements in the frame are the beam plastic moments located at beam ends followed by column plastic moments at column bases. When all non-yielding aspects of the frame are designed to be stronger than the yielding elements, the anticipated frame yield behavior is assured. Thus the shear design of beams, columns, and joints, column flexural strengths, and foundation elements are all designed to have adequate strengths to resist the anticipated flexural yield mechanism of the frame.

Table 7-13. Column axial and flexural design strengths

Column	Size (in)	ϕP_n at $M = 0$ (kips)	ϕP_b (kips)	ϕM_b (kip-ft)	ϕM_n at $P = 0$ (kip-ft)
Interior	30×36	2658	1250	1875	1700
End	30×30	2334	1000	1400	1350

9d. Design of columns for shear strength

Columns must be designed for shear strength, V_e, required by ACI 318-08 §21.6.5.1 and for the special transverse reinforcement required by ACI 318-08 §21.6.4.4. The design shear force, V_e, shall be determined from the consideration of the maximum forces that can be generated at the faces of the beam/column joints at the ends of beams framing into the joint. These joint forces are determined using one of three methods:

1. Using the maximum probable moment strengths, M_{pr}, of the column at the top and bottom between joints along with the associated factored axial loads on the column.

2. The column shear, V_e, need not exceed that determined based on the probable moment strength, M_{pr}, of the beams framing into the joint.

3. V_e shall not be less than the factored shear determined from analysis.

It is likely that the second method described above will control the shear design of the column since strong column behavior of the frame will force plastic hinges to form in the beams. At the columns in the first story, the controlling case is from column top moments based on M_{pr} of beams and column bottom moments based on M_{pr} of the column calculated with associated axial loads.

For the interior column, 30×36, at stories one and two, the maximum shear needs to be determined from maximum shear that can be transferred from beam strength, M_{pr}, as shown below.

Interior column at first story

Clear height of column = 14 ft − 3 ft = 11 ft

M_{pr} of beams framing into top of column is based on negative moment from one beam and positive moment from the other beam.

$$\Sigma M_{pr} = 1194 \text{ kip-ft} + 607 \text{ kip-ft} = 1801 \text{ kip-ft}$$

Distribution of beam moments to columns is in proportion of $4EI/L$ of columns below and above the joint. Since columns are continuous, $4EI$ is constant, and moments are distributed based on $1/L$ of columns. The lower column has a height of 14 ft, 0 in and the upper column has a height of 12 ft, 0 in. The lower column will have a moment at its top, determined as follows.

$$M = 1801 \text{ kip-ft}\left(\frac{\frac{1}{14 \text{ ft}}}{\frac{1}{14 \text{ ft}}+\frac{1}{12 \text{ ft}}}\right) = 1801 \text{ kip-ft}\left(\frac{12 \text{ ft}}{26 \text{ ft}}\right) = 831 \text{ kip-ft}$$

The lower column could develop a maximum of M_{pr} at its base. The moment, M_{pr}, for the column is determined using a reinforcement yield strength of $1.25F_y$ or 75 ksi. M_{pr} determined for an axial load of 1200 kips is approximately 2631 kip-ft.

The shear, V_e, is determined as follows based on clear column height

$$V_e = \frac{(2631 \text{ kip-ft} + 831 \text{ kip-ft})}{11 \text{ ft}} = 315 \text{ kips}$$

This value is compared with frame analysis, $V_u = 104$ kips, thus V_e controls.

Interior column at second story

Clear height of column = 12 ft – 3 ft = 9 ft

M_{pr} of beams framing into top and bottom of column is based on negative moment from one beam and positive moment from the other beam.

$$\Sigma M_{pr,\text{ above}} = 1194 \text{ kip-ft} + 607 \text{ kip-ft} = 1801 \text{ kip-ft}$$

$$\Sigma M_{pr,\text{ below}} = 1194 \text{ kip-ft} + 607 \text{ kip-ft} = 1801 \text{ kip-ft}$$

The second story column will have moments of

$$M_{\text{top}} = 1801 \text{ kip-ft}\left(\frac{12 \text{ ft}}{24 \text{ ft}}\right) = 901 \text{ kip-ft}$$

$$M_{\text{bottom}} = 1801 \text{ kip-ft}\left(\frac{14 \text{ ft}}{26 \text{ ft}}\right) = 970 \text{ kip-ft}$$

$$\Sigma M_{\text{col}} = 901 \text{ kip-ft} + 970 \text{ kip-ft} = 1871 \text{ kip-ft}$$

thus column shear, V_e, is determined as follows based on clear column height

$$V_e = \frac{1871 \text{ kip-ft}}{9 \text{ ft}} = 208 \text{ kips}$$

This value is compared with frame analysis $V_u = 104$ kips, thus V_e controls.

The tabulated calculation of column shears is shown in Table 7-14.

Table 7-14. Calculation of column shear forces, V_e

Col. at Grid Lines	Level/ Story	Col. clear height (ft)	$-M_{pr}$ (joint above) (kip-ft)	$+M_{pr}$ (joint above) (kip-ft)	ΣM_{pr} at joint	Dist. ΣM_{pr} to col.	M at col. top (kip-ft)	$-M_{pr}$ (joint below) (kip- ft)	$+M_{pr}$ (joint below) (kip- ft)	ΣM_{pr} at joint	Dist. ΣM_{pr} to col.	M at col. bot. (kip- ft)	ΣM (kip-ft)	V_e at col. (kips)[a,b]
1, 6	1	11	1194	0	1194	0.462	552	0	0	0	1	2000	2552	232
	2	9	1194	0	1194	0.5	597	1194	0	1194	0.538	642	1239	138
	3	9	1194	0	1194	0.5	597	1194	0	1194	0.5	597	1194	133
	4	9.17	1114	0	1114	0.5	557	1194	0	1194	0.5	597	1154	126
	5	9.5	614	0	614	0.5	307	1114	0	1114	0.5	557	864	91
	6	9.5	614	0	614	0.5	307	614	0	614	0.5	307	614	65
	7	9.5	495	0	495	1	495	614	0	614	0.5	307	802	84
2,3,4,5	1	11	1194	607	1801	0.462	832	0	0	0	1	2631	3463	315
	2	9	1194	607	1801	0.5	901	1194	607	1801	0.538	970	1871	208
	3	9	1194	607	1801	0.5	901	1194	607	1801	0.5	901	1802	200
	4	9.17	1114	569	1703	0.5	852	1194	607	1801	0.5	901	1753	191
	5	9.5	614	147	718	0.5	359	1114	569	1683	0.5	842	1560	164
	6	9.5	614	104	718	0.5	359	614	147	761	0.5	381	740	104
	7	9.5	495	104	599	1	599	614	104	718	0.5	359	958	202

Notes:

a. V_u appears to govern at floors 5 and 6.

b. V_u may govern at floor 6.

Special transverse reinforcement per ACI 318-08 §21.6.4

The total cross-section area of rectangular hoop reinforcement shall not be less than that required by Equations 21-4 and 21-5.

$$A_{sh} = 0.3(sb_c f'_c / f_{yt})[(A_g / A_{ch}) - 1] \qquad \text{ACI 318-08 (21-4)}$$

$$A_{sh} = 0.09(sb_c f'_c / f_{yt}) \qquad \text{ACI 318-08 (21-5)}$$

Transverse reinforcement shall be spaced at distances not exceeding (1) one quarter minimum member dimension, (2) six times the diameter of the longitudinal reinforcement, and (3) s_x as defined by Equation 21-2.

$$s_o = 4 + \left(\frac{14 - h_x}{3}\right); \quad s_o = 4 + \left(\frac{14 - 6}{3}\right) = 6.67 \text{ in}$$

The transverse reinforcement should extend beyond any joint face a distance l_o equal to the larger of (1) one column member depth, (2) one sixth of the clear column span, or (3) 18 inches. Spacing between transverse reinforcement should not exceed six bar diameters of the longitudinal steel or 6 inches (ACI 318-08 §21.6.4.1).

Table 7-15 shows calculations for special transverse reinforcement.

Table 7-15. Special transverse reinforcement in columns

Col. Size	Eq.	b	d	b_c trans	b_c long	f'_c	f_{yt}	A_g	A_{ch}	s	A_{sh}	No. legs	Size bars
30×36	21-4	30	36		26	4000	60,000	1080	891	6	0.662		
	21-5	30	36		26	4000	60,000			6	0.936	6	#4
	21-4	30	36	32		4000	60,000	1080	891	6	0.815		
	21-5	30	36	32		4000	60,000			6	1.152	6	#4
30×30	21-4	30	30		26	4000	60,000	900	729	6	0.732		
	21-5	30	30		26	4000	60,000			6	0.936	6	#4
	21-4	30	30	26		4000	60,000	900	729	6	0.732		
	21-5	30	30	26		4000	60,000			6	0.936	6	#4

Calculations for the required shear steel are shown in Table 7-16.

Table 7-16 Shear strength

Col.	Shear V_u (kips)	Shear V_e (kips)	b (in)	d (in)	f'_c (psi)	f_y (psi)	ϕV_c (kips)	A_v (sq in)	s (in)	ϕV_s (kips)	ϕV_n (kips)	DCR
30×36	104	315	30	33	4000	60,000	94	1.2	6	297	391	0.81
30×30	65	232	30	27	4000	60,000	77	1.2	6	245	322	0.72

The final column design at the first level is summarized in Table 7-17.

Table 7-17. Final column design at first level

Column	Longitudinal Reinforcement	Long. Stirrups Within Yielding Zones, l_o	Long. Stirrups Beyond Yielding Zones, l_o	Trans. Stirrups Within Yielding Zones, l_o	Trans. Stirrups Beyond Yielding Zones, l_o
30×36	20 #10	6 #4@6 inches	6 #4@6 inches	6 #4@6 inches	6 #4@6 inches
30×30	20 #10	6 #4@6 inches	6 #4@6 inches	6 #4@6 inches	6 #4@6 inches

The column design may be used for the full height columns or the reinforcement can be reduced slightly at the upper portion of the frame. Since the longitudinal reinforcement is 2.8 percent, the longitudinal reinforcement could be reduced to as little as 1 percent in upper stories.

Figures 7-11 and 7-12 show the column cross-section with dimensions and reinforcement indicated.

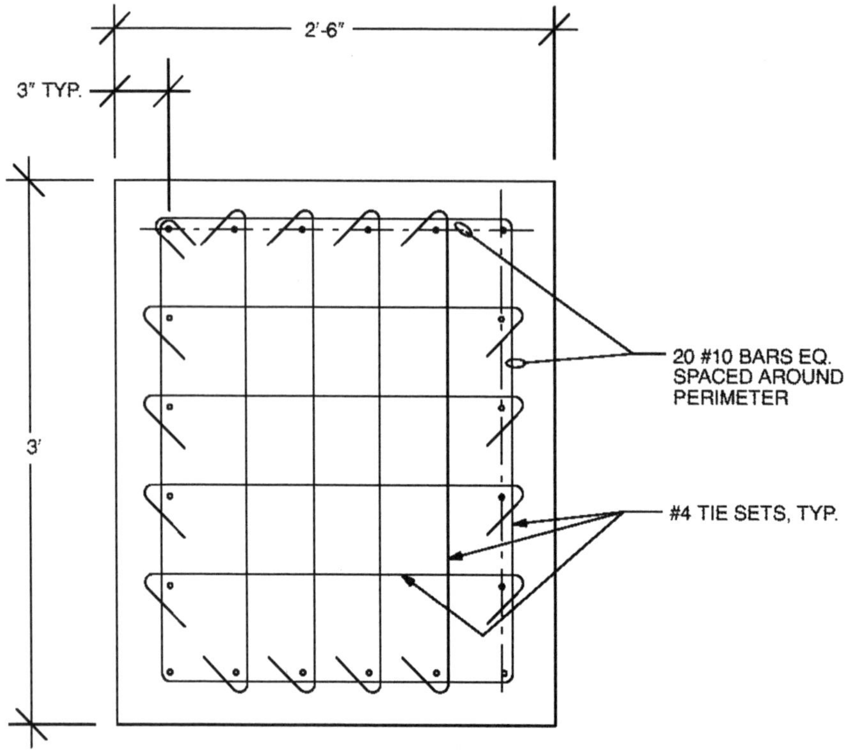

Figure 7-11. 30×36 column

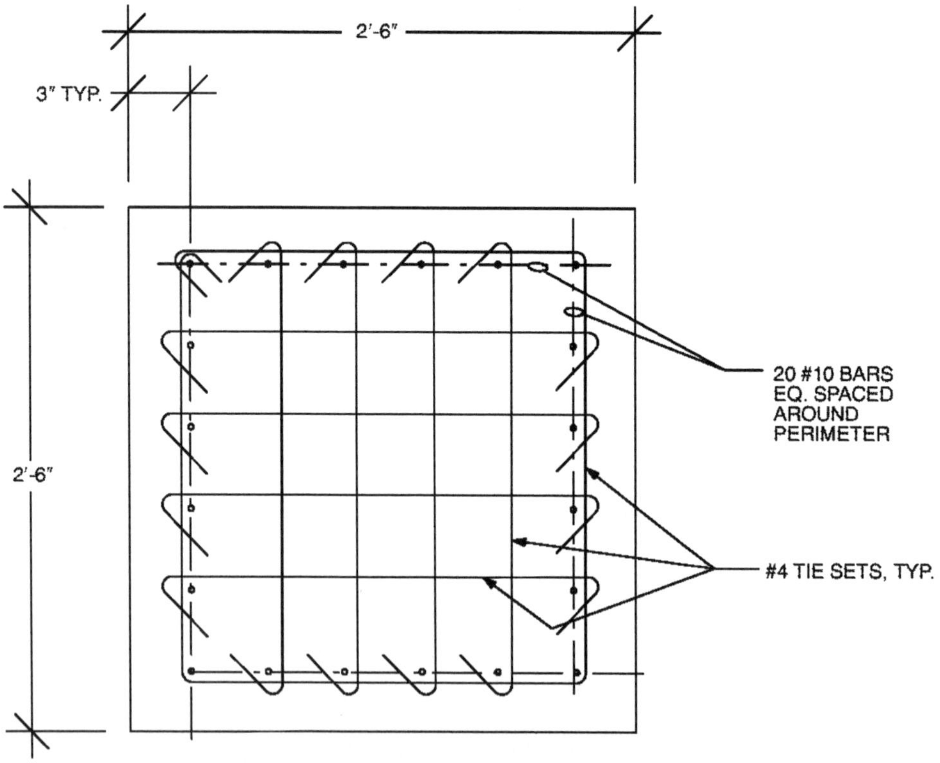

Figure 7-12. 30×30 column

10. Joint shear analysis

Beam-column joints of frames must be analyzed for joint shear in accordance with ACI 318-08 §21.5. The shear forces from analysis and the joint strength are calculated in Table 7-20. $\phi = 0.85$.

Table 7-20. Joint shear analysis

Element	Location	Shear from analysis (V_u) (kips)	V_{pr}, plastic shear (V_e) (kips)	Nominal Shear Stress	A_j (in^2)	Joint strength (kips)	Results
Int. Beam	Level 3	93	132	$\phi 15\sqrt{f'_c}A_j$	864	615	OK
End Beam	Level 3	93	132	$\phi 12\sqrt{f'_c}A_j$	720	410	OK
Int. Column	Level 2	104	315	$\phi 15\sqrt{f'_c}A_j$	864	615	OK
End Column	Level 2	67	232	$\phi 12\sqrt{f'_c}A_j$	720	410	OK

11. Detailing of beams and columns

11a. Beam reinforcement

Beams should be detailed with top, bottom, and side reinforcement as shown in Figure 7-13. In accordance with ACI 318-08, beam shear reinforcement, which meets the spacing requirements of ACI 318-08 §21.5.3.2, should be provided over a distance $2d$ from the faces of columns. The tie spacing shall not exceed: (1) $d/4$; (2) $8d_b$ of minimum beam longitudinal bar diameters; (3) $24d_b$ of stirrup bars; and (4) 12 inches. These requirements result in a 9-inch-maximum tie spacing, however, from analysis, ties required are #4 ties spaced at 6-inch centers. For ties between beam hinge regions, ties are required at $d/2$ spacing, however, based on analysis, #4 ties at 8-inch spacing are adequate across the remaining length of the beam (outside the hinge areas at each end).

Longitudinal beam bars should be spliced away from the beam-column joints and a minimum distance of $2h$ from the face of the columns, per ACI 318-08 §21.5.2.3. At the Level 2 beams for this example, the beam clear spans are approximately 21 feet and $2h$ is 2(36 inches) = 6 feet, 0 inches. The designer might consider splicing beam longitudinal reinforcement at the quarter-, one third-, or half-span locations (Figure 7-14). In this case the quarter-span locations would not be away from hinge regions, however, the one-third or mid-span locations would be okay. Increased shear reinforcement is required at the lap splice locations per ACI 318-08 §21.5.2.3. The maximum spacing of ties in these regions shall not exceed $d/4$ or 4 inches. In this case, the beam mid-point is the best place to locate the lap splices, which for the #10 top bars at Class B splices would have a splice length of 100 inches or 8 ft, 4 inches. The lap splice length for #7 bottom bars at a Class B lap splice is 54 inches or 4 ft, 6 inches. Longitudinal reinforcement can be shipped in 60-foot lengths on trucks, thus two locations of longitudinal beam lap splices would be required in the frame along Line A, conceivably on the two interior spans. Figure 7-15 shows beam-column joint reinforcement at an exterior span and beam reinforcement at interior spans is shown in Figure 7-16. A beam-column corner joint at the roof is presented in Figure 7-17.

11b. Column reinforcement

Column splices should occur at column mid-story heights (or within the center half of the column heights) per ACI 318-08 §21.6.3.2. Special transverse reinforcement is required per ACI 318-08 §21.6.4 over a length l_o above and below beams at spacing not greater than (1) the column depth, (2) one sixth the column clear span, or (3) a maximum of 18 inches. For this example, the column depth (which is either 42 inches or 44 inches) would control depending on the column. For column sections between the locations where special transverse reinforcement is required, the spacing requirements of ACI 318-08 §7.10.5.2 apply where ties should be spaced a maximum of 16 longitudinal bar diameters, 48 tie-bar diameters or the least dimension of the column. This would require ties at 20 inches; however for this example, it is recommended not to space column tie bars greater than six times the diameter of column longitudinal bars or 6 inches per ACI 318-08 §21.6.4.6 and 4 inches at lap splices.

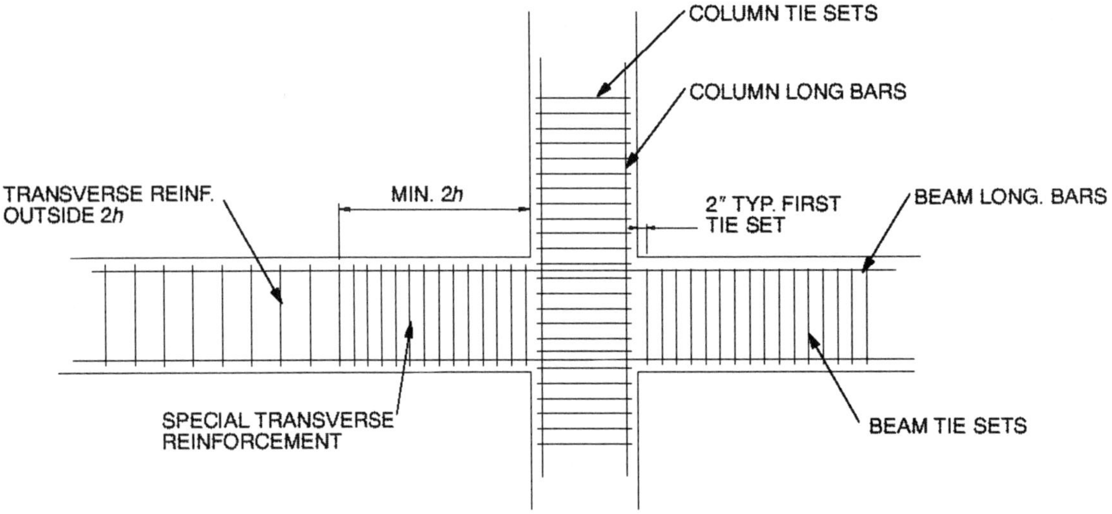

Figure 7-13. Beam-column joint

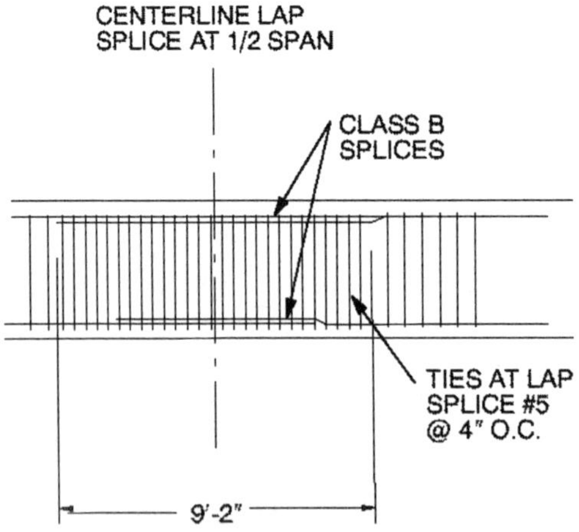

Figure 7-14. Beam reinforcement lap splice

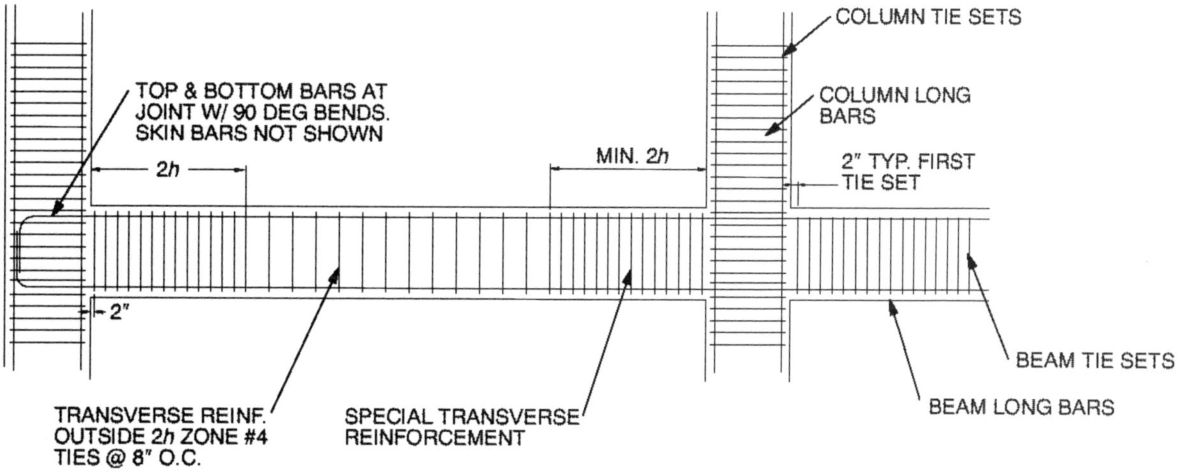

Figure 7-15. Beam-column joint reinforcement at exterior span

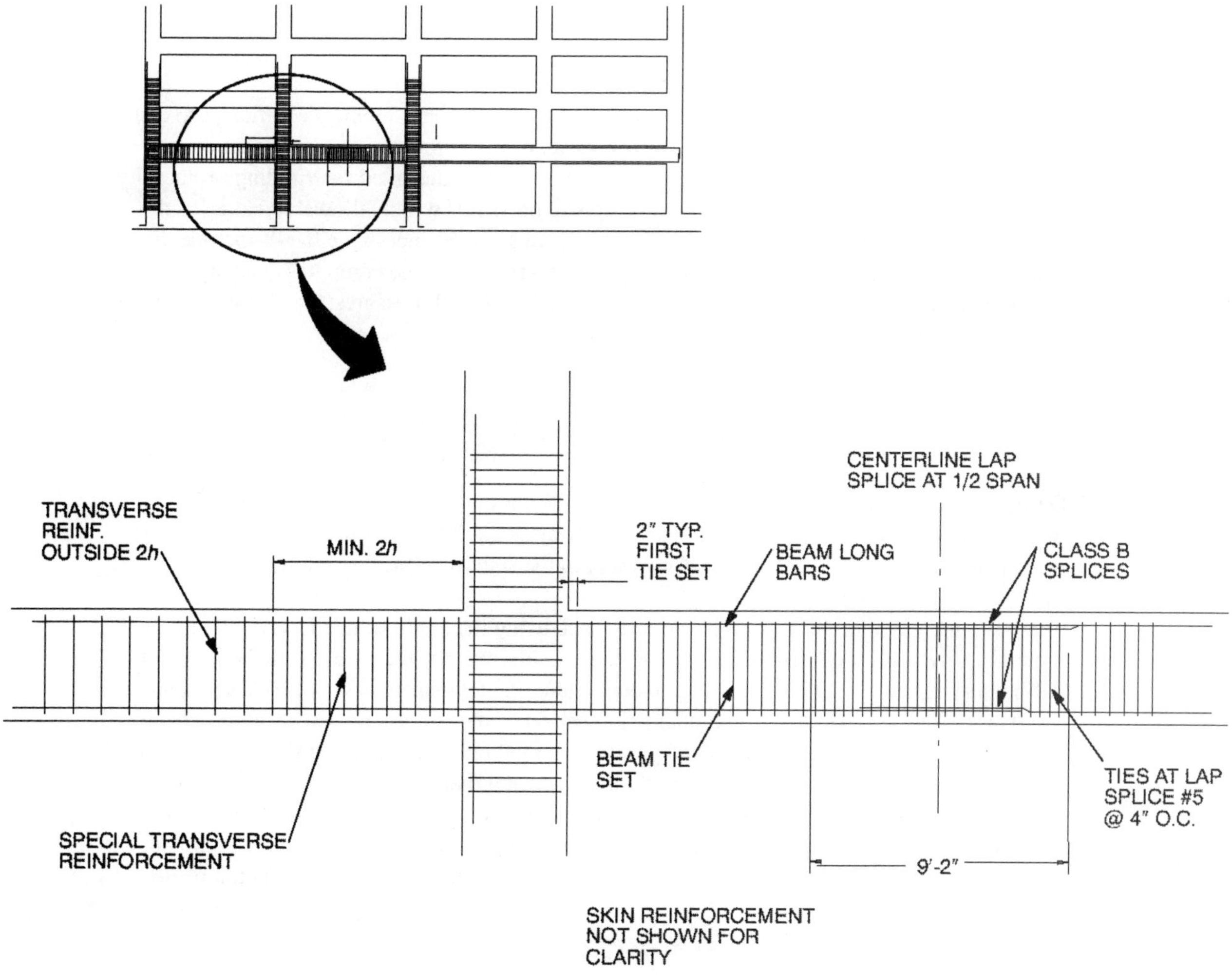

Figure 7-16. Beam reinforcement at interior spans

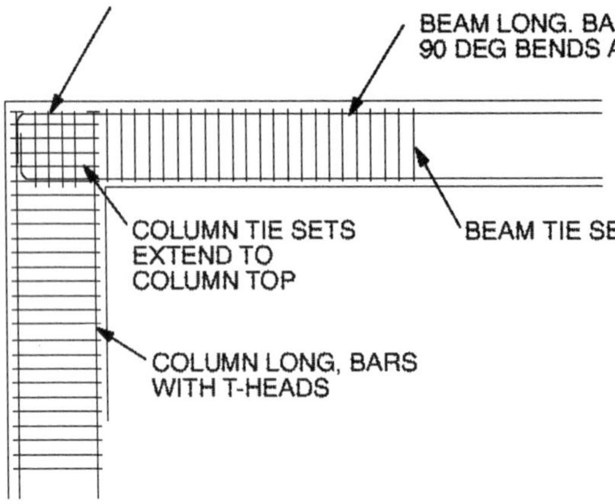

Figure 7-17. Beam-column corner joint at roof

12. Foundation considerations

The foundation system should be capable of resisting column base moments sufficient to cause plastic hinges to be located in the beams and column bases. If the plastic hinge location is forced into the columns, the foundation elements need not be designed for yielding or ductility. The foundation should also be adequate to keep soil pressures within allowable values and adequate for frame overturning stability. For this analysis, a 60-inch-wide by 48-inch-deep grade beam was used and cracked beam properties were used in the computer analysis. Note that ASD combinations of loads are used for calculation of soil pressures. The actual design of foundation elements is not performed in this example.

Commentary

Deformation compatibility should be checked at interior columns due to seismic drifts, *k*. This will lead to a conservative design for punching shear at slab/column joints. These joints may require drop panels or shear head reinforcement in the slab over interior columns.

The building period in this example was calculated using the approximate fundamental period method. A dynamic analysis may be performed considering the structure geometry, element stiffnesses, and story masses to determine eigenvalues and eigenvectors. The dynamic analysis results may be used as long as the fundamental period of the structure is not greater than 120 percent of the approximate fundamental period. If a mathematical model of the building is used to determine the period, the designer should keep in mind that non-seismic elements can add stiffness in the building and thus cause the fundamental period of the structure to be shorter. Thus, interior non-seismic columns or other important stiffening elements should be included in structural modeling and period calculations to assure conservative period calculation results.

Reinforced concrete SMRF frames can provide very ductile seismic systems for buildings with highly desirable performance characteristics. The yielding mechanisms can be predicted and the seismic performance will be ductile and not brittle or sudden. Care should be taken to assure adequate shear strength at beams, columns, and joints to assure that ductile flexural yielding will occur as anticipated. Care should also be taken with lap splices and detailing of reinforcement and with specified couplers. Longitudinal reinforcement should be ASTM A706, which has more ductile performance characteristics than ASTM A615 reinforcement.

References

ACI 318-08. *Building Code Requirements and Commentary*, American Concrete Institute, Farmington Hills, Michigan

A.H. Nilson and G. Winter. Tenth Edition, *Design of Concrete Structures*, McGraw-Hill Book Company, New York, 1986.

ASCE/SEI-7-05. *Minimum Design Loads for Buildings and Other Structures*, Structural Engineering Institute of the American Society of Civil Engineers, Reston, Virginia, 2005.

ATC-40. *Seismic Evaluation and Retrofit of Concrete Buildings*, Applied Technology Council, Redwood City, California, 1996.

IBC, *International Building Code*, 2009 Edition, International Code Council, Washington, D.C.

J.C. McCormac. Third Edition, *Design of Reinforced Concrete*, Harper Collins College Publishers, New York

J.G. MacGregor. Second Edition, *Reinforced Concrete Mechanics and Design*, Prentice Hall, New Jersey, 1992.

Paulay, T. and Priestley, N.J. *Seismic Design of Reinforced Concrete and Masonry Buildings*, John Wiley and Sons, New York, 1992.

R. Park and T. Paulay. *Reinforced Concrete Structures*, John Wiley and Sons, New York, 1975.

SEAOC, Seventh Edition, *Recommended Lateral Force Requirements and Commentary*, Structural Engineers Association of California, Sacramento, California, 1999.